SURFACE MODIFIED NANOMATERIALS FOR APPLICATIONS IN CATALYSIS

Micro and Nano Technologies Series

SURFACE MODIFIED NANOMATERIALS FOR APPLICATIONS IN CATALYSIS

Fundamentals, Methods and Applications

Edited by

MANOJ B. GAWANDE
Department of Industrial and Engineering Chemistry, Institute of Chemical Technology, Mumbai-Marathwada Campus, Jalna, Maharashtra, India

CHAUDHERY MUSTANSAR HUSSAIN
Department of Chemistry and Environmental Sciences, New Jersey Institute of Technology (NJIT), Newark, New Jersey, United States

YUSUKE YAMAUCHI
School of Chemical Engineering and Australian Institute for Bioengineering and Nanotechnology (AIBN), The University of Queensland, Brisbane, QLD, Australia

ELSEVIER

Elsevier
Radarweg 29, PO Box 211, 1000 AE Amsterdam, Netherlands
The Boulevard, Langford Lane, Kidlington, Oxford OX5 1GB, United Kingdom
50 Hampshire Street, 5th Floor, Cambridge, MA 02139, United States

Copyright © 2022 Elsevier Inc. All rights reserved.

No part of this publication may be reproduced or transmitted in any form or by any means, electronic or mechanical, including photocopying, recording, or any information storage and retrieval system, without permission in writing from the publisher. Details on how to seek permission, further information about the Publisher's permissions policies and our arrangements with organizations such as the Copyright Clearance Center and the Copyright Licensing Agency, can be found at our website: www.elsevier.com/permissions.

This book and the individual contributions contained in it are protected under copyright by the Publisher (other than as may be noted herein).

Notices

Knowledge and best practice in this field are constantly changing. As new research and experience broaden our understanding, changes in research methods, professional practices, or medical treatment may become necessary.

Practitioners and researchers must always rely on their own experience and knowledge in evaluating and using any information, methods, compounds, or experiments described herein. In using such information or methods they should be mindful of their own safety and the safety of others, including parties for whom they have a professional responsibility.

To the fullest extent of the law, neither the Publisher nor the authors, contributors, or editors, assume any liability for any injury and/or damage to persons or property as a matter of products liability, negligence or otherwise, or from any use or operation of any methods, products, instructions, or ideas contained in the material herein.

ISBN: 978-0-12-823386-3

For information on all Elsevier publications visit our website at https://www.elsevier.com/books-and-journals

Publisher: Matthew Deans
Acquisitions Editor: Kayla Dos Santos
Editorial Project Manager: Sara Valentino
Production Project Manager: Kamesh Ramajogi
Cover Designer: Mark Rogers

Typeset by TNQ Technologies

Contents

Contributors　xi
About the editors　xv
Introduction to surface-modified nanomaterials　xvii

1. New frontiers for heterogeneous catalysis: surface modification of nanomaterials　1
Anil Kumar Nallajarla, Shajeeya Amren Shaik and Anandarup Goswami

1. Introduction: catalysts, nanomaterials (NMs) and nanocatalysts (NCs)　1
2. Synthetic strategies of NCs　4
3. Need for surface functionalization　5
4. Types of NMs and their functionalization procedures　6
5. Selective applications of surface-functionalized NCs　9
6. Characterization of surface-functionalized NCs　14
7. Topics to be covered in this book　16
8. Summary: present status and future direction　17
Abbreviations　20
References　20

2. Fundamental concepts on surface chemistry for nanoparticle modifications　29
Ankush V. Biradar, Saravanan Subramanian, Amravati S. Singh, Dhanaji R. Naikwadi, Krishnan Ravi and Jacky H. Advani

1. Introduction　29
2. Esterification reaction method　46
3. Phosphate ester method　48
4. In situ modification method　48
5. Conclusions　50
References　50

3. Synthesis of surface-modified nanomaterials　53
Gianvito Vilé

1. Introduction　53
2. Nanomaterials surface chemistry and Zeta potential　55
3. Surface modification of catalytic nanoparticles by thermal and plasma treatment　58
4. Silane chemical treatment　59

vi Contents

    5. Ligand immobilization techniques to modify catalytic nanoparticles   63
    6. Surface modification via single-atom anchoring   66
    7. Industrial-scale utilization of synthetic methods to prepare surface-modified nanomaterials   69
    8. Challenges, future perspective, and conclusions   70
    References   70

4. Surface modification of nano-based catalytic materials for enhanced water treatment applications   **73**

Eleni Petala, Amaresh C. Pradhan and Jan Filip

    1. Nanomaterials for water treatment   73
    2. Advancement on modification of nanomaterials   73
    3. Examples of nanoscale modified catalytic materials   75
    4. Conclusions   94
    Acknowledgment   94
    References   94

5. Surface-modified nanomaterials-based catalytic materials for water purification, hydrocarbon production, and pollutant remediation   **103**

Ragib Shakil, Md. Mahamudul Hasan Rumon, Yeasin Arafat Tarek, Chanchal Kumar Roy, Al-Nakib Chowdhury and Rasel Das

    1. Introduction   103
    2. Nanocatalyst materials   104
    3. Wastewater treatment   105
    4. Nanocatalysts in hydrocarbon production   114
    5. Recent advancement and real-time utilization of nanocatalysts   119
    6. Conclusion   122
    Acknowledgments   122
    References   122

6. Surface-modified nanomaterial-based catalytic materials for the production of liquid fuels   **131**

Indrajeet R. Warkad, Hanumant B. Kale and Manoj B. Gawande

    1. Introduction   131
    2. Surface modified nanomaterials (SMNs) for biomass conversion to liquid fuels   133
    3. Surface-modified nanomaterials for the transformation of carbon dioxide to liquid fuels   152
    4. Future perspectives and conclusion   160

Acknowledgments	161
References	161

7. SMN-based catalytic membranes for environmental catalysis 171
Nilesh R. Manwar and Manoj B. Gawande

1. Introduction	171
2. Challenges in SMNs and catalytic MRs	173
3. Types of catalytic membrane reactors (MRs)	174
4. Basic overview of polymeric MRs	178
5. Incorporation of SMNs into polymeric membranes	179
6. SMNs based polymeric membrane-assisted catalysis	187
7. Summary and future perspectives	191
Abbreviations	192
References	192

8. Semiconductor catalysts based on surface-modified nanomaterials (SMNs) for sensors 197
E. Kuna, P. Pieta, R. Nowakowski and I.S. Pieta

1. Introduction	197
2. Zero-dimensional (0D) nanomaterials	199
3. One-dimensional (1D) nanomaterials	206
4. Two-dimensional (2D) nanomaterials	209
5. Tree-dimensional (3D) nanomaterials	214
6. Conclusion	215
Acknowledgments	216
List of resources	217
References	217

9. Surface-modified carbonaceous nanomaterials for CO_2 hydrogenation and fixation 223
Hushan Chand, Priyanka Choudhary and Venkata Krishnan

1. Introduction	223
2. Basic concepts of CO_2 sequestration (hydrogenation and fixation)	227
3. Heterogeneous catalyst in CO_2 hydrogenation	227
4. Heterogeneous catalyst in CO_2 fixation	233
5. Summary and perspectives	245
References	246

10. Surface-modified nanomaterials for synthesis of pharmaceuticals — 251
Kishore Natte and Rajenahally V. Jagadeesh

1. Introduction — 251
2. Noble metal-based nanoparticles for the synthesis of pharmaceuticals — 252
3. Nonnoble metal-based nanoparticles for the synthesis of pharmaceuticals — 254
4. Conclusions — 265
References — 265

11. Surface-modified nanomaterial-based catalytic materials for modern industry applications — 267
Priti Sharma and Manoj B. Gawande

1. Introduction — 267
2. Scope of the book chapter — 269
3. Active role of surface-modified nanomaterials in industry — 269
4. Conclusion — 284
References — 284

12. Assessment of health, safety, and economics of surface-modified nanomaterials for catalytic applications: a review — 289
Sushil R. Kanel, Mallikarjuna N. Nadagouda, Amita Nakarmi, Arindam Malakar, Chittaranjan Ray and Lok R. Pokhrel

1. Introduction — 289
2. Human health and safety consequences of SMNs — 306
3. Economic aspects of NMs used as catalysts — 308
4. Conclusions and future perspectives — 310
Acknowledgments — 311
References — 311

13. Future of SMNs catalysts for industry applications — 319
Ajaysing S. Nimbalkar, Dipali P. Upare, Nitin P. Lad and Pravin P. Upare

1. Introduction — 319
2. Nanoparticles catalysts — 321

3. Catalytic applications of nanomaterial	323
4. Shape and size dependent catalysts and reactions	325
5. Metal-isolated single atoms	334
6. Conclusion and outlook	338
References	339

Index   *347*

Contributors

Jacky H. Advani
Inorganic Materials and Catalysis Division, CSIR-Central Salt and Marine Chemicals Research Institute, Bhavnagar, Gujarat, India; Academy of Scientific and Innovative Research (AcSIR), Ghaziabad, Uttar Pradesh, India

Ankush V. Biradar
Inorganic Materials and Catalysis Division, CSIR-Central Salt and Marine Chemicals Research Institute, Bhavnagar, Gujarat, India; Academy of Scientific and Innovative Research (AcSIR), Ghaziabad, Uttar Pradesh, India

Hushan Chand
School of Basic Sciences and Advanced Materials Research Center, Indian Institute of Technology Mandi, Kamand, Himachal Pradesh, India

Priyanka Choudhary
School of Basic Sciences and Advanced Materials Research Center, Indian Institute of Technology Mandi, Kamand, Himachal Pradesh, India

Al-Nakib Chowdhury
Department of Chemistry, Bangladesh University of Engineering and Technology (BUET), Dhaka, Bangladesh

Rasel Das
Department of Chemistry, Stony Brook University, Stony Brook, NY, United States

Jan Filip
Regional Centre of Advanced Technologies and Materials, Czech Advanced Technology and Research Institute, Palacký University Olomouc, Olomouc, Czech Republic

Manoj B. Gawande
Regional Centre of Advanced Technologies and Materials, Czech Advanced Technology and Research Institute, Palacký University, Olomouc, Czech Republic; Department of Industrial and Engineering Chemistry, Institute of Chemical Technology, Mumbai Marathwada Campus, Jalna, Maharashtra, India

Anandarup Goswami
Division of Chemistry, Department of Sciences and Humanities, Vignan's Foundation for Science, Technology and Research (VFSTR, Deemed to be University), Guntur, Andhra Pradesh, India

Rajenahally V. Jagadeesh
Leibniz Institute for Catalysis, Rostock, Germany

Hanumant B. Kale
Department of Industrial and Engineering Chemistry, Institute of Chemical Technology, Mumbai Marathwada Campus, Jalna, Maharashtra, India

Sushil R. Kanel
Department of Chemistry, Wright State University, Dayton, OH, United States

Venkata Krishnan
School of Basic Sciences and Advanced Materials Research Center, Indian Institute of Technology Mandi, Kamand, Himachal Pradesh, India

E. Kuna
Institute of Physical Chemistry Polish Academy of Sciences, Warsaw, Poland

Nitin P. Lad
Organic Chemistry Research Centre, Department of Chemistry, A.M. Science College, Nashik, Maharashtra, India

Arindam Malakar
Nebraska Water Center, Part of the Robert B. Daugherty Water for Food Global Institute, University of Nebraska, Lincoln, NE, United States

Nilesh R. Manwar
Department of Industrial and Engineering Chemistry, Institute of Chemical Technology, Mumbai Marathwada Campus, Jalna, Maharashtra, India

Mallikarjuna N. Nadagouda
Department of Chemistry, Wright State University, Dayton, OH, United States; Department of Mechanical and Materials Engineering, Wright State University, Dayton, OH, United States

Dhanaji R. Naikwadi
Inorganic Materials and Catalysis Division, CSIR-Central Salt and Marine Chemicals Research Institute, Bhavnagar, Gujarat, India; Academy of Scientific and Innovative Research (AcSIR), Ghaziabad, Uttar Pradesh, India

Amita Nakarmi
Department of Chemistry, University of Arkansas at Little Rock, Little Rock, AR, United States

Anil Kumar Nallajarla
Division of Chemistry, Department of Sciences and Humanities, Vignan's Foundation for Science, Technology and Research (VFSTR, Deemed to be University), Guntur, Andhra Pradesh, India

Kishore Natte
Chemical and Material Sciences Division, CSIR-Indian Institute of Petroleum (CSIR-IIP), Mohkampur, Dehradun, India; Academy of Scientific and Innovative Research (AcSIR), Ghaziabad, Uttar Pradesh, India

Ajaysing S. Nimbalkar
Green Chemistry Division, University of Science of Technology, Daejeon, South Korea; Green Carbon Catalysis Research Group, Korea Research Institute of Chemical Technology, Daejeon, South Korea

R. Nowakowski
Institute of Physical Chemistry Polish Academy of Sciences, Warsaw, Poland

Eleni Petala
Regional Centre of Advanced Technologies and Materials, Czech Advanced Technology and Research Institute, Palacký University Olomouc, Olomouc, Czech Republic

P. Pieta
Institute of Physical Chemistry Polish Academy of Sciences, Warsaw, Poland

I.S. Pieta
Institute of Physical Chemistry Polish Academy of Sciences, Warsaw, Poland

Lok R. Pokhrel
Department of Public Health, The Brody School of Medicine, East Carolina University, Greenville, NC, United States

Amaresh C. Pradhan
Regional Centre of Advanced Technologies and Materials, Czech Advanced Technology and Research Institute, Palacký University Olomouc, Olomouc, Czech Republic

Krishnan Ravi
Inorganic Materials and Catalysis Division, CSIR-Central Salt and Marine Chemicals Research Institute, Bhavnagar, Gujarat, India; Academy of Scientific and Innovative Research (AcSIR), Ghaziabad, Uttar Pradesh, India

Chittaranjan Ray
Nebraska Water Center, Part of the Robert B. Daugherty Water for Food Global Institute, University of Nebraska, Lincoln, NE, United States

Chanchal Kumar Roy
Department of Chemistry, Bangladesh University of Engineering and Technology (BUET), Dhaka, Bangladesh

Md. Mahamudul Hasan Rumon
Department of Chemistry, Bangladesh University of Engineering and Technology (BUET), Dhaka, Bangladesh

Shajeeya Amren Shaik
Division of Chemistry, Department of Sciences and Humanities, Vignan's Foundation for Science, Technology and Research (VFSTR, Deemed to be University), Guntur, Andhra Pradesh, India

Ragib Shakil
Department of Chemistry, Bangladesh University of Engineering and Technology (BUET), Dhaka, Bangladesh

Priti Sharma
Regional Centre of Advanced Technologies and Materials, Czech Advanced Technology and Research Institute, Palacký University, Olomouc, Czech Republic

Amravati S. Singh
Inorganic Materials and Catalysis Division, CSIR-Central Salt and Marine Chemicals Research Institute, Bhavnagar, Gujarat, India; Academy of Scientific and Innovative Research (AcSIR), Ghaziabad, Uttar Pradesh, India

Saravanan Subramanian
Inorganic Materials and Catalysis Division, CSIR-Central Salt and Marine Chemicals Research Institute, Bhavnagar, Gujarat, India; Academy of Scientific and Innovative Research (AcSIR), Ghaziabad, Uttar Pradesh, India

Yeasin Arafat Tarek
Department of Chemistry, Bangladesh University of Engineering and Technology (BUET), Dhaka, Bangladesh

Dipali P. Upare
Green Chemistry Division, University of Science of Technology, Daejeon, South Korea

Pravin P. Upare
Green Carbon Catalysis Research Group, Korea Research Institute of Chemical Technology, Daejeon, South Korea

Gianvito Vilé
Department of Chemistry, Materials, and Chemical Engineering "Giulio Natta", Politecnico di Milano, Milan, Italy

Indrajeet R. Warkad
Department of Industrial and Engineering Chemistry, Institute of Chemical Technology, Mumbai Marathwada Campus, Jalna, Maharashtra, India

About the editors

Manoj B. Gawande

Manoj B. Gawande received his PhD in 2008 from the Institute of Chemical Technology, Mumbai, India, and then undertook several research stints in Germany, South Korea, Portugal, Czech Republic, the United States, and the United Kingdom. He also worked as a Visiting Professor at CBC-SPMS, Nanyang Technological University, Singapore, in 2013. Presently, he is Associate Professor at the Institute of Chemical Technology-Mumbai Marathwada Campus, Jalna, India and Visiting Professor at RCPTM-CATRIN, Palacky University. His research interests focus on SACs, advanced nanomaterials, sustainable technologies, and cutting-edge catalysis and energy applications. He has published over 140 international scientific papers.

Chaudhery Mustansar Hussain

Chaudhery Mustansar Hussain, PhD, is an Adjunct Professor and Director of Labs in the Department of Chemistry and Environmental Sciences at the New Jersey Institute of

Technology (NJIT), Newark, New Jersey, United States. His research is focused on the applications of nanotechnology, advanced technologies and materials, analytical chemistry, environmental management, and various industries. Dr. Hussain is the author of numerous papers in peer-reviewed journals as well as a prolific author and editor of more than 60 scientific monographs and handbooks in his research areas, published with Elsevier, Royal Society of Chemistry, John Wiley and Sons, CRC, Springer, and so forth.

Yusuke Yamauchi

Yusuke Yamauchi received his bachelor's (2003), master's (2004), and PhD (2007) degrees from the Waseda University, Japan. After receiving his PhD, he joined the National Institute of Materials Science (NIMS), Japan, to start his own research group. In 2017, he moved to The University of Queensland (UQ). Presently, he is a senior group leader at AIBN and a full professor at School of Chemical Engineering. He concurrently serves as an honorary group leader at NIMS, an associate editor of the Journal of Materials Chemistry A published by the Royal Society of Chemistry (RSC). He has published more than 800 papers in international refereed journals with ~50,000 citations (h-index > 115). He has been selected as one of the highly cited researchers (chemistry in 2016–2020 and materials science in 2020).

Introduction to surface-modified nanomaterials

Hanumant B. Kale, Manoj B. Gawande
Department of Industrial and Engineering Chemistry, Institute of Chemical Technology, Mumbai-Marathwada Campus, Jalna, Maharashtra, India

In the 21st century, various branches of science and technology have emerged due to the investigation and exploration of advanced technology in the field of not only electronics but also in catalysis as sustainable technological development.[1] Many of these branches have the potential to grow in the future. Nanotechnology is one such popular branch that has provided solutions to many of the problems that arise due to lack/gaps in material science. For example, it is proved that the bulk counterpart, when compared to the nanomaterials' (NMs) high surface area, abundant active sites, and porosity outperforms its properties manyfold.[2–4] So, it is very true that NMs are the cornerstone of advanced science and technology. Nanomaterials have been the subject of extensive research over the decades due to their high surface-to-volume ratio and mechanical strength. Because of the advancement of nanotechnology, it is now possible to alter the physical and chemical properties of NMs for molecular recognition along with their catalytic functionality.[5,6] Such efforts have resulted in a large number of catalytic platforms for a wide range of analytes, including metal ions, support precursors, molecular compounds, ionic liquids, and biomolecules like nucleic acids and proteins. The development of nanotechnology has transformed research into advanced catalysts with high intrinsic activity. To date, several applications of NMs documented, and many of them are utilized for the commercialization of industrial processes including automobiles, marine coatings, aerospace, and construction. Furthermore, because of their vast range of applications in catalysis, imaging, photonics, nanoelectronics, sensors, biomaterials, and biomedicine, the significance of NMs in modern technologies is growing.

At the moment, one of the major concerns of the global economic vision is the ability to meet the challenges of energy and sustainability. So, catalytic innovation is crucial for energy, the synthesis process, and the environment. For instance, catalysis is important in modern chemical industries because at least one catalytic process is used in approximately 80%–90% of manufactured chemicals.[7] Since it was noticed that catalysts must be involved in the development of sustainable processes, the relationship between catalysts and NMs has grown stronger to provide the desired solutions for the development of stable, recyclable, efficient, and selective nanocatalytic systems. In this regard, NMs play a significant role in the fabrication of many devices and modified nanostructures because of their unique structural properties. Many researchers around the world are interested in the design and synthesis of organic–inorganic hybrid NMs, as well as their catalytic applications in pharmaceutical, biomedical, optical, fuel cells, and environmental technology, as they enable environmentally friendly and benign catalytic processes.[8] In recent years, novel approaches to designing highly selective nanostructured catalysts have been developed by controlling the interaction between the active catalytic species and the support materials. Nanomaterials enable simple surface modification by tailoring electronic properties, conductivity, catalytic properties, and response to physical events.[9] The functionalization of NMs enables the inclusion of functional groups and species across the surface, resulting in a better structure with novel properties.

Why is it necessary to modified the surface of nanomaterials?

Both bare nanocomposites and surface-modified nanocomposites are the foundation of nanoscience and bioscience. Because of the unique size- and shape-dependent properties and higher surface area of nanostructured materials, as well as the possibility of controlling their surface properties employing simple surface modifications techniques; these surface-modified nanomaterials (SMNs) have revolutionized the field of catalysis particularly heterogeneous catalysis.[10] As a result, appropriate surface modifications can either improve catalytic activity and selectivity, allow for the incorporation of catalytically active centers on nanosupport, and improve their stability and compatibility during synthesis as well as during reaction conditions.[11–13] However, directly prepared pristine NMs show many disadvantages like irregular pore size distribution, low binding capacity, and low metal selectivity because of their heterogeneous structure; thus, it is essential to develop NMs with highly active catalytic functionality. The surface of NMs can be chemically tailored to increase the adsorption capacity of reactants involved in catalytic transformations and subsequently performance of heterogeneous catalysts.[14] The NM surface is the interface between the bulk materials and the external environment. The surface of the NM is likely to alter because of the interaction with the environment. This is essential because changes in the surface characteristics of the NMs can modify the chemical compositions, ruggedness, hydrophilicity, crystallinity, biocompatibility, conductivity, permeability, frozenness, lubrication, and the interlinked density of materials.

To date, various synthetic methods for preparing different classes of SMNs like chemical adsorption techniques, solvothermal method, a one-pot synthetic method involving surfactants, co-precipitation, sol-gel method, polyol synthesis, are documented to achieve fast reaction kinetics, high binding capacity, selectivity, stability, and reusability. The ideal SMNs should have all of these fascinating properties including (1) well-designed with appropriate pore size and geometry, as well as an open pore structure, (2) exposed adsorption sites that are easily accessible and have suitable surface properties, (3) high selectivity of targeted metal in the presence of other chemical functionality, (4) easy synthesis and required reagents are inexpensive, especially when considering not only the cost of material synthesis but also the cost of equipment which utilized in fabrication, (5) good operational stability at high temperature and pressure, and (6) reusability for several catalytic cycles.

The utilization of NMs in catalysis has been used in numerous research endeavors throughout the world to develop innovative and greener approaches.[15] The most often documented various SMNs as a catalytic material owing to their high catalytic efficacy, selectivity, and reusability with abundant intrinsic active sites.[16] Also, different metal nanoparticles (NPs) with changes in surface structures can be utilized as catalysts or mediators. Besides their small size and enlarged area, NMs have surfaced as an appealing candidate for homogeneous and heterogeneous catalyst interfaces. Heterogeneous catalysis is extremely important for chemical synthesis, as well as in the energy sector for fuel production. Additionally, pollutant remediation activities in environmental catalysis are driving extensive research efforts in optimizing the reactivity and selectivity of catalyst materials. For developing cost-effective and environmentally benign chemical processes, heterogeneous catalysts with low metal use but well-defined catalytic sites are greatly desired. "Never modify a running system," as the saying goes. New methods, on the other hand, have the potential to be considerably superior to earlier ones. While catalytic materials of hundreds of atoms have been employed to speed up chemical reactions in the past, while nowadays reducing particle size, usually down to the nanoscale, and thus increasing surface metal atom exposure and chemical functionality with surface modulation has been a common and efficient technique for improving performance and atom utilization efficiency of nanocatalysts.[17,18] In this context, the usage of SMNs with different chemical functionality on the surface along with metal NPs acts as an intrinsic active site of catalytic material could bring a new method to catalysis.

The last decade has witnessed an exponential rise in the interest pertaining to the synthesis and application of SMNs which has been driven tremendously by their potential to offer integrated benefits of catalytic activity, selectivity, stability, and recovery. This has led to a change in the complete surface chemistry of NMs. Fig. 1 clearly indicates the importance and growth of research in the field of SMNs in the scientific community. This citation index figure is taken from the Web of Science with the keyword search "surface-modified nanomaterials." Surface modification of NMs has been accomplished using a variety of techniques and researchers are currently working on developing a new and industrially efficient process for NMs functionalization. The more prevalent and effective approaches for the functionalization of NMs are biochemical treatment, polymer grafting, silane grafting, and surfactant-assisted modification. Apart from these chemical methods, other physical and mechanical methods including grinding, polishing, mechanochemical treatment, spraying, plasma treatment, ion implantation and deposition also documented. Today, if we have a glimpse at the literature reports, we find that SMNs have been proficiently exploited not just in biomedicines but also in material science and catalysis. In fact, they are proven to be very promising materials in diverse catalytic applications ranging from industrial organic catalysis to biomass scale-up, nanomedicines, sensors, electrocatalysis, energy conversions and storage, photooxidation technologies for water treatment, C—H activation and catalytic remediation of toxic

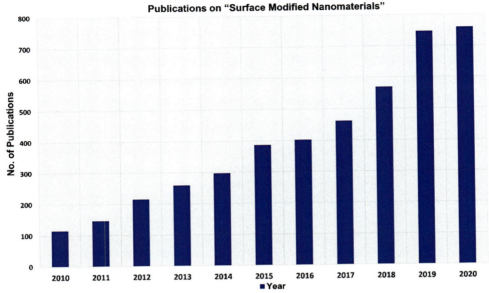

Figure 1 The number of publications on the topic "Surface-modified nanomaterials," timespan: 2010—2020, as found in Web of Science (date: September 20, 2021). This citation index figure is taken from the Web of Science with the topic search "surface-modified nanomaterials" from the year 2010—2020.

metals (environmental remediation). Their potential activity, selectivity and long-term stability has been accomplished by appropriate surface engineering and supporting/immobilizing metal NPs on different solid matrixes. The resulting surface-engineered materials can be successfully employed as catalysts for many challenging reactions. The anchoring of active catalytic species on support also plays an indispensable role by preventing the product stream from getting contaminated with the catalyst residues. Today, theoretical studies with the aid of modern computational tools are also being applied for the investigation of the mechanism responsible for the observed catalysis of many materials, including SMNs. With the help of theory, and computational tools, researchers are gaining insights into how wonderfully the surface of different metal NPs and nanostructures can be tailored for the effective and selective catalytic applications.

Role of surface-modified nanomaterials

SMNs differ from pristine NMs in absorption, mechanical, electrical, and optical properties. Notably, through surface modification, functionalization improves the qualities and features of metal NPs as well as nanostructures, allowing them to play an important role in catalysis. Increasing and tailoring the surface functionality and intrinsic activity of active sites of catalysts is a common strategy for improving catalytic efficacy. As a result, carbon nanostructures, quantum dots, silica, chalcogenides, and metal oxides-based NMs should be designed to fully expose active sites to increase the number of surface defects on nanocatalysts. The electronic structure of NMs should be modified using molecular adsorbates or ionic liquids to boost inherent catalytic efficiency.[19] The strong structure sensitivity is observed in several NMs during catalytic transformations because of the nature of reactants and solvents, so it is necessary to modulate NMs structures in the form of size, facet, boundary, composition, interface, and ligand modification to manifest the catalytic performance. For example, the number of coordination sites and surface chemisorption influence the size and shape of metal NPs. The chemical binding of reactive intermediates on the catalyst surface may be altered by the geometric and electrical impacts of metallic alloy NMs. SMNs may form interfacial active sites, which will regulate the adsorption and stability of essential reactive species due to their electrostatic interaction with other components such as metals, metal NPs, complexes, and ligands.[20] Apart from this, the SMNs with well-defined interfaces are derived from a fundamental understanding of the boundary between two phases, i.e., interfacial interaction between two surfaces and their structural properties. The integration of various NMs with different chemical moieties enables the generation of unique and/or enhanced functionalities that do not exist in the individual components of NMs. The design and development of integrated nanocatalyst systems,[21] i.e., SMNs is the solution for the availability of highly specific nanocatalysts which easily combined with reagents and reacted at very optimum reaction conditions.

Surface chemistry is very important for the engineering and fabrication of SMNs, apart from this the computational studies are also used to establish requirements for catalytic sites for certain chemical processes.[22] Since the last two decades NMs can be modulated in different specific aspects like size, shape, facet, grain boundary, composition (percent of metal loading, doping of metals like N, O, and S in support interface), ligand modification, and crystal phase to increases the catalytic activity, stability, and selectivity with respect to bare NMs or metal NPs.[23] Furthermore, the functionalized NPs have excellent physical features, including anticorrosion, antiagglomeration, and noninvasive qualities.

1. **Geometric and composition of NMs**: Geometric adjustments of NPs include morphology, pore size, shell thickness, profile shape, porosity, crystallinity, etc. The morphology of NPs, i.e., cubic, rhombic, tetrahedral directly affected on certain active facets and low-coordinated sites, which are subsequently responsible for their performance in chemical reactions. At the nanoscale or subnanometer scale, NPs exhibited unique physicochemical properties and by modifying the surface, various catalytic sites are accessible to reactants. The ratio of corner, edge, and terrace atoms varies depending on the size of the metal nanostructure. The atoms in the corner or edge have a different coordination number and chemical interaction energy, which could have a significant impact on catalytic activity and selectivity toward several catalytic applications.

Composition modification refers to the addition of an additive element to form an alloy or mixed cationic/anionic materials. The change in the composition of NMs or metal nanocatalysts is efficiently increased the catalytic performance. Bimetallic phase or metal alloying in nanocatalyst fabrication can efficiently control the binding strength of key intermediates to alter the pathway in chemical conversion. For instance, colloidal, reverse, and dendrimers chemistry is utilized to modulate surface atomic assemblies in active metals and metal oxide NPs with well-defined sizes, shapes, and compositions for selective catalysis.[22]

2. **Crystallographic facets and grain boundary (GB)**: The crystallographic facets in NPs are nothing but the orientation of atoms, which directly affect the metal bond energy and the adsorption and binding energy of the intermediates, the periodic arrangement of the atomic structure, and subsequently the output of catalytic reactions. For example, the Cu (100) facilitates the formation of selectively C_2 product and Cu (111) resulted into C_1/C_2 products subsequently via carbon dioxide reduction.[24,25]
Bulk defects in metals like GBs creates a vacancy and increases the surface strain.[26] Chorkendorff and coworkers proved that strong CO-binding sites assisted by grain boundaries are the intrinsic active center for electrocatalytic CO reduction on oxide-derived Cu (OD-Cu).[27] Notably, characterization studies verified that the OD-Cu is made up of an interlinked network of nanocrystallites with arbitrarily aligned nanograin structures at the crystallite interfaces. Disordered surfaces at grain boundaries and defect terminations can produce metastable active sites, which are stabilized by the cascade of the nanocrystalline network.

3. **Crystal phase in NMs**: The phase engineering in NMs has been emerged as a new technique to tune the intrinsic catalytic efficiency of nanocatalysts. The latest evidence has shown that the crystal phase of metal nanostructures, arising from the different arrangement of atoms and alter the electronic structure can drastically change their characteristic properties.[28] Based on the different phases, NMs can be divided into three categories such as amorphous NMs, amorphous—crystalline, and crystal phase-based heteronanostructures, which is helpful to determine the properties and functionalities of NMs.[29]

4. **Metal-support interactions**: The metal NPs anchored on various supports like N-doped carbon, silica, carbon nanotubes, graphene oxide, graphitic carbon nitride, and metal oxides are typically stabilized by ionic or covalent interactions with an adjacent atom. The electron transfer process across the metal NPs and supports is very important for the well-functioning of NMs. The geometric and electronic behavior of NMs is heavily influenced by their microenvironment, i.e., supporting medium, coordination sites, and precise control on the dispersion of metal atoms, which determines their catalytic efficiency.[15] Surprisingly strong metal-support interaction enhanced the stability of nanocatalysts and also affected the kinetics of the

catalytic reaction. The supporting materials also reduce the chances of uncontrolled growth of nanoparticle aggregation by stabilizing the metal NPs on support.[11] Along with this, the supports and metal NPs have synergistically activated the substrates during the catalytic transformations.[17]

5. **Functionalization of NMs with molecular ligands**: This is the most common and efficient technique utilized for the modification of NMs surface. The use of molecular ligands to modify the surface of NMs is a benefit for improving the catalytic performance of NMs. Herein ligands act as a coordination medium with metal NPs to enhance the chemical functionality on the surface like —NH, —OH, and —SH. This technique can alter the microenvironment around the metal nanoparticle surface, resulting in improved interfacial electron transfer and stabilized key intermediates. The surface of NPs such as silica, graphene, titania, magnetic NPs, and metal oxides are chemically modified via various ligands such as —OH, —NH$_2$, —Cl, —SH, —COOH, —PO$_4$, organic complexes (cysteamine, amino acids), and polymers.[30] Apart from that, researchers have used a variety of other surface-modification techniques to make more accessible sites for various chemical reactions, including the integration of metal NPs with acidic and basic functional groups, other species such as zeolites, MOFs, fullerenes, click modification, mercaptopropyl, and anchoring substrates such as carboxylic acid, thiols, and thioethers. Carbon nanostructures have been identified as one of the most promising reinforcement agents for increasing the properties of polymer composites due to their exceptional features such as high mechanical stability, improved electrical conductivity, and recyclability.

Organic ligands are utilized as templating reagents, pore direct reagents, or stabilizing reagents in the synthesis of SMNs such as P123 (mixture of polyethylene oxide and polypropylene oxide), CTAB (cetyltrimethylammonium bromide), and PANI (polyaniline). These organic compounds may obstruct catalytic sites and mass transfer pathways for reactants and products in catalytic reactions. When a particular structure is formed, these species are usually eliminated by a further thermal or mechanical process like calcination/extraction. However, surface modification of organic molecules on NPs to create organic/inorganic interfaces is beneficial to catalytic processes in various research fields.[31–33] The effectiveness of functionalization is determined by the affinity of functional head groups for the surface of NPs. In different NPs assemblies, i.e., metal NPs, quantum dots, core-shells, etc. both noncovalent and covalent interactions are important for modulating surface characteristics and constructing SMNs. Noncovalent interatomic interactions are helpful for regulating assembly/disassembly reversibility, whereas covalent interactions make it easier to assemble stable nanostructures for fine-tuning the sensitivity of chemical sensing.[34]

As previously stated, many catalytic applications required NMs with specialized functionality; nevertheless, the usage of SMNs demonstrated good catalytic performance in a variety of applications. For instance, biologically modified NMs utilized in nanomedicines as well as carbon materials modified with acidic/basic functionality are efficiently utilized in catalysis. Thus, the journey and rise of the SMNs are inspiring the next generation of chemists to tailor and design surface modulated NMs that could effectively combat the environmental challenges on the horizon of catalysis.

This book covers the various fabrication methods utilized for modifying the surface of NMs along with their applications in different fields of catalysis. In Fig. 2, schematically represented the overall outline of this book. As one of the initial chapters of the book, Chapter 1 focused on many introductory themes in connection with nanotechnology, catalysis, and changes in the surface of NMs and gives the readers important background information for a smooth transition to the rest of the chapters on the SMNs and their applications in catalysis.

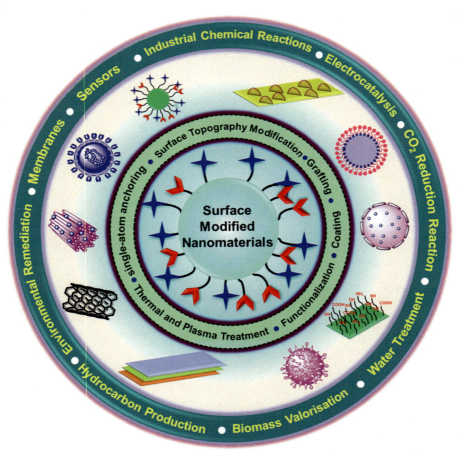

Figure 2 The schematic representation of synthesis strategies of surface modulated nanomaterials and their applications in catalysis.

The support plays an important role in surface modifications, stabilization, and the activities of NPs. Biradar and coworkers discussed the advanced methods of surface chemistry employed in NMs synthesis in Chapter 2. The next Chapter 3 deals with the synthesis of SMN. In this chapter different syntheses strategies, including thermal and plasma treatment, silane chemical methods, ligand immobilization via grafting techniques, and atom anchoring for single-atom incorporation are introduced by Vilé et al.

The zero-valent iron NPs (NZVI) rank among the best candidates for water treatment applications, demonstrating high surface activity and remarkable environmental uptake capacity. In Chapter 4, Petala et al. summarized the most frequent and successful NZVI modification strategies that can enhance multifunctional improvement of diverse NMs qualities. The NZVI reacts quickly with water as well as oxygen and has a high tendency to agglomerate, apart from this lack of stability, rapid passivation, and limited particle mobility, all of which can result in a significant decrease in catalytic activity. Surface modification to NZVI can remove disadvantages and barriers to this technology's real-field applications, which can provide new pathways for water remediation. In line with this, the surface-modified nanocatalysts display specific and efficient catalytic activities toward the removal of pollutants from wastewater. Specifically, various organic pollutants like dyes, pathogens (virus, bacteria, algae, fungi, etc.), and other inorganic impurities, especially the metal ions are being removed by a range of nanocatalysts. In Chapter 5, Das and coworkers have reviewed the extensive pieces of evidence reported for NMs modulations, which are catalytically important to drive the water treatment and hydrocarbon production processes.

Researchers are looking for a more sustainable way for the production of fuels to fulfill their energy demands. To overcome the two major concerns, i.e., diminishing the fossil fuels and raising concerns about global climate change. In this regard scientists are working to expand pathways for fuel production that rely on renewable resources, such as liquid fuels synthesis from bioderived precursors for transportation and energy source, so the cost of these fuels is lower than that of petroleum fuels. In Chapter 6, Gawande and coworkers discussed important protocols for the synthesis and applications of SMNs utilized in the production of liquid fuels. These fuels include alcohols and polyols i.e., 1,2-propanediol, 1,4-butanediol, ethylene glycol, glycerol, 1,3-propanediol, also furanics compounds (furfural and 5-hydroxymethylfurfural), biodiesel (fatty acid methyl esters, free fatty acid, etc.) from natural occurring organic biomass as a precursor. This chapter also summarized the synthesis of energy fuels like methanol, ethanol, propanol, and formic acid from hydrogenation and electrocatalytic carbon dioxide (CO_2) reduction in presence of SMNs.

Hybridizing SMNs with catalytic membrane reactors (MRs) will be innovative to develop environmental catalysis. In Chapter 7, Manwar et al. discussed the context of

SMNs and engineering catalytic MR for environmental industries. Special attention has been paid toward the environmental catalytic applications of functionalized polymeric membranes.

Semiconductor catalysts are a group of materials that possess a preferable combination of electronic structure, superb charge transport characteristics, excellent optical properties, and increased adsorption capacity, making possible its application in the electronics industry. Therefore, semiconducting NMs are of great interest in sensing-based devices such as electrochemical sensors, optical sensors, or gas adsorptive sensors. In Chapter 8, different fabrication strategies of semiconductor NMs (0D, 1D, 2D, and 3D) in terms of their sensing properties are discussed by Kuna et al.

As we all know, CO_2 is the biggest contributor to global warming, hence the scientific community has great challenges in reducing CO_2. The utilization of CO_2 is one of the most tangible solutions to the global warming problem arising from the growing level of CO_2 in the atmosphere. At present, the scientific community has garnered enormous interest in CO_2 capture and conversion into usable fuels and chemicals. Due to the abundance of CO_2, low prices, and nontoxicity of the natural source of CO_2, other industrially important chemicals might also be synthesized. In Chapter 9, Krishnan and coworkers reviewed the different strategies for the conversion of CO_2 to carbon-based valuable products using surface-modified carbonaceous NMs as catalysts.

Metal nanoparticle-based nanostructures are extremely important for the production of essential chemicals due to their ease of recycling and reusability, as well as their high activity. In Chapter 10, Natte et al. summarized the latest developments on the synthesis of supported NPs and their applications for the chemical processes involving the synthesis of pharmaceuticals and related compounds. Particularly the authors discussed the metal NPs catalyzed hydrogenations and aminations for the synthesis and functionalization of pharmaceuticals and their intermediates.

Today, NMs are being effectively commercialized in every sector around the world, slowly but steadily, and are emerging as future-based protocol technologies. Because NMs and engineered NMs with specific functional groups are the next generation smart materials with vast potential applications in future industrially important reactions such as oxidation, isomerization, polymerization and reduction, hydrogenation, and so on. In Chapter 11, Sharma et al. elucidated the present scientific objectives for successful functionalization of NMs, as well as their potential industrial catalytic applications employing scientific perceptiveness and wide scientific divination. In this chapter, the authors described the various surface-modification technique that have been reported for NMs such as graphene, silica, TiO_2, magnetic-modified surface, and such combinations could be further explored for industrially important chemical transformations.

Besides the several fabrication techniques and catalytic applications of SMNs, it is very necessary to understand the potential health and safety of these SMNs catalysts. It is very critical as the SMNs could be released into the workplace and environmental matrixes such as water and soil, potentially exposing humans and other ecological organisms and thus may present a health hazard. In Chapter 12, Kanel and coworkers reviewed the state of the knowledge on the health, safety, and economic aspects of various SMNs that are purportedly used in catalytic applications.

SMN have emerged as a new class of nanostructures with promising performance improvements over outdated micro and macro materials. As a result, it is expected to advance in the field of catalytic and chemistry-related applications. The last Chapter 13 of this book, Upare and coworkers highlighted the status of surface modification of NPs, their established industrial applications, evolving applications in the chemical transformations, and shed a light on future research prospects with respect to SMNs as industrial catalysis.

Considering the wide applications of SMNs, this book highlights the design, synthesis, and development of SMNs concerning its historical development to recent progress in the field. Also, their respective applications in catalysis including industrial catalysis, energy production, sensor, environmental remediation, photo-, and electrocatalysis. Along with this, the health, safety, and economic aspects of various SMNs that are purportedly used in catalytic applications are also discussed in this book. It gives the reader a comprehensive and cohesive picture of practically all relevant up-to-date improvements by providing an all-encompassing overview of SMNs and their applications. Surface-modulating techniques and processes are described here to enhance the efficacy of NMs that can significantly alter the performance of already used procedures and deliver fascinating consumer products to match the requirement of catalysis. This book will be useful to advanced undergraduate and graduate students, as well as industrial professionals, as a source of knowledge and a guide for their studies and research. Due to the integral nature of the topics, it will be of interest to a wide range of audiences, including industrial scientists, industrial engineers, nanotechnologists, materials scientists, physicists, chemists, pharmacists, chemical engineers, biologists, and all those involved and interested in the future frontiers of NMs. The goal of this book is to illuminate the scientific community with fabrication strategy and utilization of SMNs in the aforementioned field along with challenges associated with them and propose alternative answers for the future. To the best of our knowledge, this book has been covered various important topics in depth, and we believe that this book will be a valuable resource for readers who want to learn about the past, present, and future of surface-modification strategies for NMs with a focus on their catalytic properties and applications in catalysis, and further they can explore them to advanced and industrial catalytic applications.

References

1. Polshettiwar V, Luque R, Fihri A, Zhu H, Bouhrara M, Basset JM. Magnetically recoverable nanocatalysts. *Chem Rev* 2011;**111**(5):3036—75.
2. Jeevanandam J, Barhoum A, Chan YS, Dufresne A, Danquah MK. Review on nanoparticles and nanostructured materials: history, sources, toxicity and regulations. *Beilstein J Nanotechnol* 2018;**9**:1050—74.
3. Baig N, Kammakakam I, Falath W. Nanomaterials: a review of synthesis methods, properties, recent progress, and challenges. *Mater Adv* 2021;**2**(6):1821—71.
4. Bayda S, Adeel M, Tuccinardi T, Cordani M, Rizzolio F. The history of nanoscience and nanotechnology: from chemical-physical applications to nanomedicine. *Molecules* 2019;**25**(1):112.
5. Varma RS. Greener approach to nanomaterials and their sustainable applications. *Curr Opin Chem Eng* 2012;**1**(2):123—8.
6. Nasrollahzadeh M, Sajjadi M, Iravani S, Varma RS. Green-synthesized nanocatalysts and nanomaterials for water treatment: current challenges and future perspectives. *J Hazard Mater* 2021;**401**:123401.
7. Thomas JM. The enduring relevance and academic fascination of catalysis. *Nat Catal* 2018;**1**(1):2—5.
8. Sharma RK, Sharma S, Dutta S, Zboril R, Gawande MB. Silica-nanosphere-based organic—inorganic hybrid nanomaterials: synthesis, functionalization and applications in catalysis. *Green Chem* 2015;**17**(6):3207—30.
9. Sudarsanam P, Zhong R, Van den Bosch S, Coman SM, Parvulescu VI, Sels BF. Functionalised heterogeneous catalysts for sustainable biomass valorisation. *Chem Soc Rev* 2018;**47**(22):8349—402.
10. Gawande MB, Goswami A, Felpin FX, Asefa T, Huang X, Silva R, Zou X, Zboril R, Varma RS. Cu and Cu-based nanoparticles: synthesis and applications in catalysis. *Chem Rev* 2016;**116**(6):3722—811.
11. Xie H, Wang T, Liang J, Li Q, Sun S. Cu-based nanocatalysts for electrochemical reduction of CO_2. *Nano Today* 2018;**21**:41—54.
12. Liu L, Corma A. Metal catalysts for heterogeneous catalysis: from single atoms to nanoclusters and nanoparticles. *Chem Rev* 2018;**118**(10):4981—5079.
13. Bai ST, De Smet G, Liao Y, Sun R, Zhou C, Beller M, Maes BUW, Sels BF. Homogeneous and heterogeneous catalysts for hydrogenation of CO_2 to methanol under mild conditions. *Chem Soc Rev* 2021;**50**(7):4259—98.
14. Wieszczycka K, Staszak K, Woźniak-Budych MJ, Litowczenko J, Maciejewska BM, Jurga S. Surface functionalization — The way for advanced applications of smart materials. *Coord Chem Rev* 2021;**436**:213846.
15. Sharma RK, Yadav S, Dutta S, Kale HB, Warkad IR, Zbořil R, Varma RS, Gawande MB. Silver nanomaterials: synthesis and (electro/photo) catalytic applications. *Chem Soc Rev* 2021. https://doi.org/10.1039/D0CS00912A.
16. Varma RS. Greener and sustainable trends in synthesis of organics and nanomaterials. *ACS Sustainable Chem Eng* 2016;**4**(11):5866—78.
17. Astruc D. Introduction: nanoparticles in catalysis. *Chem Rev* 2020;**120**(2):461—3.
18. Imaoka T, Kitazawa H, Chun WJ, Yamamoto K. Finding the most catalytically active platinum clusters with low atomicity. *Angew Chem Int Ed* 2015;**54**(34):9810—5.
19. Ross MB, De Luna P, Li Y, Dinh CT, Kim D, Yang P, Sargent EH. Designing materials for electrochemical carbon dioxide recycling. *Nat Catal* 2019;**2**(8):648—58.
20. Zhang ZC, Xu B, Wang X. Engineering nanointerfaces for nanocatalysis. *Chem Soc Rev* 2014;**43**(22):7870—86.
21. Burgener S, Luo S, McLean R, Miller TE, Erb TJ. A roadmap towards integrated catalytic systems of the future. *Nat Catal* 2020;**3**(3):186—92.
22. Zaera F. Nanostructured materials for applications in heterogeneous catalysis. *Chem Soc Rev* 2013;**42**(7):2746—62.
23. Yang CH, Nosheen F, Zhang ZC. Recent progress in structural modulation of metal nanomaterials for electrocatalytic CO_2 reduction. *Rare Metals* 2020;**40**(6):1412—30.
24. Schouten KJP, Kwon Y, van der Ham CJM, Qin Z, Koper MTM. A new mechanism for the selectivity to C_1 and C_2 species in the electrochemical reduction of carbon dioxide on copper electrodes. *Chem Sci* 2011;**2**(10):1902—9.

25. Luo W, Nie X, Janik MJ, Asthagiri A. Facet dependence of CO_2 reduction paths on Cu electrodes. *ACS Catal* 2016;**6**(1):219–29.
26. Shih YJ, Wu ZL, Lin CY, Huang YH, Huang CP. Manipulating the crystalline morphology and facet orientation of copper and copper-palladium nanocatalysts supported on stainless steel mesh with the aid of cationic surfactant to improve the electrochemical reduction of nitrate and N_2 selectivity. *Appl Catal B* 2020;**273**:119053.
27. Verdaguer-Casadevall A, Li CW, Johansson TP, Scott SB, McKeown JT, Kumar M, Stephens IEL, Kanan MW, Chorkendorff I. Probing the active surface sites for CO reduction on oxide-derived copper electrocatalysts. *J Am Chem Soc* 2015;**137**(31):9808–11.
28. Cheng H, Yang N, Lu Q, Zhang Z, Zhang H. Syntheses and properties of metal nanomaterials with novel crystal phases. *Adv Mater* 2018;**30**(26). 1707189.
29. Chen Y, Lai Z, Zhang X, Fan Z, He Q, Tan C, Zhang H. Phase engineering of nanomaterials. *Nat Rev Chem* 2020;**4**(5):243–56.
30. Georgakilas V, Tiwari JN, Kemp KC, Perman JA, Bourlinos AB, Kim KS, Zboril R. Noncovalent functionalization of graphene and graphene oxide for energy materials, biosensing, catalytic, and biomedical applications. *Chem Rev* 2016;**116**(9):5464–519.
31. Kwon SG, Krylova G, Sumer A, Schwartz MM, Bunel EE, Marshall CL, Chattopadhyay S, Lee B, Jellinek J, Shevchenko EV. Capping ligands as selectivity switchers in hydrogenation reactions. *Nano Lett* 2012;**12**(10):5382–8.
32. Niu Z, Li Y. Removal and utilization of capping agents in nanocatalysis. *Chem Mater* 2014;**26**(1):72–83.
33. Zhu W, Chen Z, Pan Y, Dai R, Wu Y, Zhuang Z, Wang D, Peng Q, Chen C, Li Y. Functionalization of hollow nanomaterials for catalytic applications: nanoreactor construction. *Adv Mater* 2019;**31**(38): 1800426.
34. Lim SI, Zhong CJ. Molecularly mediated processing and assembly of nanoparticles: exploring the interparticle interactions and structures. *Acc Chem Res* 2009;**42**(6):798–808.

CHAPTER 1

New frontiers for heterogeneous catalysis: surface modification of nanomaterials

Anil Kumar Nallajarla, Shajeeya Amren Shaik and Anandarup Goswami

Division of Chemistry, Department of Sciences and Humanities, Vignan's Foundation for Science, Technology and Research (VFSTR, Deemed to be University), Guntur, Andhra Pradesh, India

1. Introduction: catalysts, nanomaterials (NMs) and nanocatalysts (NCs)

Traditionally, "catalysts" can be defined as a species capable of altering the rate of the reaction without being converted during the process.[1] The concept of catalysis was first introduced by Swedish chemist J.J. Benzelius in his annual report in 1836 in which he proposed the presence of a "catalytic force" to act on substrates to bring out homogeneous or heterogeneous transformations in the presence of a chemical entity, the characteristics and properties of which remain intact after the transformation.[2] Since then, the field of catalysis has witnessed a very sharp growth and catalysis has now became an integral part of research in both academia and industries.[3,4] While the conventional chemical transformations utilizing metal-based catalytic systems have dominated the research efforts,[1,5] the non-metal-based catalytic systems have also gained significant interest in recent times.[6,7]

Primarily, the catalysts are divided into two categories depending on the state of catalyst and substrate: (1) homogeneous and (2) heterogeneous. The homogeneous catalysts are those which remain in the same phase as substrates (e.g., soluble metal complexes, etc.) whereas heterogeneous catalysis involves different phases of catalysts and substrates. As far as the historical developments are concerned, probably the first theory about homogeneous catalysis was reported by Charles Benned and Nicolas Clement when they developed intermediate compound theory based on their observation on catalytic effects of nitrogen oxides for the synthesis of sulfuric acid.[8,9] In case of heterogeneous catalysts, the first observation of a heterogeneous metal-catalyzed chemical reaction of two gases was reported by Sir Humphry Davy who quoted the observation as "discovery of a new and curious series of phenomena." While all these experimental findings helped in developing the field of catalysis, the possibility of applying these for industrial applications was first recognized in 1831 by Phillips, Jr. during manufacturing of sulfuric acid.[10] Since then the path of catalysis has come a long way and based on the recent survey, the global catalyst market size was around USD 33.9 billion in 2019 and is estimated to grow at a growth rate of 4.4% from 2020 to 2027.[11]

Though significant achievements were observed in both classes, the respective fields of catalysis have their own advantages and disadvantages. For example, while homogeneous catalysts exhibit higher activity and solubility, the separation of catalyst and recyclability limits their wider applications.[12] On the other hand heterogeneous catalysts could often be recovered and recycled but their poorer activity and selectivity (especially under harsh reaction conditions) poses significant challenges. Initially, the development of different catalytic systems was concentrated on metal-based systems (primarily transition metals) and how to improve their reactivity and selectivity for specific chemical transformations. In that context, several homogeneous and heterogeneous catalytic systems were developed and utilized in academia and industry. However, when the concept of sustainability and green chemistry started being associated with catalysis, more active efforts were engaged to develop greener variants of the catalytic systems with environmentally benign protocols.[13] Since the realization that the involvement of catalysts is essential in order to develop sustainable processes, the relationship between catalysts and nanomaterials (NMs) has become stronger to provide the desirable solutions to the development of stable, recyclable, efficient, and selective nanomaterial-based catalytic systems.

The use of NMs for catalytic purposes is definitely considered as one of the biggest triumphs in the area of catalysis.[14] By definition, NMs are a class of materials, size of which ranges from 1 to 100 nm in at least one dimension. While the existence of the NMs can be traced back to origin of the universe, probably the first use of artificial NMs in the form of carbon-soot could be associated when human invented fire.[15] Since then, the progress about the utilization of NMs in various applications such as electronic, biomedical, optical, catalytic, and so forth has gone to an extent where it is nearly impossible not to encounter with a NMs in modern-day lives.[16,17] The exponential growth in the areas of nanoscience and nanotechnology can be attributed to the size- and shape-dependent properties, high surface area to volume ratio, quantum confinement effect, and last but not the least, the tunability of the surface properties of the NMs.[18,19] In addition, the fundamental understanding about the origin of the unique properties and the structure-activity relationships has also improved significantly in last two decades due to the advancement of sophisticated instrumentations and computer power.[20–22] These unique properties coupled with modern facilities opened up new avenues for the synthesis and applications of NMs and in that context, the catalytic entities using NMs, often termed as "nanocatalysts" (NCs), has indeed transformed the conventional landscape of catalysis.[23]

As mentioned earlier, in case of homogeneous catalysis, both the substrate and catalyst are in the same phase whereas heterogeneous catalysis requires the substrate and catalyst to be in different phases. While the activity and the selectivity of the catalyst is generally high in case of homogeneous catalysis, the recovery of the catalyst is challenging. On the other hand, heterogeneous catalysts can potentially overcome the challenges related to reusability of the catalysts but due to phase difference between the reactants and the catalyst, the poor activity and the selectivity appears as significant concerns. Thus, any catalytic option where an appropriate balance between the two i.e., where the recyclability can be achieved without loss of the activity/selectivity, is highly desirable and encouraged among the catalytic fraternity. In that context, owing to their unique physical and chemical properties, NMs as catalysts offer the promise of fulfilling those criteria, hence making them most sought-after alternatives in recent times[24] (Fig. 1.1A).

In general, metal nanoparticles (MNPs) such as Cu, Pd, Au, Ag etc., and metal-based NPs such as metal-oxides, sulfides, carbides, and so forth are prevalent in literature as NCs for various organic reactions including reduction, oxidation, coupling, etc.[25,26] Among these systems also, the ease of synthesis, cost, stability, and compatibility of the catalytic systems along with the extent of recyclability of catalyst primarily influence the selection of the final catalytic NMs. For example, while the expensive Pd NPs are conventionally used for various coupling reactions including C−C, C−N, and C−S reactions to yield the corresponding products with high yield and selectivity, Cu NPs catalyzed processes may often be preferred because of the low cost and high abundance of the NCs.[27] The metal complexes can also be grafted on the surface of the nanosupport such as silica, titania, and so forth. In that case, the surface of the nanosupport can be initially modified with different organic moieties and/or ligands[28] and the metal ions are then allowed to bind to the ligands/organic moieties connected to nanosupport. While the metal complex works as an efficient catalyst, the nanosupport helps to make the removal of the

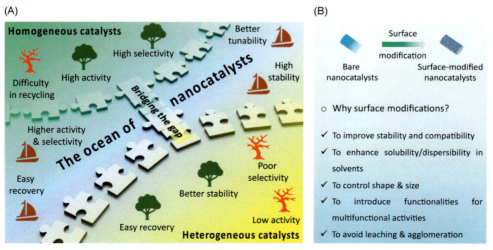

Figure 1.1 (A) Advantages of nanocatalysts in comparison to homo- and heterogeneous catalysts. (B) The need of surface functionalization of nanomaterials.

catalysts at the end of the reaction simpler and easier.[29,30] Alternatively the surface of the NMs could just simply be modified to improve their activity, stability, and compatibility.[31,32] All these surface modification techniques have become instrumental to modulate the characteristics and the activities of NCs and hence demand a special place during any discussion of NCs including this book. In addition to this brief overview on catalysis and NMs, the subsequent subsections will highlight importance of surface modifications (Fig. 1.1B), syntheses of NCs with special relevance to surface-functionalization strategies and selected applications of surface-modified NCs. At the end, the content of the books along with a concluding remarks emphasizing the present status and future direction is also presented.

2. Synthetic strategies of NCs

Since NCs are essentially the NMs which are employed for the catalytic applications, their preparation methods do not deviate much from conventional synthetic strategies of NMs (Fig. 1.2). In general, the synthesis of NMs can be classified into two categories: (1) top-down and (2) bottom-up approaches.[33–35] In case of top-down approach, the NMs are synthesized from the bulk through various "cutting" strategies, which allow fragmentation of the bulk into the materials at the nanoscale. On the other hand, the bottom-up approach involves the preparation of the NMs from their atomic or molecular precursors. Among the top-down procedures, ball-milling, lithography, and so forth are considered to be the top-choices, whereas in case of bottom-up procedures, sol-gel, and co-precipitation methods have gained popularity because of the ease of synthesis and availability of molecular precursors. The possibility of classifying these processes further

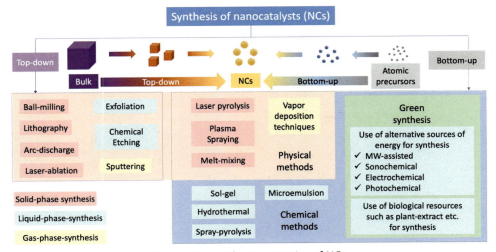

Figure 1.2 Synthetic strategies of NCs.

based on the physical and chemical methods have also garnered interest among the researchers as often they provide valuable information regarding the feasibility/success of the procedure depending on the availability of bulk/precursor materials and reaction conditions. In addition, as the development of sustainable choices are increasingly becoming essential in every sectors, several greener synthetic approaches are being adopted for the preparation of NMs/NCs. For example, various synthetic processes such as sonochemical, microwave (MW)-assisted, light-induced, electrochemical, etc. have been reported which primarily involve alternative sources of energies.[36–39] On the other hand, due to the vast availability of the green resources, several plant-derived bioresources have been utilized either as reducing agents and/or stabilizing agents for the preparation of primarily metal-based NPs with different shapes and sizes.[40–42] Moreover, in the context of reutilizing the waste for various applications, contemporary research areas also focus on the development of sustainable greener catalytic processes and/or utilization of waste-derived NMs as NCs.[43,44] While all these approaches have their own advantages and disadvantages, the primary goal remains the same, i.e., controlled synthesis of NMs with a definite shape and size and other interesting properties which enable them to be exploited for various applications.

3. Need for surface functionalization

Despite the significant advances in the syntheses of NMs/NCs, often the experimental results do not correlate well with the expected outcome. For instance, even the synthetic procedures lead to the desired NPs, the stability of the NPs often becomes an issue which raises questions on their potential for long-term applications. Though the detailed understanding about the factors affecting the stability of the NPs is yet to be fully explored, primarily the interaction between the surface atoms and the surrounding environment often alters the surface morphology and compositions, leading to changes in the stability as well as the other characteristics of the NPs.[45] In addition, the compatibility of the NMs/NCs with the surface and the other reaction conditions have also limited the practical utility.[46,47] For example, to improve solubility/dispersibility of the NCs in the solvent and/or to control their hydrophobicity/hydrophilicity, often further surface modifications become essential. Moreover, due to the sintering/agglomeration/leaching of the NCs under catalytic reaction conditions, the precise control on shape and size especially after the first cycle of the reaction has become challenging.[48] As the mechanistic details about the synthetic procedures are being understood due to the advancement of sophisticated instrumentation and collaborative interdisciplinary research endeavors, it is becoming clearer that simple alterations of the preparation procedures may not be sufficient to address those issues.[49–51] In fact, in that context, the use of surface stabilizing/modifying agents is becoming imperative, which is also reflected into the novel design and synthesis of NPs with controlled shape/sizes and stability. In view of the present scenario related to NCs, these surface modifications not only have helped to synthesize desired NPs with specific size, morphology and properties, but they have also been shown to control the reactivity and selectivity of the catalytic procedures.[52,53] In addition, the smart design and execution of these modifications can also allow incorporation of additional functionalities as surface groups (even the complimentary or orthogonal ones) so that the final materials can be utilized as multifunctional NMs.

4. Types of NMs and their functionalization procedures

The "surface modification or functionalization" refers to a certain modification of the NPs' surface by introducing various "modifying agents" to obtain them with high stability and/or controlled morphology and size, and/or additional functionalities. Depending on the preparation methods as well as the types of NPs, the strategies of surface modifications may vary. For example, they can be coupled with the original synthetic strategies in which the synthesis of the NPs is carried out in presence of surface modifying agents where the growth kinetics of the process and the thermodynamic stability of the surface-modified NPs dictate the overall success of the process.[54,55] In other case, the surface modifications have been implemented in a post-synthetic way where the surface of the NMs are functionalized with different functionalities after the synthesis of NMs.[56,57] The selection between these types of processes are primarily dictated by the types of NMs to be synthesized along with choice of precursors and procedures.

Since the types of NMs has been given the prime importance in selecting the strategies for surface modifications, the classification of NMs in relation to surface modifications seems timely (Fig. 1.3). As opposed to conventional classification of NMs which mainly divides NMs based on their dimensions, in the case of surface-modified NMs, the composition plays a major role. In that context, primarily four types of materials can be considered: (1) metal NPs, (2) metal-oxides or similar NPs, (3) carbon-based NPs, and last but not the least (4) composite NPs. In case of metal NPs, the surface of the NPs is often functionalized using small molecules to improve the stability.[58] For example, surface-modified Au NPs are often synthesized using solution-based reduction process in the presence of organic thiol groups which provides better stability and morphological control due to the strong interaction between NPs' surface and thiol groups.[59,60] In addition, a selective and optimized "etching" of these thiol groups from supported Au NPs has been shown to improve the catalytic properties of the NPs due to better availability of catalytic centers whereas, residual thiol groups on the surface has been reported to maintain the size, shape, and stability of the NPs.[61] Not only small molecules, several polymeric materials have also been utilized to protect the surface of the NPs as well as to maintain the unique structure and morphology.[62,63] In most of these cases, the synthesis is carried out in the presence of the surface-stabilizing agents which are often called as "capping agents" due to their ability to cap

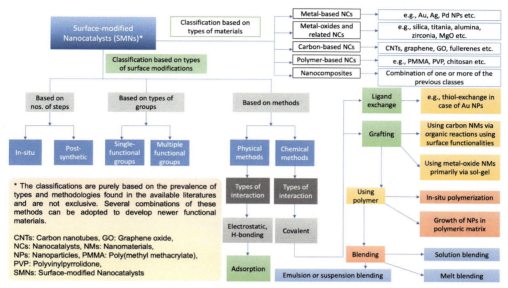

Figure 1.3 Classification of SMNs based on different parameters.

the NPs in a particular shape/size.[64,65] In addition, recently bio-derived synthetic procedures are employed, where the bio-derived precursors act as both reducing as well as capping agents for the metal salts.[66,67] In case of metal-oxides such as silica, titania, zirconia, and so forth, several sol-gel techniques have been adopted, which primarily lead to very robust and stable NMs with higher surface area.[68–70] Interestingly, these procedures generate the surfaces with abundant alkoxy or alkanol groups which can be further functionalized with various types of molecules using additional steps. In recent times, this methodology have been widely adopted to "graft" ligand molecules on these surfaces which were utilized to anchor several metal-based catalytic entities (either in the form of metal NPs or just simple catalytically active metal complexes).[71–73] Moreover, this strategy is also used extensively to prepare magnetically recyclable NCs where the magnetic core is first coated with silica-based shell which is then functionalized with the desired NCs.[74–76] Another example where these post-synthetic procedures are widely used includes the surface modifications of carbon-based NMs. In this case, the degree of unsaturation and the amount/types of surface functional groups of carbon-based NMs dictate the success of post-synthetic modifications.[77–81] It is quite evident that with suitable design, all these surface functionalization steps can potentially help to integrate more than one type of NMs or even combine NMs with the bulk materials to make composite structures. The synthesis of hybrid structures depends on the design of the connection approaches which are primarily governed by utility of these materials for specific applications. For example, covalent linkage between a carbon-based NMs and ceramic materials help to improve the strength and chemical and thermal stability of the composite.[82–84] In case of NCs, the connectivity between surface-functionalized carbon-based materials and metal NPs often offers better catalytic performance due to the synergistic effect.[85–87]

As far as the types of surface functionalization is concerned, there are different ways to classify them (Fig. 1.3). For example, based on the no. of steps involved, the functionalization could be carried out either in an in situ fashion or via multistep synthesis where a post-synthetic step is added at the end. Depending on the types of functionalities to be incorporated, the functionalization protocols could vary (e.g., to attach ligand molecule on the surface to anchor metal or metal-based NPs). In addition, any control on spatial distribution of functionalities is also preferred for the synthesis of multifunctional materials. Since the beginning of the historical development to the recent examples of the surface functionalization techniques, there are two major ways by which these strategies can be classified: (1) physical methods and (2) chemical methods (recently biological methods also have emerged as a separate category, despite being primarily a chemical method).[88] The selection of all these methods is primarily dependent on interaction (or the type of bond formation) between the bare NMs with the surface modifiers. In case of physical methods, adsorption dominates the research efforts and coulombic/electrostatic and/or H-bonding interactions are commonly reported.[89,90] One of the examples of this category is the adsorption of ionic or neutral surfactant(s) on the surface of metal NPs to improve their colloidal stability and dispersibility/ solubility in various solvents.[91] Several polymeric coating materials have also been exploited in this case.[92] In case of chemical methods, they are often preferred where covalent interactions are desired. In case of ligand-stabilized-MNPs, ligand substitution on pre-functionalized NPs with a different ligand have been reported.[93] As mentioned earlier, in case of thiol-capped Au NPs, other thiol-based ligands have been explored to replace the existing ligands on the surface of Au NPs to change the properties of ligand-capped Au NPs.[59,60] While the types of interactions are still a topic of research, the common theories center around covalent bonds between Au NPs and thiol ligands as evidenced by various spectroscopic techniques.[58,94] In case of surface functionalization of metal-oxides and other inorganic oxides like silica, several functionalities have been attached on the surface of NMs primarily via sol-gel method, which involves hydrolysis and condensation of the precursors of surface modifiers. In case of silica and other metal-oxides, the silanol groups on the surface act as anchoring sites where the guest precursor molecules can interact in order to be functionalized on the surface.[68,70,95] In fact, this strategy has been generalized for various number of inorganic oxide NPs such as Fe-oxides, titania, zirconia, and even binary ones.[96,97] Apart from the inorganic systems, carbon-based NMs are also covalently functionalized utilizing the surface groups using various kinds of organic reactions. The main difference between this type of functionalization steps in comparison to the ones for metal-oxides is that in this case, the various organic transformations can be employed depending on the functionalities on the surface whereas in case of metal-oxides, the steps of functionalization predominantly involve sol-gel method, governed by a combination of hydrolysis and condensation steps. For example, fullerenes could be functionalized using Diels−Alder reaction[98,99] and the oxygen-based functionalities on the surface of graphene oxide (GO), carbon nanotubes (CNTs) help to graft other functional moieties via various reactions namely epoxide ring opening, coupling, and so forth.[80,100−102] In terms of surface functionalization, polymeric substances are extensively used as surface coating of NPs and in this case various types of blending (solution, emulsion, melt) as well as in situ polymerization techniques have been used to coat polymeric layers on the NP surface.[103−106] In most cases the nature of interaction between NPs and polymers is designed to be covalent ones but other physical interactions are also prevalent in literature depending on the choice of precursors and NPs.[105] In fact, the suitable combination of these functionalization strategies allows development of several hybrid composite materials with novel properties. These polymeric hybrid materials show exceptional mechanical and thermal stability and other conducive properties for various application ranging from catalysis, environmental remediation, and so forth.[107−109]

Though "core-shell" NMs should technically be considered as surface-functionalized NMs, historically, it has often been classified as a separate class depending on the extent of coverage of the core by the shell and the unique interactions the core has with the shell part, often leading to improved stability and synergistic properties. In addition, the surface of the core-shell can also be further functionalized depending on the requirements.[110−112] The interested readers are advised to go to excellent review by Gawande et al. in order to explore more about this kind of materials along with their properties and catalytic applications.[113] Irrespective of the strategies and/or the types of NMs, a careful design and utilization of surface modification step has become instrumental for synthesizing NMs with improved stability, novel properties and applications. It is undeniable that the surface modification plays a pivotal role in providing the stability of the NPs by stopping them from agglomeration, leaching, and so forth. In fact, this has been proven to be more important in case of NCs where recyclability of the catalysts with unchanged composition, structural integrity and the activity becomes essential.[114,115] Keeping that in mind, some recent representative examples are presented in a systematic manner to highlight the importance of surface-functionalized NMs as NCs in the next section.

5. Selective applications of surface-functionalized NCs

The surface functionalization of NMs renders the resultant materials with unique properties and superior stability than their unfunctionalized counterparts which eventually allow them to be explored in various applications. NCs which are synthesized using surface functionalization steps are primarily explored in different areas of catalysis like catalytic organic transformations, electrocatalysis, photocatalysis, and so forth. In addition, recently, surface-functionalized NCs are widely used in environmental applications such as dye degradation, water purification, etc. In case of utilizing the functionalized NMs for catalytic organic transformations, often the surface properties of the parent support NMs are modified with suitable ligands/metal complexes, followed by installation of metal NPs and the surface-modified NCs are used in several reactions including oxidation, reduction, coupling (C—C, C—H, C—N, C—O etc.) among many others.[116,117] For example, the surface silanol groups on the MCM-41 were modified with bidentate nitrogen-based ligands using 3-(2-aminoethylamino) propyl trimethoxy silanes and subsequently treated with Cu(I) salt to incorporate Cu(I), complexed with the surface-bound ligands (Fig. 1.4A).[118] After thorough characterization using various techniques, the NC was used for C—N coupling reaction in which it showed high catalytic activity with good substrate scope. The catalysts were recycled up to 10 cycles without any significant loss of the reactivity. In another example, in order to improve the recyclability of the expensive Pd NPs, Rohleder et al. coated palladium nanocubes with mesoporous silica and used core-shell catalyst for C—C Heck coupling reactions (Fig. 1.4B).[119] The presence of mesopores on the shell allowed the reactants to interact with the Pd nanocubes at the core but the silica coating limited the aggregation and sintering and improved the recyclability. The catalysts exhibited good to excellent reactivity toward a variety of substrates and could be isolated via centrifugation before using for the next cycle. The NC could be reused up to 8 times and even after eighth cycle, both the integrity and the reactivity of the catalysts were maintained.

During last two decades, the magnetic nanaocatalysts have gained immense popularity primarily due to the recovery of the catalytic materials using external magnet, thereby

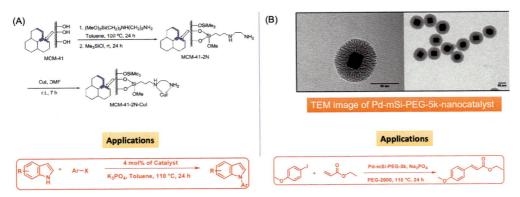

Figure 1.4 (A) Synthesis of surface-functionalized amine-grafted Cu(I) nanocatalysts for the C—N coupling reaction. (B) TEM images of the mesoporous-silica-coated Pd-nanocubes (top) and the C—C coupling reactions catalyzed by them in PEG medium. ((A) Adopted from Ref. 118 with permission, 2013 Elsevier. (B) Adopted from the Ref. 119 with permission, 2020, CC-BY RSC).

improving its recyclability.[74,115,120,121] While in most cases magnetically separable Fe-oxides are used as support, grafting of a catalytic species directly onto Fe-oxides is often found challenging. To circumvent that issue, coating of a silica and/or carbon-based materials has been proven to be successful. In that regard, Chitosan-bound pico-naldehyde Cu complex on Fe_3O_4 support was synthesized using multiple steps which included (1) the synthesis of iron-oxide NPs with surface $-OH$ groups, (2) grafting of bioderived carbon-based chitosan onto the surface to incorporate amine functionalities, (3) imine formation between the surface amines and the 4-picolinaldehyde, and finally (4) complexation of Cu(II) precursors with surface ligands (Fig. 1.5A).[122] The NCs were thoroughly characterized using various material characterization techniques and applied in the catalytic oxidation of sulfides into sulfoxides. The catalytic materials after the completion of reaction were simply removed from the mixture using external magnet and the recovered materials were recycled up to four times. The latest addition into the field of surface-functionalized NCs is "single-atom" catalysts where the size of the active catalytic species can be reduced to atomic level to achieve the highest reactivity.[123,124] Nearly for all the reported cases, they have been supported on various NMs. For instance, recently Zboril and coworkers designed a cyanographene support where the protruded

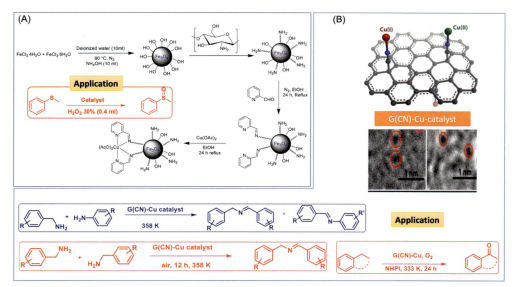

Figure 1.5 (A) Grafting of a Cu-catalyst onto the chitosan-coated magnetic NPs via piconaldehyde ligand and its use in the oxidation of sulfides and (B) Schematic structure and HRTEM image of Cu(I)-single-atom catalysts on cyanographene and their applications in oxidative coupling and benzylic oxidation. ((A) Adopted from Ref. 122 with permission, 2018, AIChE. (B) Adopted from Ref. 125 with permission, 2019 Wiley).

dangling cyano groups could be utilized to anchor Cu(I) and Cu(II) single atoms on the surface and the combination was effectively used for oxidative homocoupling of substituted benzylamine as well as for oxidation of benzylic C—H bonds with excellent yields and selectivity (Fig. 1.5B).[125] Based on the available reports[126–128] in this category, the field of single-atom catalysis is expected to explode in the coming years, where more surface modifications will be explored to stabilize the single atoms onto the support materials.

In addition to catalytic organic transformations, the NMs also have been widely exploited for electro- and photocatalytic applications. In fact, the progress of NCs in these sectors has revolutionized the field of sustainable development beyond the conventional organic transformations. In case of electrocatalysis, the surface-modified carbon-based NMs (such as graphene, CNTs etc.) have been extensively used for various electrocatalytic applications including hydrogen evolution reaction (HER), oxygen evolution reaction (OER), and so forth.[129] The surface modifications on these nanosystems often exhibit higher electrocatalytic performance. For instance, Pt NPs were anchored on the surface of nanocarbon films and employed them in electrocatalytic and supercapacitor applications.[130] The presence of Pt NPs on the surface helped to increase the catalytic performance where the extended unsaturated networks improved the electron transport. Among other examples, both noble-metal-based and noble-metal-free core-shell NMs have been synthesized and used for electrocatalytic water splitting where synergistic effect stemming from electronic modulation due to unique core-shell interaction is attributed for their enhanced catalytic performance.[131] In case of photocatalytic transformations, TiO_2 NMs have been extensively explored as photocatalysts for water splitting and other environmental applications (such as dye and pollutant degradation, etc.).[132,133] The recent examples also showed that TiO_2 surface can be functionalized with various other species and the resultant material exhibits improved catalytic performance. For example, hematite-decorated TiO_2, synthesized using simple hydrothermal process could be used in photocatalytic water splitting.[134] This catalysts helped to overcome the certain disadvantages of TiO_2 nanoparticles (such as low absorption in the visible region, fast electron-hole recombination rate etc.). Moreover, additional metal doping on this combination with various surface dopants were also developed which displayed superior photocatalytic properties. For example, Huerta-Flores et al. developed a series of heterostructured α-Fe_2O_3—TiO_2:X (X = Co, Cu, Bi) photoanodes and studied the effect of metal doping on the photocatalytic performance for water splitting (Fig. 1.6A).[135] The study showed that the performance changed in the order Bi > Cu > Co and the improvement in photoconversion and charge separation were attributed to high conductivity and carrier concentration leading to enhancement in the electron-hole transport. In addition to TiO_2, there are other surface-modified, photoactive NMs which were also explored for other

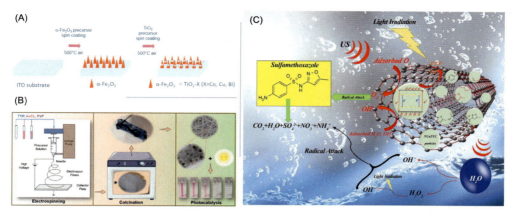

Figure 1.6 (A) Preparation of metal-doped hematite-TiO₂ nanocomposites and their superior photocatalytic behavior in water splitting, (B) Synthesis of Au NPs-decorated TiO₂ and its effective utilization in photocatalytic degradation of organic dyes and (C) An overview of sono-photo-catalytic wastewater treatment using a nanocomposite of SWCNTs and N,Cu-doped-TiO₂. ((A) Adopted from Ref. 135 with permission, 2021 Elsevier. (B) Adopted from Ref. 145 with permission, 2020 Elsevier. (C) Adopted from Ref. 147 with permission, 2020 Elsevier).

photocatalyzed reactions. Among them, graphitic carbon nitride (g-C₃N₄) has emerged as a unique photoactive materials with band-gap conducive for various photocatalytic reactions.[136–138] Due to their structural similarity with other carbon-based NMs, the material has been used in conjunction with other carbon-based NMs like graphene,[139] CNTs[140,141] and so forth for photocatalytic water splitting and environmental remediation and the results revealed higher activity of the composite material due to the unique band-gap alignment and a facile electron transfer mechanism.

Owing to this unique photocatalytic activity of various photoactive NMs, expectedly, their applications have been extended to several environmental applications such wastewater treatment, dye degradation, and so forth.[142–144] Among the different processes of wastewater treatment, degradation of dyes/pollutants using photocatalysis has opened up newer opportunities to explore the surface-functionalized NMs. For instance, pairing of titania with plasmonic Au NPs significantly improved photocatalytic decomposition of rhodamine B and methylene blue dyes under solar simulator irradiation, overcoming the disadvantages of pristine titania (Fig. 1.6B).[145] These were synthesized by electrospinning technique, followed by calcination at 500 °C. The presence of Au NPs played an important role to enhance photocatalytic activity in presence of UV irradiation. In another example, a composite material ZSM-5/TiO₂/Ni was fabricated by immobilizing TiO₂ onto ZMS-5 zeolite surface by sol-gel method, followed by incorporating nickel NPs on the surface to enhance the photocatalytic activity.[146] Methylene blue dye was chosen as model for wastewater pollutant and UV irradiation assisted photocatalytic degradation of dye removal was studied. A composite of functionalized single-walled carbon nanotubes (SWCNTs) and N,Cu-codoped-titania was reported to act as a sono-photocatalyst for wastewater treatment (Fig. 1.6C).[147] Sulfamethoxazole was taken as a model contaminant of waste water released from pharma industries and in comparison to other reference samples, N, Cu codoped TiO$_2$@functionalized SWCNTs showed significant improvement in their degradation activity when employed as a heterogenous sono-photocatalyst. The results indicated that the nanocomposite had excellent catalytic ability under UV-coupled ultrasonic irradiation for the decomposition of non-biodegradable compounds and pharmaceutical industrial effluents. In another instance, bio-derived polymeric material chitosan was functionalized with zero-valent metal Fe(0) and ZnO and the combination has also been used for adsorption of toxic metal ions from wastewater.[148] All these representative methods are essentially only a fraction of the reported literature to clearly indicate the potential of surface-functionalized NMs for various catalytic applications.

6. Characterization of surface-functionalized NCs

While the major focus of the present chapter/book has been given to the surface-functionalization strategies of the NCs and their applications, without a brief mention of the instrumental techniques, used in order to characterize the NMs, the discussion seems incomplete. After the synthesis of the expected catalytic materials, the information about shape, size and morphology of the NMs can be obtained using various microscopic techniques among which scanning electron microscopy (SEM) and transmission electron microscopy (TEM) remain the primary ones.[149,150] The selection criteria regarding the type of microscopic techniques to be used, depend on the resolution, dimension (2D vs. 3D), stability of NPs during measurement as well as the information, expected to obtain from the techniques. Due to the recent sophistication of the instrumental techniques, advanced microscopic techniques are also used to couple with various other facilities such as energy dispersive X-ray spectroscopy (EDX or EDS) for determining the presence of elements, high-resolution TEM (HRTEM), and high-angle annular dark-field scanning transmission electron microscopy (HAADF-STEM) for elemental mapping to find out precise size and location of the atoms/clusters even at subnanometer level, selected area (electron) diffraction (SAED) for recognizing particular crystal planes, to name a few.[151] In addition, atomic force microscopy (AFM) has been instrumental in finding out the surface pattern and has been extensively used.[152,153] In order to identify particular crystal structures, level of crystallinity, size of the crystallite and also possible defects primarily powder X-ray diffraction (PXRD) has now routinely been used.[154,155] In this case, it is worth mentioning that the ordering in mesoporous silica like MCM-41, SBA-15 could be identified using small-angle X-ray scattering (SAXS).[156] In the present context of surface-functionalized NCs, other properties such as distribution of surface charge, size of the agglomerated particles can often be obtained from static- and/or dynamic light scattering (SLS or DLS) experiments.[157] The surface area of the particles is one of the major important factors in the context of its catalytic activity and N_2-adsorption-desorption experiments are widely used to calculate the surface area, pore-sizes, pore-volumes using Brunauer–Emmett–Teller (BET) isotherm and the associated Barrett–Joyner–Halenda (BJH) methods.[158,159] The chemical nature and bonding connectivity of the surface atoms prove to be very important in establishing the understanding of surface behavior and in this regard, X-ray photoelectron spectroscopy (XPS) has contributed immensely to provide both qualitative and quantitative information regarding amount of dopants/ligands/metals, oxidation number of metals, the nature of connectivity for both metal and nonmetals etc.[160,161] Moreover, Fourier-transform infrared (FT-IR) and Raman spectroscopy have been extensively used to identify the functional groups on the surface (especially where the surface is modified via covalent attachments) and the degree of unsaturation/defects in case of carbon-based NMs respectively.[162–164] The recent advancement in solid-sate nuclear magnetic resonance (ss-NMR) has also enabled the researchers to identify the nature of functionality and can be used for nonmagnetic samples, preferably powdered ones.[165] In case of polymeric functionality, the degradation behavior as well as the stability of the polymeric scaffolds are routinely assessed using thermogravimetric analysis (TGA) and differential scanning calorimetry (DSC) measurements.[166] TGA can also be extended to metal-oxide and carbon-based NMs to ascertain their stability and integrity.[167] As far as the specific interaction between support and catalytic entity and/or catalytic species with substrates are concerned, often various computational techniques for theoretical calculations have been used. In addition, the use of artificial intelligence (AI), machine learning (ML) approach, and so forth have started being utilized to find out the optimized performance of NCs in context of reaction parameters. While this list provides just a glimpse of the routine techniques that are being employed for the characterization of NCs, sophisticated advanced features among these techniques are presently exploited to provide complementarity among the techniques. From the catalysis standpoint, the characterization parameters regarding the reactions such as measuring the conversion at the different time interval, identification of products, and so forth have been also routinely carried out using various techniques among which gas chromatography (GC), gas chromatography mass-spectrometry (GC-MS), high performance liquid chromatography (HPLC), FT-IR, 1D, and 2D solution NMRs are worth mentioning.

7. Topics to be covered in this book

In the begining, Gawande et al. beautifully describes the importance of surface modification with respect to the applications of nanoparticles as a prelude. As Chapter 1 of the book, it aims to set a platform by providing a glance on the introductory topics to motivate the readers to go deeper into the next part of the book. While highlighting the topics, a special care has been taken to minimize the overlap with the remaining chapters. Hence, rather than having a detailed discussion on each topic, only a brief idea has been presented with the anticipation that the readers will gain a thorough knowledge on the topics in subsequent chapters. Since the focus of the book has primarily been on the development of surface modifications strategies for catalytic applications, the subsequent chapters are concentrated mainly on the two themes, namely, (1) surface modification strategies and (2) catalytic applications of the surface-modified systems. Following that thought, in Chapter 2 of the book, historical developments as well as the present status of the fundamental concepts of surface chemistry, which are essential to identify the various surface interactions as well as the modification strategies, is provided. Chapter 3 aims to extend those knowledge to identify the synthetic strategies of the surface-modified NMs in a more systematic and elaborative way. These two chapters coupled with the present chapter are expected to build the necessary foundation for the readers to allow them to have a smooth journey for the remainder of the book where catalytic applications are highlighted. Rather than discussing every possible NC applications, a special emphasis has been given to the contemporary research areas while choosing the utility of the surface-functionalized NMs. In relation to that, Chapter 4 summarizes the present concerns and challenges regarding removal of toxic effluents using NCs and how surface-modified NMs (specifically zero-valent iron nanoparticles, NZVI) can be utilized as a catalytic materials to address those issues. In Chapter 5, the contemporary approaches to show the effectiveness of the surface-functionalized NCs for water purification, hydrocarbon production and pollutant remediation is described. As the dependence of NMs for the development of various energy-related catalytic applications seems undeniable, it indeed requires special mention in this book. In that context, Chapter 6 emphasizes on the importance of surface-functionalized nanocatalysts for the production of liquid fuels. As a natural progression of that idea, Chapter 7 highlights some of the recent advances in the field of membrane technology involving surface-modified NCs with a special emphasis on environmental catalysis. The superiority of the NCs in comparison to their bulk counterpart has led to the discovery of the several novel applications, among which surface-modified NCs for sensing and electrocatalytic and photocatalytic applications (such water splitting, artificial photosynthesis, CO_2 reduction, etc.) are surely considered as front runners. All these advanced topics will be thoroughly discussed in Chapters 8 and 9. As one of the initial challenges in the area of nanocatalysis has been the uncertainty related to the exploitation of NCs for the industrial applications, current research efforts are being directed toward addressing those issues. In that respect, in Chapter 10, Rajenahally et al. summarize the recent studies where the options of NCs are explored for the syntheses of pharmaceutical candidates. On the other hand, several modern industrial applications which involve surface-modified NCs are highlighted in the Chapter 11. Since the initial days of nanoscience/nanotechnology, the primary focus has always been the development of NMs with novel structures and properties for various applications and the area of nanocatalysis is also no exception in that. However, with the realization that sustainable developments are imperative for any processes to be successful in long run, the investigations on the effects of NMs on health and environment have started to attract significant attraction. In Chapter 12, Nadagouda et al. describe the protocols and associated tools for assessing the toxicological effects of NMs on health and environment and emphasize on their safety issues, both short- and long-termsocioeconomic impacts specially in the context of sustainability. Lastly, various interdisciplinary topics are discussed in the Chapter 13 primarily focusing on the future of surface-functionalized NCs for industrial and commercial applications. As the uniqueness of the book lies in choosing the contemporary research topics and the competent expert authors related to NCs, we strongly believe that the readers especially who want to work in the area of surface-modified NCs will be surely benefitted from not only the topics discussed in the book but also from the approach by which each chapter has been written.

8. Summary: present status and future direction

Given the present status of functional NMs, it is clear that the surface functionalization of NMs plays a very significant part in determining the final outcome. The surface-functionalized products often are expected to exhibit superior structural and functional characteristics than their unmodified counterparts. The strategies chosen to modify the surface are primarily dependent on the types of surface and its interaction/compatibility with the surface modifiers. In addition, the choice and the availability of precursors, reaction conditions and also the expectation from the final surface-modified products (in terms of properties and applications) influence the selection procedures. In case of metal NPs, functionalization is mostly done using the organic ligands which can interact strongly with the metallic counterparts. As an alternating option, the biological synthesis of NMs involve the formation of metal/metal-based NPs from their precursors where the biological entities act both as reducing agents and capping agents. While in case of metal-oxide NPs, sol-gel procedures are preferred for both synthesis and functionalization steps, the utilization of unsaturation and the existing functional groups on the surface of carbon-based NMs prove beneficial to synthesize the desired surface-functionalized

NMs. In case of polymer embedded or polymer adorned NPs, polymers could be synthesized from the monomers using various procedures in the presence of NPs without disturbing the morphology, size and properties of nanoparticles. Depending on the utilization of surface-modified NMs for specific applications, more than one types of functionalities can be incorporated in a single nanosystem and various surface modification procedures can be exploited to improve the co-operativity among the individual components. Due to recent advancements of alternating energy resources, most of such functionalization strategies/methodologies are also being explored using MW-based or sono-/photo-/electrochemical methods.[37,168–170] All these strategies have expanded the scope of NMs to be used in various applications in a more sustainable way. For example, the stability, the colloidal dispersibility and compatibility of NMs in a particular reaction medium have often been questioned for long-term applications and surface modifications strategies have surely helped them improving the situations. In addition, they can often be used to install additional surface functionalities which enable the materials to have multifunctional behavior. With these advantages in hand, in the present context, the resultant NMs are widely used in various catalytic applications in the form of NCs.

The emergence of NCs can primarily be traced back to their ability in bridge the gap between homogeneous and heterogeneous catalysts by introducing the advantages of both the catalytic systems in one, thereby eliminating the drawbacks of each. While the surface modification strategies of NMs leads to improved stability, compatibility, and so forth, it can also help to alter the reactivity and selectivity of a particular reaction. It was reported that the organic transformations catalyzed by metal NPs often suffer from sintering and agglomeration problems after the first cycle. This can be controlled by the use of suitable surface-functionalized capping agents/ligands to protect the structural integrity of NPs. Moreover, they also modulate the reactivity and selectivity of the product formation and increase the possibility of recycling. In addition, the successful installation of various catalytic functionalities on the support materials also allows them to be employed for multicatalytic transformations which are otherwise difficult without the suitable surface modifications strategies. Apart from nanocatalytic organic transformations, the areas of electrocatalytic and photocatalytic transformations have been revolutionized by the advent of NMs. In addition to increasing the stability of NPs, the introduction of defects and dopants into the NMs' surface render them with unique catalytic properties including higher stability, superior activity/selectivity, and possible recyclability, due to conducive surface interactions leading to better electron transport, and lower work functions. Due to the improved catalytic properties, the applications of these surface-modified NMs are extended in various energy and environmental applications including water splitting, pollutant removal, and some of these materials have also been considered for industrial commercialization.[171]

While the above summary highlights the triumphs of nanocatalysis especially with the surface-modified ones, the field still experiences certain challenges which become the central theme of ongoing and future research works and hence are highlighted below:

(1) Most of the synthetic procedures and the subsequent surface functionalization of NMs involve bottom-up approach because of the ease in controlling the product outcome. Thus, despite having significant progress in the areas of top-down approaches, the surface functionalization steps are mainly carried out in a post-synthetic manner using the NMs synthesized from top-down approach and hence more efforts are indeed needed in order to exploit top-down approaches of surface functionalization strategies.

(2) As far as the chemistry of functionalization goes, the understanding about the control and selectivity of the functionalization and the rationale behind any observed preference of functionalization (e.g., site selectivity, etc.) still remain unclear. This has become especially important when the biomolecules are used as surface modifiers and hence require more systematic investigations.

(3) In the early days of nanocatalysis, surface modifiers and support materials were considered to be just spectators in catalysis and able to provide stability and anchoring points to the actual catalytically active species. With the advent of sophisticated instrumentation and computation power, their role in tuning the catalytic activity/selectivity has started to be identified,[172,173] but more efforts such combining the machine-learning approach with operando spectroscopic techniques[174,175] and so forth are still needed.

(4) As opposed to homogeneous catalytic systems, the batch dependency of the synthesis as well as the functionalization steps of the nanocatalytic systems pose imminent challenge toward reproducing the catalytic materials, thereby hindering the smooth transition from a lab-scale synthesis to an industrial-scale preparation.

(5) Last but not least, the evaluation of the safety and nanotoxicological effects of these nanocatalysts[176,177] deserves a much more systematic studies to move toward more sustainable design of these protocols.

This book aims to shed lights on the above-mentioned challenges in the respective chapters and offer potential solutions for the future. To the best our knowledge, until now no books are available covering these topics in detail, and we believe this book will surely be an asset to the readers if they want to know the past, present, and future of surface modification strategies of the NMs with special relevance to their catalytic properties.

Abbreviations

Abbreviations	Full form
AI	Artificial intelligence
AFM	Atomic force microscopy
BET	Brunauer—Emmett—Teller
BJH	Barrett—Joyner—Halenda
CNTs	Carbon nanotubes
DLS	Dynamic light scattering
DSC	Differential scanning microscopy
EDS	Energy dispersive X-ray spectroscopy
FT-IR	Fourier-transform infrared
GC	Gas chromatography
GO	Graphene oxide
HER	Hydrogen evolution reaction
HPLC	High performance liquid chromatography
IR	Infrared
ML	Machine learning
MCM-41	Mobil composite matter-41
MNPs	Metal nanoparticles
MS	Mass spectroscopy
MW	Microwave
nm	Nanometer
NMs	Nanomaterials
NPs	Nanoparticles
nZVI	Nanoscale zero-valent iron
OER	Oxygen evolution reaction
PMMA	Poly(methyl methacrylate)
PVA	Polyvinyl alcohol
PXRD	Powder X-ray diffraction
SEM	Scanning electron microscopy
SLS	Static light scattering
SS-NMR	Solid state nuclear magnetic resonance
SWCNTs	Single-walled carbon nanotubes
TGA	Thermogravimetric analysis
XPS	X-ray photoelectron spectroscopy
XRD	X-ray diffraction
ZMS-5	Zeolite socony mobil-5

References

1. Védrine JC. Heterogeneous catalysis on metal oxides. *Catalysts* 2017;**7**:341.
2. Robertson AJB. The early history of catalysis. *Platin Met Rev* 1975;**19**:64.
3. Anslyn EV, James TD, Sessler JL. Catalyst: academia and industry, continually blurring research roles. *Chem* 2016;**1**:173.
4. Hagen J. Future development of catalysis. In: *Industrial catalysis; A practical approach*. John Wiley & Sons, Ltd; 2015. p. 463.

5. Iwasawa Y. *Tailored metal catalysts; catalysis by metal complexes.* Springer Netherlands; 2012.
6. Bertelsen S, Jørgensen KA. Organocatalysis—after the gold rush. *Chem Soc Rev* 2009;**38**:2178.
7. Gupta S, Banu R, Ameta C, Ameta R, Punjabi PB. Emerging trends in the syntheses of heterocycles using graphene-based carbocatalysts: an update. *Top Curr Chem* 2019;**377**:13.
8. Clément N, Desormes CB. Mémoire Sur l'Outremer. *Ann Chim* 1806;**57**:317.
9. Wisniak J. Nicolas clément. *Educ Quím* 2011;**22**:254.
10. Phillips P. Manufacture of sulphuric acid. *Br Pat* 1831:6096.
11. *Catalyst market size & share, industry report*; n.d. 2020—2027. https://www.grandviewresearch.com/industry-analysis/catalyst-market [Accessed 13 February 2021].
12. de Vries JG, Jackson SD. Homogeneous and heterogeneous catalysis in industry. *Catal Sci Technol* 2012;**2**:2009.
13. North M, editor. *Sustainable catalysis: with non-endangered metals, part 1. Green chemistry series.* The Royal Society of Chemistry; 2016.
14. Polshettiwar V, Asefa T. Introduction to nanocatalysis. In: *Nanocatalysis synthesis and applications.* John Wiley & Sons, Ltd; 2013. p. 1—9.
15. Bayda S, Adeel M, Tuccinardi T, Cordani M, Rizzolio F. The history of nanoscience and nanotechnology: from chemical—physical applications to nanomedicine. *Molecules* 2020;**25**:112.
16. Gupta R, Xie H. Nanoparticles in daily life: applications, toxicity and regulations. *J Environ Pathol Toxicol Oncol* 2018;**37**:209.
17. Nasrollahzadeh M, Sajadi SM, Sajjadi M, Issaabadi Z. Chapter 4 - applications of nanotechnology in daily life. In: Nasrollahzadeh M, Sajadi SM, Sajjadi M, Issaabadi Z, Atarod M, editors. *An introduction to green nanotechnology*, vol. 28. Interface Science and Technology Elsevier; 2019. p. 113.
18. Khan I, Saeed K, Khan I. Nanoparticles: properties, applications and toxicities. *Arab J Chem* 2019;**12**:908.
19. Thangadurai TD, Manjubaashini N, Thomas S, Maria HJ. Nanomaterials, properties and applications. In: *Nanostructured materials.* Cham: Springer International Publishing; 2020. p. 11.
20. Liu R, Zhang HY, Ji ZX, Rallo R, Xia T, Chang CH, et al. Development of structure—activity relationship for metal oxide nanoparticles. *Nanoscale* 2013;**5**:5644.
21. Wang X, Yao F, Xu P, Li M, Yu H, Li X. Quantitative structure—activity relationship of nanowire adsorption to SO_2 revealed by in situ TEM technique. *Nano Lett* 2021;**4**:1679.
22. Speck-Planche A, Kleandrova VV, Luan F, DS Cordeiro MN. Computational modeling in nanomedicine: prediction of multiple antibacterial profiles of nanoparticles using a quantitative structure—activity relationship perturbation model. *Nanomed* 2015;**10**:193.
23. Asefa T, Huang X. Nanocatalysis: catalysis with nanoscale materials. In: *Handbook of solid state chemistry.* American Cancer Society; 2017. p. 443.
24. Polshettiwar V, Varma RS. Green chemistry by nano-catalysis. *Green Chem* 2010;**12**:743.
25. Wang D, Astruc D. The recent development of efficient earth-abundant transition-metal nanocatalysts. *Chem Soc Rev* 2017;**46**:816.
26. Xu Y, Chen L, Wang X, Yao W, Zhang Q. Recent advances in noble metal based composite nanocatalysts: colloidal synthesis, properties, and catalytic applications. *Nanoscale* 2015;**7**:10559.
27. Gawande MB, Goswami A, Felpin F-X, Asefa T, Huang X, Silva R, et al. Cu and Cu-based nanoparticles: synthesis and applications in catalysis. *Chem Rev* 2016;**116**:3722.
28. Neouze M-A, Schubert U. Surface modification and functionalization of metal and metal oxide nanoparticles by organic ligands. *Monatshefte Für Chem - Chem Mon* 2008;**139**:183.
29. Khan SA, Vandervelden CA, Scott SL, Peters B. Grafting metal complexes onto amorphous supports: from elementary steps to catalyst site populations via kernel regression. *React Chem Eng* 2020;**5**:66.
30. Averseng F, Vennat M, Che M. Grafting and anchoring of transition metal complexes to inorganic oxides. In: *Handbook of heterogeneous catalysis.* American Cancer Society; 2008. p. 522.
31. Wilhelm S, Kaiser M, Würth C, Heiland J, Carrillo-Carrion C, Muhr V, et al. Water dispersible upconverting nanoparticles: effects of surface modification on their luminescence and colloidal stability. *Nanoscale* 2015;**7**:1403.
32. Cheah P, Brown P, Qu J, Tian B, Patton DL, Zhao Y. Versatile surface functionalization of water-dispersible iron oxide nanoparticles with precisely controlled sizes. *Langmuir* 2021;**37**:1279.

33. Ghoranneviss M, Soni A, Talebitaher A, Aslan N. Nanomaterial synthesis, characterization, and application. *J Nanomater* 2015;**2015**:892542.
34. Kolahalam LA, Viswanath IVK, Diwakar BS, Govindh B, Reddy V, Murthy YLN. Review on nanomaterials: synthesis and applications. *Mater Today Proc* 2019;**18**:2182.
35. Kumar N, Kumbhat S. Nanomaterials: general synthetic approaches. In: *Essentials in nanoscience and nanotechnology*. John Wiley & Sons, Ltd; 2016. p. 29.
36. Bilecka I, Niederberger M. Microwave chemistry for inorganic nanomaterials synthesis. *Nanoscale* 2010;**2**(8):1358.
37. Xu H, Zeiger BW, Suslick KS. Sonochemical synthesis of nanomaterials. *Chem Soc Rev* 2013;**42**:2555.
38. Li G-R, Xu H, Lu X-F, Feng J-X, Tong Y-X, Su C-Y. Electrochemical synthesis of nanostructured materials for electrochemical energy conversion and storage. *Nanoscale* 2013;**5**:4056.
39. Bárta J, Procházková L, Vaněček V, Kuzár M, Nikl M, Čuba V. Photochemical synthesis of nano- and micro-crystalline particles in aqueous solutions. *Appl Surf Sci* 2019;**479**:506.
40. Das RK, Pachapur VL, Lonappan L, Naghdi M, Pulicharla R, Maiti S, et al. Biological synthesis of metallic nanoparticles: plants, animals and microbial aspects. *Nanotechnol Environ Eng* 2017;**2**:18.
41. Mohanpuria P, Rana NK, Yadav SK. Biosynthesis of nanoparticles: technological concepts and future applications. *J Nanoparticle Res* 2008;**10**:507.
42. Shaik SA, Sengupta S, Varma RS, Gawande MB, Goswami A. Syntheses of N-doped carbon quantum dots (NCQDs) from bioderived precursors: a timely update. *ACS Sustain Chem Eng* 2021;**9**:3.
43. Rana A, Yadav K, Jagadevan S. A comprehensive review on green synthesis of nature-inspired metal nanoparticles: mechanism, application and toxicity. *J Clean Prod* 2020:122880.
44. Salem SS, Fouda A. Green synthesis of metallic nanoparticles and their prospective biotechnological applications: an overview. *Biol Trace Elem Res* 2020:1.
45. Xu L, Liang HW, Yang Y, Yu SH. Stability and reactivity: positive and negative aspects for nanoparticle processing. *Chem Rev* 2018;**118**:3209.
46. Mukhopadhyay S, Veroniaina H, Chimombe T, Han L, Zhenghong W, Xiaole Q. Synthesis and compatibility evaluation of versatile mesoporous silica nanoparticles with red blood cells: an overview. *RSC Adv* 2019;**9**:35566.
47. Rajabi F, Karimi N, Saidi MR, Primo A, Varma RS, Luque R. Unprecedented selective oxidation of styrene derivatives using a supported iron oxide nanocatalyst in aqueous medium. *Adv Synth Catal* 2012;**354**:1707.
48. Alayoglu S. Chapter 3 - achievements, present status, and grand challenges of controlled model nanocatalysts. In: Fornasiero P, Cargnello M, editors. *Morphological, compositional, and shape control of materials for catalysis. Studies in surface science and catalysis*, vol. 177. Elsevier; 2017. p. 85.
49. Alayoglu S, Somorjai GA. Nanocatalysis II: in situ surface probes of nano-catalysts and correlative structure—reactivity studies. *Catal Lett* 2015;**145**:249.
50. Weng Z, Zaera F. Increase in activity and selectivity in catalysis via surface modification with self-assembled monolayers. *J Phys Chem C* 2014;**118**:3672.
51. Zhang H, Wang C, Sun HL, Fu G, Chen S, Zhang YJ, et al. In situ dynamic tracking of heterogeneous nanocatalytic processes by shell-isolated nanoparticle-enhanced Raman spectroscopy. *Nat Commun* 2017;**8**:15447.
52. Chung I, Song B, Kim J, Yun Y. Enhancing effect of residual capping agents in heterogeneous enantioselective hydrogenation of α-keto esters over polymer-capped Pt/Al$_2$O$_3$. *ACS Catal* 2021;**11**:31.
53. Nasrallah HO, Min Y, Lerayer E, Nguyen T-A, Poinsot D, Roger J, et al. Nanocatalysts for high selectivity enyne cyclization: oxidative surface reorganization of gold sub-2-nm nanoparticle networks. *J Am Chem Soc Au* 2021;**1**:187.
54. Ben-Sasson M, Lu X, Nejati S, Jaramillo H, Elimelech M. In situ surface functionalization of reverse osmosis membranes with biocidal copper nanoparticles. *Desalination* 2016;**388**:1.
55. Qu H, Ma H, Zhou W, O'Connor CJ. In situ surface functionalization of magnetic nanoparticles with hydrophilic natural amino acids. *Inorganica Chim Acta* 2012;**389**:60.
56. Bohara RA, Thorat ND, Pawar SH. Role of functionalization: strategies to explore potential nano-bio applications of magnetic nanoparticles. *RSC Adv* 2016;**6**:43989.

57. Ha J-M, Solovyov A, Katz A. Postsynthetic modification of gold nanoparticles with calix[4]Arene enantiomers: origin of chiral surface plasmon resonance. *Langmuir* 2009;**25**:153.
58. Rossi LM, Fiorio JL, Garcia MAS, Ferraz CP. The role and fate of capping ligands in colloidally prepared metal nanoparticle catalysts. *Dalton Trans* 2018;**47**:5889.
59. Kluenker M, Mondeshki M, Nawaz Tahir M, Tremel W. Monitoring thiol−ligand exchange on Au nanoparticle surfaces. *Langmuir* 2018;**34**:1700.
60. Caragheorgheopol A, Chechik V. Mechanistic aspects of ligand exchange in Au nanoparticles. *Phys Chem Chem Phys* 2008;**10**:5029.
61. Das S, Goswami A, Hesari M, Al-Sharab JF, Mikmeková E, Maran F, et al. Reductive deprotection of monolayer protected nanoclusters: an efficient route to supported ultrasmall Au nanocatalysts for selective oxidation. *Small* 2014;**10**:1473.
62. Bhol P, Mohanty M, Mohanty PS. Polymer-matrix stabilized metal nanoparticles: synthesis, characterizations and insight into molecular interactions between metal ions, atoms and polymer moieties. *J Mol Liq* 2021;**325**:115135.
63. Králik M, Biffis A. Catalysis by metal nanoparticles supported on functional organic polymers. *J Mol Catal Chem* 2001;**177**:113.
64. Kinge S, Bönnemann H. One-pot dual size- and shape selective synthesis of tetrahedral Pt nanoparticles. *Appl Organomet Chem* 2006;**20**:784.
65. Campisi S, Schiavoni M, Chan-Thaw CE, Villa A. Untangling the role of the capping agent in nanocatalysis: recent advances and perspectives. *Catalysts* 2016;**6**:185.
66. Saravanan A, Kumar PS, Karishma S, Vo D-VN, Jeevanantham S, Yaashikaa PR, et al. A review on biosynthesis of metal nanoparticles and its environmental applications. *Chemosphere* 2021;**264**:128580.
67. Patil S, Chandrasekaran R. Biogenic nanoparticles: a comprehensive perspective in synthesis, characterization, application and its challenges. *J Genet Eng Biotechnol* 2020;**18**:67.
68. Parashar M, Shukla VK, Singh R. Metal oxides nanoparticles via sol−gel method: a review on synthesis, characterization and applications. *J Mater Sci Mater Electron* 2020;**31**:3729.
69. Sui R, Charpentier P. Synthesis of metal oxide nanostructures by direct sol−gel chemistry in supercritical fluids. *Chem Rev* 2012;**112**:3057.
70. Esposito S. "Traditional" sol-gel chemistry as a powerful tool for the preparation of supported metal and metal oxide catalysts. *Materials* 2019;**12**:668.
71. Mallakpour S, Madani M. A review of current coupling agents for modification of metal oxide nanoparticles. *Prog Org Coat* 2015;**86**:194.
72. Deshmukh R, Niederberger M. Mechanistic aspects in the formation, growth and surface functionalization of metal oxide nanoparticles in organic solvents. *Chem − Eur J* 2017;**23**:8542.
73. Rajh T, Chen LX, Lukas K, Liu T, Thurnauer MC, Tiede DM. Surface restructuring of nanoparticles: an efficient route for ligand−metal oxide crosstalk. *J Phys Chem B* 2002;**106**:10543.
74. Gawande MB, Monga Y, Zboril R, Sharma RK. Silica-decorated magnetic nanocomposites for catalytic applications. *Coord Chem Rev* 2015;**288**:118.
75. Fatahi Y, Ghaempanah A, Ma'mani L, Mahdavi M, Bahadorikhalili S. Palladium supported aminobenzamide modified silica coated superparamagnetic iron oxide as an applicable nanocatalyst for Heck cross-coupling reaction. *J Organomet Chem* 2021;**936**:121711.
76. Esfandiari N, Kashefi M, Mirjalili M, Afsharnezhad S. Role of silica mid-layer in thermal and chemical stability of hierarchical Fe_3O_4-SiO_2-TiO_2 nanoparticles for improvement of lead adsorption: kinetics, thermodynamic and deep XPS investigation. *Mater Sci Eng B* 2020;**262**:114690.
77. Zydziak N, Yameen B, Barner-Kowollik C. Diels−alder reactions for carbon material synthesis and surface functionalization. *Polym Chem* 2013;**4**:4072.
78. Munirasu S, Albuerne J, Boschetti-de-Fierro A, Abetz V. Functionalization of carbon materials using the diels-alder reaction. *Macromol Rapid Commun* 2010;**31**:574.
79. Zhang X, Hou L, Samorì P. Coupling carbon nanomaterials with photochromic molecules for the generation of optically responsive materials. *Nat Commun* 2016;**7**:11118.
80. Punetha VD, Rana S, Yoo HJ, Chaurasia A, McLeskey JT, Ramasamy MS, et al. Functionalization of carbon nanomaterials for advanced polymer nanocomposites: a comparison study between CNT and graphene. *Prog Polym Sci* 2017;**67**:1.

81. Mallakpour S, Soltanian S. Surface functionalization of carbon nanotubes: fabrication and applications. *RSC Adv* 2016;**6**:109916.
82. Ma M, Li H, Xiong Y, Dong F. Rational design, synthesis, and application of silica/graphene-based nanocomposite: a review. *Mater Des* 2021;**198**:109367.
83. Bhanja P, Das SK, Patra AK, Bhaumik A. Functionalized graphene oxide as an efficient adsorbent for CO_2 capture and support for heterogeneous catalysis. *RSC Adv* 2016;**6**:72055.
84. Dreyer DR, Park S, Bielawski CW, Ruoff RS. The chemistry of graphene oxide. *Chem Soc Rev* 2010;**39**:228.
85. Reyes-Cruzaley AP, Félix-Navarro RM, Trujillo-Navarrete B, Silva-Carrillo C, Zapata-Fernández JR, Romo-Herrera JM, et al. Synthesis of novel Pd NP-PTH-CNTs hybrid material as catalyst for H_2O_2 generation. *Electrochimica Acta* 2019;**296**:575.
86. Shi X, Yu H, Gao S, Li X, Fang H, Li R, et al. Synergistic effect of nitrogen-doped carbon-nanotube-supported Cu–Fe catalyst for the synthesis of higher alcohols from syngas. *Fuel* 2017;**210**:241.
87. Shi W, Zhang J, Li X, Zhou J, Pan Y, Liu Q, et al. Synergistic effect between surface anhydride group and carbon–metal species during catalytic reduction of nitric oxide. *Energy Fuels* 2017;**31**:11258.
88. Dhand C, Dwivedi N, Loh XJ, Jie Ying AN, Verma NK, Beuerman RW, et al. Methods and strategies for the synthesis of diverse nanoparticles and their applications: a comprehensive overview. *RSC Adv* 2015;**5**:105003.
89. Ijaz I, Gilani E, Nazir A, Bukhari A. Detail review on chemical, physical and green synthesis, classification, characterizations and applications of nanoparticles. *Green Chem Lett Rev* 2020;**13**:223.
90. Jamkhande PG, Ghule NW, Bamer AH, Kalaskar MG. Metal nanoparticles synthesis: an overview on methods of preparation, advantages and disadvantages, and applications. *J Drug Deliv Sci Technol* 2019;**53**:101174.
91. Dixit SG, Mahadeshwar AR, Haram SK. Some aspects of the role of surfactants in the formation of nanoparticles. *Colloids Surf Physicochem Eng Asp* 1998;**133**:69.
92. Quarta A, Curcio A, Kakwere H, Pellegrino T. Polymer coated inorganic nanoparticles: tailoring the nanocrystal surface for designing nanoprobes with biological implications. *Nanoscale* 2012;**4**:3319.
93. Rambukwella M, Sakthivel NA, Delcamp JH, Sementa L, Fortunelli A, Dass A. Ligand structure determines nanoparticles' atomic structure, metal-ligand interface and properties. *Front Chem* 2018;**6**:330.
94. Wang X, Wang X, Bai X, Yan L, Liu T, Wang M, et al. Nanoparticle ligand exchange and its effects at the nanoparticle–cell membrane interface. *Nano Lett* 2019;**19**:8.
95. Kankala RK, Han YH, Na J, Lee CH, Sun Z, Wang SB, et al. Nanoarchitectured structure and surface biofunctionality of mesoporous silica nanoparticles. *Adv Mater* 2020;**32**:1907035.
96. Yarbrough R, Davis K, Dawood S, Rathnayake H. A sol–gel synthesis to prepare size and shape-controlled mesoporous nanostructures of binary (II–VI) metal oxides. *RSC Adv* 2020;**10**:14134.
97. Thiagarajan S, Sanmugam A, Vikraman D. Facile methodology of sol-gel synthesis for metal oxide nanostructures. In: Chandra U, editor. *Recent applications in sol-gel synthesis*. Rijeka: IntechOpen; 2017.
98. Kräutler B, Maynollo J. Diels-alder reactions of the [60] fullerene functionalizing a carbon sphere with flexibly and with rigidly bound addends. *Tetrahedron* 1996;**52**:5033.
99. Śliwa W. Diels-alder reactions of fullerenes. *Fuller Sci Technol* 1997;**5**:1133.
100. Wang X, Shi G. An introduction to the chemistry of graphene. *Phys Chem Chem Phys* 2015;**17**:28484.
101. Amirov RR, Shayimova J, Nasirova Z, Dimiev AM. Chemistry of graphene oxide. Reactions with transition metal cations. *Carbon* 2017;**116**:356.
102. Tasis D, Tagmatarchis N, Bianco A, Prato M. Chemistry of carbon nanotubes. *Chem Rev* 2006;**106**:1105.
103. Hore MJA. Polymers on nanoparticles: structure & dynamics. *Soft Matter* 2019;**15**:1120.
104. Balazs AC, Emrick T, Russell TP. Nanoparticle polymer composites: where two small worlds meet. *Science* 2006;**314**:1107.
105. Grubbs RB. Roles of polymer ligands in nanoparticle stabilization. *Polym Rev* 2007;**47**:197.
106. Mackay ME, Tuteja A, Duxbury PM, Hawker CJ, Van Horn B, Guan Z, et al. General strategies for nanoparticle dispersion. *Science* 2006;**311**:1740.

107. Tjong SC. Structural and mechanical properties of polymer nanocomposites. *Mater Sci Eng R Rep* 2006;**53**:73.
108. Fischer H. Polymer nanocomposites: from fundamental research to specific applications. *Mater Sci Eng C* 2003;**23**:763.
109. Fu S, Sun Z, Huang P, Li Y, Hu N. Some basic aspects of polymer nanocomposites: a critical review. *Nano Mater Sci* 2019;**1**:2.
110. Kumar KS, Kumar VB, Paik P. Recent advancement in functional core-shell nanoparticles of polymers: synthesis, physical properties, and applications in medical biotechnology. *J Nanoparticles* 2013; **2013**:672059.
111. Heuzé K, Rosario-Amorin D, Nlate S, Gaboyard M, Bouter A, Clérac R. Efficient strategy to increase the surface functionalization of core—shell superparamagnetic nanoparticles using dendron grafting. *New J Chem* 2008;**32**:383.
112. Khatami M, Alijani HQ, Nejad MS, Varma RS. Core@shell nanoparticles: greener synthesis using natural plant products. *Appl Sci* 2018;**8**:411.
113. Gawande MB, Goswami A, Asefa T, Guo H, Biradar AV, Peng DL, et al. Core—shell nanoparticles: synthesis and applications in catalysis and electrocatalysis. *Chem Soc Rev* 2015;**44**:7540.
114. Zhang S, Zhuo H, Li S, Bao Z, Deng S, Zhuang G, et al. Effects of surface functionalization of mxene-based nanocatalysts on hydrogen evolution reaction performance. *Catal Today* 2020. https://doi.org/10.1016/j.cattod.2020.02.002.
115. Taheri-Ledari R, Maleki A. Magnetic nanocatalysts utilized in the synthesis of aromatic pharmaceutical ingredients. *New J Chem* 2021;**45**:4135.
116. Chng LL, Erathodiyil N, Ying JY. Nanostructured catalysts for organic transformations. *Acc Chem Res* 2013;**46**:1825.
117. Shakil Hussain SM, Kamal MS, Hossain MK. Recent developments in nanostructured palladium and other metal catalysts for organic transformation. *J Nanomater* 2019:1562130.
118. Xiao R, Zhao H, Cai M. MCM-41-Immobilized bidentate nitrogen copper(I) complex: a highly efficient and recyclable catalyst for Buchwald N-arylation of indoles. *Tetrahedron* 2013;**69**:5444.
119. Rohleder D, Vana P. Mesoporous-silica-coated palladium-nanocubes as recyclable nanocatalyst in C—C-coupling reaction — a green approach. *RSC Adv* 2020;**10**:26504.
120. Nehlig E, Motte L, Guénin E. Magnetic nano-organocatalysts: impact of surface functionalization on catalytic activity. *RSC Adv* 2015;**5**:104688.
121. Magnetic nanocatalysts: supported metal nanoparticles for catalytic applications. *Nanotechnol Rev* 2013; **2**:597.
122. Fakhri A, Naghipour A. Fe_3O_4@chitosan-Bound picolinaldehyde Cu complex as the magnetically reusable nanocatalyst for adjustable oxidation of sulfides. *Environ Prog Sustain Energy* 2018;**37**:1626.
123. Cheng N, Zhang L, Doyle-Davis K, Sun X. Single-atom catalysts: from design to application. *Electrochem Energy Rev* 2019;**2**:539.
124. Mitchell S, Pérez-Ramírez J. Single atom catalysis: a decade of stunning progress and the promise for a bright future. *Nat Commun* 2020;**11**:4302.
125. Bakandritsos A, Kadam RG, Kumar P, Zoppellaro G, Medved' M, Tuček J, et al. Mixed-valence single-atom catalyst derived from functionalized graphene. *Adv Mater* 2019;**31**:1900323.
126. Datye AK, Guo H. Single atom catalysis poised to transition from an academic curiosity to an industrially relevant technology. *Nat Commun* 2021;**12**:895.
127. Xue Q, Zhang Z, Ng BKY, Zhao P, Lo BTW. Recent advances in the engineering of single-atom catalysts through metal—organic frameworks. *Top Curr Chem* 2021;**379**:11.
128. Gawande MB, Fornasiero P, Zbořil R. Carbon-based single-atom catalysts for advanced applications. *ACS Catal* 2020;**10**:2231.
129. Meng Y, Huang X, Lin H, Zhang P, Gao Q, Li W. Carbon-based nanomaterials as sustainable noble-metal-free electrocatalysts. *Front Chem* 2019;**7**:759.
130. Es-Souni M, Schopf D. Modified nanocarbon surfaces for high performance supercapacitor and electrocatalysis applications. *Chem Commun* 2015;**51**:13650.
131. Yin X, Yang L, Gao Q. Core—shell nanostructured electrocatalysts for water splitting. *Nanoscale* 2020; **12**:15944.

132. Nakata K, Fujishima A. TiO$_2$ photocatalysis: design and applications. *J Photochem Photobiol C Photochem Rev* 2012;**13**:169.
133. Guo Q, Zhou C, Ma Z, Yang X. Fundamentals of TiO$_2$ photocatalysis: concepts, mechanisms, and challenges. *Adv Mater* 2019;**31**:1901997.
134. Liŭ D, Li Z, Wang W, Wang G, Liú D. Hematite doped magnetic TiO$_2$ nanocomposites with improved photocatalytic activity. *J Alloys Compd* 2016;**654**:491.
135. Huerta-Flores AM, Chávez-Angulo G, Carrasco-Jaim OA, Torres-Martínez LM, Garza-Navarro MA. Enhanced photoelectrochemical water splitting on heterostructured α-Fe$_2$O$_3$-TiO$_2$: X (X = Co, Cu, Bi) photoanodes: role of metal doping on charge carrier dynamics improvement. *J Photochem Photobiol Chem* 2021;**410**:113077.
136. Qi K, Liu S, Zada A. Graphitic carbon nitride, a polymer photocatalyst. *J Taiwan Inst Chem Eng* 2020;**109**:111.
137. Wang L, Wang K, He T, Zhao Y, Song H, Wang H. Graphitic carbon nitride-based photocatalytic materials: preparation strategy and application. *ACS Sustain Chem Eng* 2020;**8**:16048.
138. Wang X, Blechert S, Antonietti M. Polymeric graphitic carbon nitride for heterogeneous photocatalysis. *ACS Catal* 2012;**2**:1596.
139. Song L, Guo C, Li T, Zhang S. C60/graphene/g-C$_3$N$_4$ composite photocatalyst and mutually-reinforcing synergy to improve hydrogen production in splitting water under visible light radiation. *Ceram Int* 2017;**43**:7901.
140. Christoforidis KC, Syrgiannis Z, Parola VL, Montini T, Petit C, Stathatos E, et al. Metal-free dual-phase full organic carbon nanotubes/g-C$_3$N$_4$ heteroarchitectures for photocatalytic hydrogen production. *Nano Energy* 2018;**50**:468.
141. Wu Y, Liao H, Li M. CNTs modified graphitic C$_3$N$_4$ with enhanced visible-light photocatalytic activity for the degradation of organic pollutants. *Micro Nano Lett* 2018;**13**:752.
142. Padmanabhan NT, Thomas N, Louis J, Mathew DT, Ganguly P, John H, et al. Graphene coupled TiO$_2$ photocatalysts for environmental applications: a review. *Chemosphere* 2021;**271**:129506.
143. Paola AD, García-López E, Marcì G, Palmisano L. A survey of photocatalytic materials for environmental remediation. *J Hazard Mater* 2012;**211**:3.
144. Li X, Xie J, Jiang C, Yu J, Zhang P. Review on design and evaluation of environmental photocatalysts. *Front Environ Sci Eng* 2018;**12**:14.
145. Tang KY, Chen JX, Legaspi EDR, Owh C, Lin M, Tee ISY, et al. Gold-decorated TiO$_2$ nanofibrous hybrid for improved solar-driven photocatalytic pollutant degradation. *Chemosphere* 2021;**265**:129114.
146. Badvi K, Javanbakht V. Enhanced photocatalytic degradation of dye contaminants with TiO$_2$ immobilized on ZSM-5 zeolite modified with nickel nanoparticles. *J Clean Prod* 2021;**280**:124518.
147. Isari AA, Hayati F, Kakavandi B, Rostami M, Motevassel M, Dehghanifard EN. Cu Co-doped TiO$_2$@functionalized SWCNT photocatalyst coupled with Ultrasound and visible-light: an effective sono-photocatalysis process for pharmaceutical wastewaters treatment. *Chem Eng J* 2020;**392**:123685.
148. Saad AHA, Azzam AM, El-Wakeel ST, Mostafa BB, El-latif MBA. Removal of toxic metal ions from wastewater using ZnO@chitosan core-shell nanocomposite. *Environ Nanotechnol Monit Manag* 2018;**9**:67.
149. Su D. Advanced electron microscopy characterization of nanomaterials for catalysis. *Green Energy Environ* 2017;**2**:70.
150. Mao C. Introduction: nanomaterials characterization using microscopy. *Microsc Res Tech* 2004;**64**:345.
151. Smith DJ. Chapter 1 characterization of nanomaterials using transmission electron microscopy. In: *Nanocharacterisation (2)*, vols. 1—29. The Royal Society of Chemistry; 2015.
152. Li TD, Chiu HC, Ortiz-Young D, Riedo E. Nanorheology by atomic force microscopy. *Rev Sci Instrum* 2014;**85**:123707.
153. Ikai A, Afrin R, Saito M, Watanabe-Nakayama T. Atomic force microscope as a nano- and micrometer scale biological manipulator: a short review. *Semin Cell Dev Biol* 2018;**73**:132.
154. Mourdikoudis S, Pallares RM, Thanh NTK. Characterization techniques for nanoparticles: comparison and complementarity upon studying nanoparticle properties. *Nanoscale* 2018;**10**:12871.

155. Whitfield P, Mitchell L. X-ray diffraction analysis of nanoparticles: recent developments, potential problems and some solutions. *Int J Nanosci* 2004;**3**:757.
156. Zienkiewicz-Strzalka M, Skibińska M, Pikus S. Small-angle X-ray scattering (SAXS) studies of the structure of mesoporous silicas. *Nucl Instrum Methods Phys Res B* 2017;**411**:72.
157. Kaszuba M, McKnight D, Connah MT, McNeil-Watson FK, Nobbmann U. Measuring sub nanometre sizes using dynamic light scattering. *J Nanoparticle Res* 2008;**10**:823.
158. Bardestani R, Patience GS, Kaliaguine S. Experimental methods in chemical engineering: specific surface area and pore size distribution measurements—BET, BJH, and DFT. *Can J Chem Eng* 2019;**97**:2781.
159. Song D, Li J. Effect of catalyst pore size on the catalytic performance of silica supported cobalt fischer—tropsch catalysts. *J Mol Catal Chem* 2006;**247**:206.
160. Baer DR. Guide to making XPS measurements on nanoparticles. *J Vac Sci Technol A* 2020;**38**:031201.
161. Korin E, Froumin N, Cohen S. Surface Analysis of nanocomplexes by X-ray photoelectron spectroscopy (XPS). *ACS Biomater Sci Eng* 2017;**3**:882.
162. Katumba G, Mwakikunga B, Mothibinyane T. FTIR and Raman spectroscopy of carbon nanoparticles in SiO_2, ZnO and NiO matrices. *Nanoscale Res Lett* 2008;**3**:421.
163. Mester L, Govyadinov AA, Chen S, Goikoetxea M, Hillenbrand R. Subsurface chemical nanoidentification by nano-FTIR spectroscopy. *Nat Commun* 2020;**11**:3359.
164. Capeletti LB, Zimnoch JH. Fourier transform infrared and raman characterization of silica-based materials. In: Stauffer MT, editor. *Applications of molecular spectroscopy to current research in the chemical and biological Sciences*. Rijeka: IntechOpen; 2016.
165. Hirsh DA. *Solid-state NMR of complex nano- and microcrystalline materials* [Ph.D. Thesis]. Canada: University of Windsor; 2017.
166. Schindler A, Doedt M, Gezgin S, Menzel J, Schmölzer S. Identification of polymers by means of DSC, TG, STA and computer-assisted database search. *J Therm Anal Calorim* 2017;**129**:833.
167. Corcione CE, Frigione M. Characterization of nanocomposites by thermal analysis. *Materials* 2012;**5**:2960.
168. Kumar A, Kuang Y, Liang Z, Sun X. Microwave chemistry, recent advancements, and eco-friendly microwave-assisted synthesis of nanoarchitectures and their applications: a review. *Mater Today Nano* 2020;**11**:100076.
169. McGilvray KL, Decan MR, Wang D, Scaiano JC. Facile photochemical synthesis of unprotected aqueous gold nanoparticles. *J Am Chem Soc* 2006;**128**:15980.
170. Yanilkin VV, Nasretdinova GR, Kokorekin VA. Mediated electrochemical synthesis of metal nanoparticles. *Russ Chem Rev* 2018;**87**:1080.
171. Olveira S, Forster SP, Seeger S. Nanocatalysis: academic discipline and industrial realities. *J Nanotechnol* 2014;**2014**:324089.
172. Zhang Y, Fu D, Xu X, Sheng Y, Xu J, Han Y. Application of operando spectroscopy on catalytic reactions. *Curr Opin Chem Eng* 2016;**12**:1.
173. Song B, Yang TT, Yuan Y, Sharifi-Asl S, Cheng M, Saidi WA, et al. Revealing sintering kinetics of MoS_2-supported metal nanocatalysts in atmospheric gas environments via operando transmission electron microscopy. *ACS Nano* 2020;**14**:4074.
174. Timoshenko J, Jeon HS, Sinev I, Haase FT, Herzog A, Roldan Cuenya B. Linking the evolution of catalytic properties and structural changes in copper—zinc nanocatalysts using operando EXAFS and neural-networks. *Chem Sci* 2020;**11**:3727.
175. Rück M, Garlyyev B, Mayr F, Bandarenka AS, Gagliardi A. Oxygen reduction activities of strained platinum core—shell electrocatalysts predicted by machine learning. *J Phys Chem Lett* 2020;**11**:1773.
176. Ma C, White JC, Dhankher OP, Xing B. Metal-based nanotoxicity and detoxification pathways in higher plants. *Environ Sci Technol* 2015;**49**:7109.
177. Ong KJ, MacCormack TJ, Clark RJ, Ede JD, Ortega VA, Felix LC, et al. Widespread nanoparticle-assay interference: implications for nanotoxicity testing. *Plos One* 2014;**9**:1.

CHAPTER 2

Fundamental concepts on surface chemistry for nanoparticle modifications

Ankush V. Biradar[1,2], Saravanan Subramanian[1,2], Amravati S. Singh[1,2], Dhanaji R. Naikwadi[1,2], Krishnan Ravi[1,2] and Jacky H. Advani[1,2]
[1]Inorganic Materials and Catalysis Division, CSIR-Central Salt and Marine Chemicals Research Institute, Bhavnagar, Gujarat, India; [2]Academy of Scientific and Innovative Research (AcSIR), Ghaziabad, Uttar Pradesh, India

1. Introduction

Humans have been creative and often invent new concepts in science and technologies for the betterment of socioeconomic life. The term nano is not new in both nature and science, but the recent scientific surmounts provided the researchers with new tools better to understand the physical, chemical, and biological phenomena better; that naturally occur at nanoscales. Understanding the concepts of nanoparticles and their interactions may considerably improve and strengthen the academia and industrial sectors. Nanoparticles (NPs) and their diverse products are ubiquitous and form an integral part of many consumer products and lifestyles. The size of the nanoparticles is significantly smaller and lies between the world of everyday matters, described by Newton's laws of motion, and atom/single molecules governed by quantum mechanics. In modern science, nanotechnology is one of the most promising areas of research to customize the traditional bulk material for its most exemplary applications. The material properties of nanoparticles are mainly dependent on the shape, size, morphology, and distribution. NPs are commonly classified as zero-dimensional (0D), one-dimensional (1D), two-dimensional (2D), and three-dimensional (3D) (Fig. 2.1) based on their structural features. NPs have unique physicochemical properties and provide a versatile scaffold for diverse applications.[1] Depending on the physical and chemical characteristics, these NPs can be classified as shown in Fig. 2.1.

In recent years, enormous efforts are being made globally to develop ecofriendly nanoparticle synthesis, which produces environmentally benign products using green nanotechnology. The synthetic method has a significant impact on the physical, chemical, and biological properties of the nanoparticles. The synthesis of stable nanoparticles with appropriate dimensionality in a single step remains valuable and quite challenging.

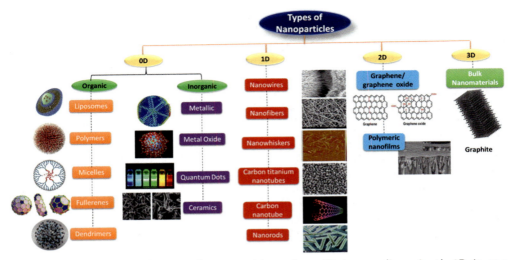

Figure 2.1 Classification of types of nanoparticles; where 0D is zero-dimensional, 1D is one-dimensional, 2D is two-dimensional, and 3D is three-dimensional.

Thus, many efforts have been devoted toward eliminating or minimizing the superfluous processing steps and allow the synthesis to proceed at controlled physiological conditions such as temperature, pH, pressure, particle shape, pore size, environment, and proximity.

A variety of methods are known for the NP synthesis, broadly classified into two different approaches, i.e., (1) Bottom-up and (2) Top-down approach.[2,3] The top-down approach, as the name suggests, uses bulk materials that are reduced to produce the NPs via chemical, physical or mechanical methods. While in the case of the bottom-up approach, atoms or molecules generally grow to produce the NPs. These methods are further subclassified based on the reaction conditions, operations, and adopted procedures. Although there are several ways accessible for NPs synthesis, single-step environmentally benign alternative methods are highly desired for several applications. The use of algae, microorganisms, enzymes, fungus, living cells, tissue, and plants or plant extracts serves this purpose. Such environmentally friendly synthetic strategies have paved the way for an important branch in nanotechnology.[4-6] Nanoparticles find excessive application in medicine, pharmaceuticals, food, and agroindustries.[6-9] Hence, developing greener and more convenient methods that produce ecofriendly, nontoxic, and ecofriendly nanoparticles is focused. Several factors influence the quality and quantity of the synthesized nanoparticles for their potential use in different applications. The synthesized NPs generally a undergo the purification process (till a satisfactory level is reached) in order to perform desired surface modifications. The surface modification influences on to (1) stabilize the aggregative NPs in a solvent medium, (2) passivate the reactive NP, and (3) promote the assembly of NP. A wide range of surface types like silica, a carbon-based material, polymer, and others materials with different functionalities are used for surface modification.[10–16] The stabilization of the nanoparticles through surface modifications helps to avoid the leaching of the nanoparticles. On the other hand, the high surface area and surface energies of the naked NPs make them kinetically unstable. Thus, the stabilization of these NPs using a surfactant is highly desirable.

The stabilization of the NPs can be achieved using several strategies as shown in Fig. 2.2. Among the diverse methodologies, one of the effective methods is electrostatic stabilization.[17–19]

The stabilization could be achieved by the anions or cations from the precursor, which associates with the nanoparticles and forms a neutral electrical double layer. The neutrality of the double layer arises due to the arrangement of oppositely charged layers in parallel to each other. The first layer, also known as the Stern layer, is generated from the formal charges on the surface of the particles, while the second one is formed by the Coulombic attraction of the charged ions to the surface charge. This double layer formed on the surface of the nanoparticles develops the same surface charge, which prevents aggregation via the Coulombic repulsion. The use of polymers and surfactants inhibits the aggregation, thus resulting in the steric stabilization of the nanoparticles. This combined instruction of the steric and electrostatic effects is known as electrostatic stabilization.

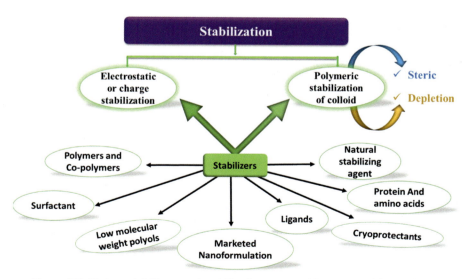

Figure 2.2 Various stabilizers and stabilization are used in nanoparticle synthesis.

Polymers are well-known steric stabilizers for nanoparticle synthesis. A repulsive force neutralizes the attraction between the particles owing to the van der Waals force.[16] Polymeric ligands are classified into two major groups: (1) polymers with multiple stabilizing functionalities, and (2) a single stabilizing functional group located either at the initial or terminal position of the polymer chain. Both direct polymerization and living polymerization techniques have been practiced to prepare polymers bearing the functional groups. Along with the adapted nanoparticle synthetic method, the choice of a specific polymeric ligand also dramatically controls the size and shape of the resulting NPs.[16] The polymers stabilize the nanoparticles either via steric stabilization or via depletion stabilization. The effectiveness of stabilization is known by its protective value. For instance, the protective values of poly(N-vinyl-2-pyrrolidone) (PVP), poly(vinyl alcohol), poly(acrylamide), poly(acrylic acid), and poly(ethyleneimine) are 50.0, 5.0, 1.3, 0.07, and 0.04, respectively. Polymer ligands also displayed a control on retarding the oxidation of NPs, which is a major issue with different metallic NPs.

1.1 Steric stabilization

Steric stabilization is the phenomenon where the polymer or surfactant forms a coating on the nanoparticles inducing a repulsive force, thus separating the particles from each other. Generally, the polymer or the surfactant possesses two different functionalities, i.e., the hydrophilic functional group and the hydrophobic alkyl chains. The hydrophilic group reacts with the nanoparticles, while the alkyl group is projected freely toward the solution. This creates a steric hindrance that stabilizes the nanoparticles from aggregation. High salt concentration does not have any effect on such stabilization. Moreover, the stabilization is also unaffected under conditions where zeta potential is nearly zero.

1.2 Use of silica nanoparticles for the stabilization of nanoparticles

The availability of several tunable functional groups and a high surface-to-volume ratio proves to be beneficial for the synthesis of a wide variety of structural materials having a broad range of applications. For instance, the stabilization of Au NPs with varied sizes was done by reducing the Au (III) on the hemiaminal functionalized silica.[10] The size of the NPs was found to be dependent on both the concentration and the time (see Fig. 2.3). These stabilized NPs, when used as a catalyst, gave excellent oxidation of both the linear alkanes and the alkylbenzenes.

Another excellent method was used to synthesize silica encapsulated Pd nanoparticles (5 and 20 nm) decorated on monodispersed silica nanospheres of 250 nm and synthesized in a stepwise manner.[12] Wherein the Pd NPs grown on slice nanospheres were encapsulated by a silica shell to obtain a core-shell type material. The shell was then etched

Fundamental concepts on surface chemistry for nanoparticle modifications 33

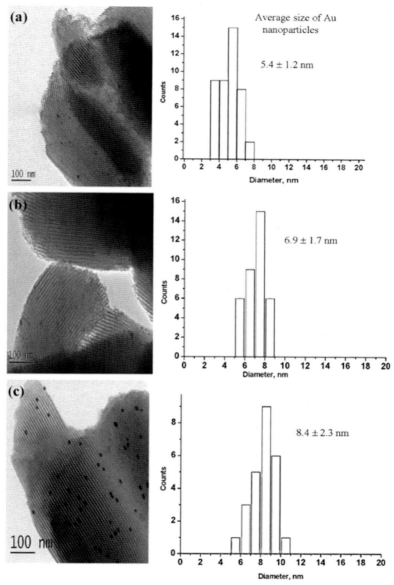

Figure 2.3 Transmission electron microscope (TEM) images and Au nanoparticle size distribution of Au/SBA-15 catalysts (A) 0.01 mM, (B) 0.01 mM and (C) 1.0 mM of HAuCl$_4$ solutions. *(Reprinted with permission from Ref. 10. Copyright 2012 Elsevier Ltd).*

to produce a nanoporous shell which allows the reactants to diffuse toward the nanoparticles. This core-shell structure was used as a catalyst for room-temperature hydrogenation of various substrates with high turnover numbers. The NPs also catalyzed the C—C coupling reactions effectively with negligible leaching and aggregation yielded high recyclability.

Further, catalytically active porous Pt NPs were synthesized via a simple transmetallation reaction. The simplest method for the synthesis of these porous was carried out by Pt NPs by transmetallation reactions between sacrificial nickel nanoparticles and chloroplatinic acid (H_2PtCl_6) in solution and in the constrained environment at the air-water interface.[11] Overall, the Pt porous nanoparticles exploiting the replacement reaction between Ni nanoparticles and H_2PtCl_6 causes a newly modified surface (Fig. 2.4). These porous Pt NPs were used as the catalyst for the hydrogenation reactions. Furthermore, SiO_2 sphere-supported Pd NPs were synthesized (Fig. 2.5), then coated with a hollow nanoporous zirconia shell.[12] The nanoporous shell allowed the easy passage of the reactants toward the active sites while the shell itself stabilized the NPs aggregation/sintering

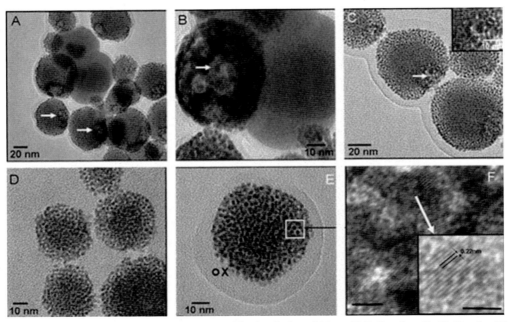

Figure 2.4 HRTEM images of a monolayer of hydrophobized Ni on H_2PtCl_6 after (A, B) 10 min, (C) 30 min, and (D) 45 min of transmetallation. HRTEM images of a monolayer of hydrophobized Ni on H_2PtCl_6 after 45 min. The highlighted region in (E) is shown in (F), which shows the fringes representing Pt NPs. *(Reprinted with permission from Ref. 11. Copyright 2009 Wiley-VCH GmbH, Weinheim).*

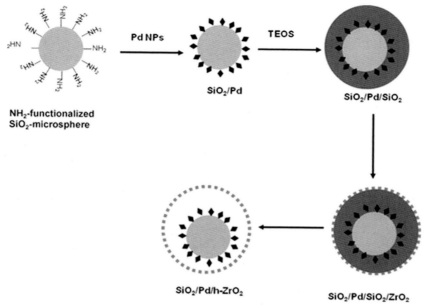

Figure 2.5 Schematic representation of the synthesis of $SiO_2/Pd/h$-ZrO_2 core-shell-shell nanostructures. *(Reprinted with permission from Ref. 12. Copyright 2011 Wiley-VCH GmbH, Weinheim).*

(Fig. 2.4). These catalytic nanoreactors were found highly efficient in reducing different olefins and nitro groups.

The acetalization of glycerols with different aldehydes over nano MoO_3 particles dispersed on the silica support was carried.[20] The MoO_3/SiO_2 catalysts with varying MoO_3 loadings were prepared by a sol-gel technique using ethyl silicate-40 and ammonium heptamolybdate as silica and molybdenum source, respectively. The finely dispersed mesoporous silica was used to synthesize molybdenum nanoparticles. Moreover, the nanosized MoO_3 particles on the mesoporous silica support was substantiated by TEM analysis Fig. 2.6, which showed the MoO_3 the size of 1–20 nm in range. The silica support helped in the stabilization of molybdenum nanoparticles even after several catalytic cycles of activity. Overall, the silica played an important role in retaining the activity and selectivity.

Further, one-pot synthesis of ultrasmall MoO_3 nanoparticles supported on SiO_2, TiO_2, and ZrO_2 nanospheres was performed, and these materials were utilized as catalysts for the epoxidation reaction. The molybdenum oxide nanoparticles supported on various (SiO_2, TiO_2, or ZrO_2) nanospheres were produced in one-pot using a reverse micelle method. The nanoparticles were uniformly protected by SiO_2, TiO_2, or ZrO_2, and activity was enhanced. Furthermore, TEM images confirmed the uniform dispersion of MoO_3 nanoparticles (1.5–4 nm) on silica surface (275 nm) as shown in Fig. 2.7.[13]

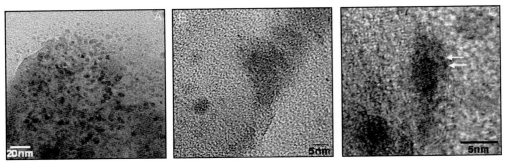

Figure 2.6 TEM micrograph of 1% MoO$_3$/SiO$_2$ prepared by sol-gel at different magnification and focused on a different area. *(Reprinted with permission from Ref. 20. Copyright 2009 Royal chemical society).*

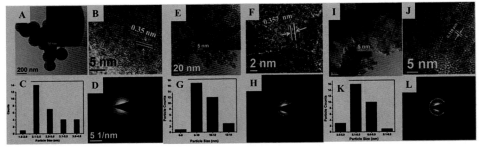

Figure 2.7 HRTEM images of (A) MoO$_3$/SiO$_2$; (B) MoO$_3$/SiO$_2$ particles showing the fringes corresponding to (021) plane of MoO$_3$; (C) histogram showing the distribution of particles; (D) SAED pattern of the MoO$_3$/SiO$_2$ catalyst; HRTEM images of (E) MoO$_3$/TiO$_2$ and (F) MoO$_3$/TiO$_2$ particles showing the fringes corresponding to (101) plane of MoO$_3$/TiO$_2$; (G) histogram showing the distribution of particles; (H) SAED pattern of the MoO$_3$/ZrO$_2$ catalyst; HRTEM images of (I) MoO$_3$/ZrO$_2$; (J) MoO$_3$/ZrO$_2$ particles showing the fringes corresponding to (101) plane of ZrO$_2$; (K) histogram showing the distribution of particles; and (L) SAED pattern of the MoO$_3$/SiO$_2$ catalyst. *(Reprinted with permission from Ref. 13. Copyright 2014 Royal chemical society).*

The bimetallic Au—Cu@SiO$_2$ catalyst was synthesized via a reverse micelle (water-in-oil) method.[21] The first step of the synthesis involved the formation of reverse micelles. The metal salts were added to these micelles, followed by hydroxide addition to producing corresponding hydroxides. To this solution, a 3-aminopropyltriethoxysilane (APTES) was added to functionalize the formed hydrated mixed metal oxide composite, followed by the addition of a silica source to produce discrete reverse micelles. The amine groups of APTES attached strongly to the mixed metal oxide surface while the organosilane part of APTES attached to the silica surface. Finally, the micelle is disrupted to produce the final silica-coated alloy nanoparticles as shown in Fig. 2.8.

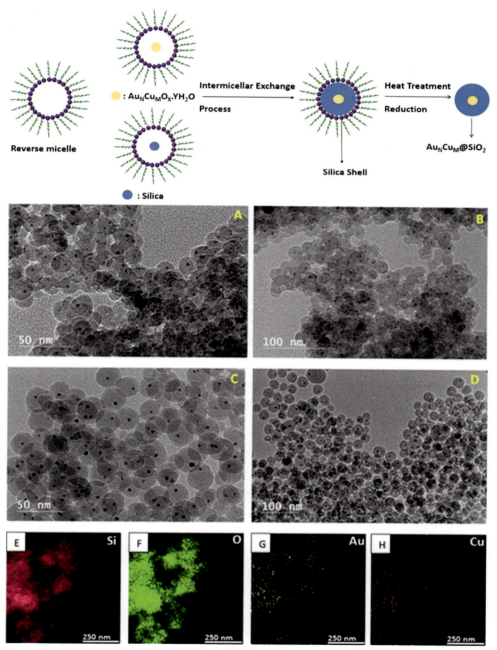

Figure 2.8 (Top) Synthesis of Au–Cu nanocomposite catalyst; (A)Au@SiO$_2$ catalyst; (B) Au@SiO$_2$ catalyst after pretreatment; (C) Au–Cu@SiO$_2$ catalyst; (D) Au–Cu@SiO$_2$ catalyst after pretreatment; and (E–H) elemental mapping of the Au–Cu@SiO$_2$ catalyst. *(Reprinted with permission from Ref. 21. Copyright 2018 American Chemical Society).*

Further, a modified Stober method is presented to synthesize Fe@SiO$_2$ nanoparticles (NPs) using 3-aminopropyltriethoxysilane (APTES) for modification to render with metal particle surface attuned with silica NPs.[22] The insulating silica layer on FeO nanoparticles is helpful to enhance the resistivity of the polymer nanocomposites. However, the completely encapsulated FeO NPs on silica is essential to acquire a high reflection loss and broad absorption bandwidth for microwave absorption. The detailed synthesis and surface modification of Fe@SiO$_2$ is illustrated in Fig. 2.9.

1.3 Carbon materials in the stabilization of nanoparticles

The stabilization of the nanoparticles on a carbon nanosphere[23] are generally performed in three different routes. The synthetic methodologies followed were as follows:

(1) The first route involved simple metal salt impregnation on the mesoporous aminophenol formaldehyde (APF) nanospheres followed by pyrolysis at high temperature to produce metal NP-loaded N-doped carbon nanospheres. (Fig. 2.10A)

(2) In the second route, the metal salt was mixed with APF solution, followed by the addition of hexadecyltrimethylammonium bromide (HDTMB), water, ethanol, and ammonium solution, and finally TEOS to produce APF@metal@silica nanospheres. These spheres were further pyrolyzed at high temperatures to produce Yolk-shell type N-doped carbon@silica nanospheres. The removal of silica using HF gave the hollow N-doped carbon nanospheres. (Fig. 2.10B)

(3) The third route involved the dispersion of APF@silica core-shell nanoparticles in aq. HCl solution to remove all the surfactants. These surfactant-free nanoparticles were then redispersed in a mixture of water, ethanol, HDTMB, and ammonium solution, followed by the addition of aminophenol and formaldehyde. This solution was again redispersed in different metal salt (based on requirement) solution to produce metal 1-APF@silica@M2-APF nanoparticles, which was further pyrolyzed to give bicoreshell type material, i.e., metal-1 doped N-doped carbon@metal-2 doped N-doped carbon (Fig. 2.10C).

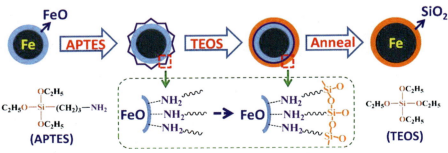

Figure 2.9 Stabilization of Fe@SiO$_2$ NPs using silica shell. *(Reprinted with permission from Ref. 22. Copyright 2011 American Chemical Society).*

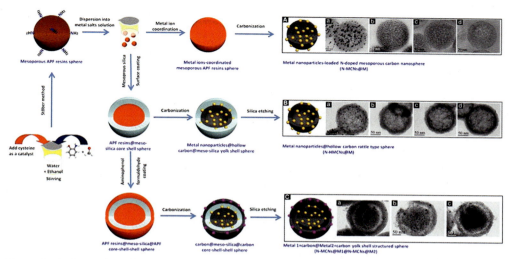

Figure 2.10 General synthetic routes for the synthesis of different carbon materials stabilized metal nanoparticles. *(Reprinted with permission from Ref. 23. Copyright 2016 American Chemical Society).*

The cost-effective and scalable Co−SCN/RGO electrocatalyst is synthesized through the modified Hummers' method. The synthesized NPs as a unique Co−(N, S)−C active site for overall water splitting in alkaline media. Remarkably, the optimized Co−SCN/RGO catalyst displayed exceptional hydrogen evolution reaction and oxygen evolution reaction activities. The stabilization of nanoparticles has also been achieved by different carbon dopants apart from nitrogen. For instance, cobalt nanoparticles stabilized by S-doped C_3N_4 were synthesized by treatment with H_2S followed by impregnation and finally, high-temperature heat treatment. This material was mixed with graphitic oxide under ethanoic conditions, followed by separation and drying to produce the final S-doped C_3N_4 over RGO (Fig. 2.11).[24]

Generally, the nanoparticles are affected by high temperature treatment that results in sintering. Most of the time, the stabilization of nanoparticles using various carbon materials face the issue of nanoparticle segregation via sintering, if not stabilized effectively. To reduce such sintering, Pt alloy nanoparticles stabilized on carbon were encapsulated in a silica shell to produce stable nanoparticles which do not undergo sintering during heat treatment.[25] The addition of the silica layer on Pt NPs provides a physical barrier to prevent direct contact of the Pt_3Co_1 core with adjacent particles. The synthetic steps involved are shown in Fig. 2.12. Initially, the PVP stabilized alloy nanoparticles are coated with silica. These coated nanoparticles are coated on the carbon support, calcined, and etched with HF to produce stable alloy nanoparticles.

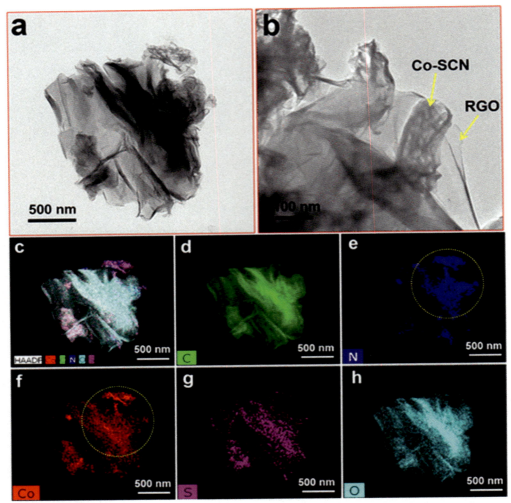

Figure 2.11 (A,B) TEM images of Co-SCN-RGO catalyst; (C–H) HAADF elemental mapping of the constituent elements. *(Reprinted with permission from Ref. 24. Copyright 2019 American Chemical Society).*

Co NPs with <15 nm particle size supported on the graphitic *N*-doped carbon were synthesized by a simple impregnation-pyrolysis strategy shown in Fig. 2.13.[14] A Co salt was impregnated on a curcubit[6] uril (CB[6]) and pyrolyzed to produce the desired catalyst. Pd (2.56 wt%) supported on CB[6] was synthesized by the simple impregnation method. This catalyst was utilized for the catalytic nitroarenes hydrogenation at milder conditions. This catalyst effectively hydrogenated different nitroarenes in the THF-water solvent at

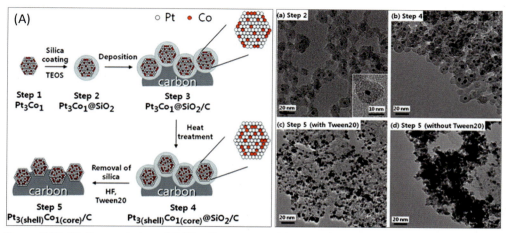

Figure 2.12 (A) Encapsulation of PdCo alloy nanoparticles via silica encapsulation process; HRTEM images of some synthetic steps: (a) PVP coated alloy nanoparticles, (b) calcined silica-coated alloy supported on carbon, (c) alloy nanoparticles after silica etching in the presence of Tween 20, and (d) alloy nanoparticles after silica etching in the absence of Tween 20).

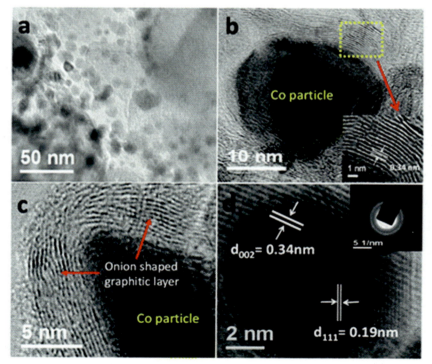

Figure 2.13 TEM images of the Co@g-C/N-800 catalyst. *(Reprinted with permission from Ref. 14. Copyright 2018 Royal Chemical Society).*

50°C with a high turnover frequency. CB⁶ showed a cuboid morphology; however, Pd(0)CB⁶ showed the uniform distribution of Pd nanoparticles as shown in Fig. 2.13.

In similar lines, Pd/Ni nanoparticles supported on the *N*-doped carbonaceous materials were synthesized using a simple pyrolysis method.[15,26] The metal salts were complexed with chitosan, followed by pyrolysis under the reducing atmosphere (700 °C). The Pd@N—C nanocatalyst was synthesized by carbonization of the Pd-chitosan complex at 800°C under a nitrogen atmosphere. However, the TEM analysis showed the uniform distribution of the Pd nanoparticles on the surface of N-doped carbon with an average size in the range of 2—6 nm Fig. 2.14. The resulting materials were found to be Pd NPs supported on the *N*-doped carbon flakes (Pd NPs@N—C), while the other was found to be Ni NPs supported on the *N*-doped CNTs. Both the catalysts were utilized for the hydrogenation reactions. The Pd NPs@N—C was found to be efficient for the room-temperature hydrogenation of various functional groups like C—C, C═C, —NO₂, epoxides, and so forth. While in the case of the Ni NPs@N-CNT, the catalyst efficiently and chemoselectively hydrogenated various nitroarenes at a lower reaction temperature. However, these materials were prepared by almost similar strategy, the carbon support formed after carbonization was found to be different in both cases. This change in the shape of the formed support was attributed to the formation of Ni nanoparticles which act as the active metal centers/nuclei for the deposition of the carbon. The migration of the carbon over the Ni nanoparticles results in the formation of *N*-doped CNT support.

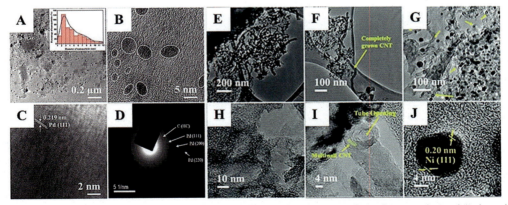

Figure 2.14 (A) TEM micrograph of palladium nanoparticles supported on chitosan derived N-doped carbon (Pd@NC)—inset shows the particle distribution of the nanoparticles; (B) HRTEM images of Pd NPs; (C) HRTEM image showing the fringe corresponding to Pd NP; (D) SAED pattern of Pd@NC catalyst; (E–G) TEM images of Ni NPs supported on *N*-doped carbon nanotubes (Ni@NCNT); (H) graphitic sheets in the synthesized catalyst; (I) HRTEM image showing the mouth opening of the multiwalled CNT; and (J) HRTEM images showing a single Ni nanoparticle with lattice fringe of 0.20 nm of Ni (111) plane. *(Reprinted with permission from Ref. 15. Copyright 2019 Elsevier Ltd., Reprinted with permission from Ref. 26. Copyright 2020 Royal Chemical Society).*

Similarly, a simple hydrothermal-pyrolysis method synthesize Ni NPs with 2.4–10 nm particle size supported on the N-doped carbon nanosheets from bagasse.[27] The as-synthesized catalyst was utilized for the reductive amination of nitroarenes with bio-derived aldehydes. Doping of nitrogen on carbon materials helps the uniform distribution of Ni nanoparticles on the carbon nanosheets and controls the formation of small nanoparticles around 2.4–10 nm are shown in Fig. 2.15. The smaller nanoparticles enhance the reduction capability of nitro compounds to corresponding imines.

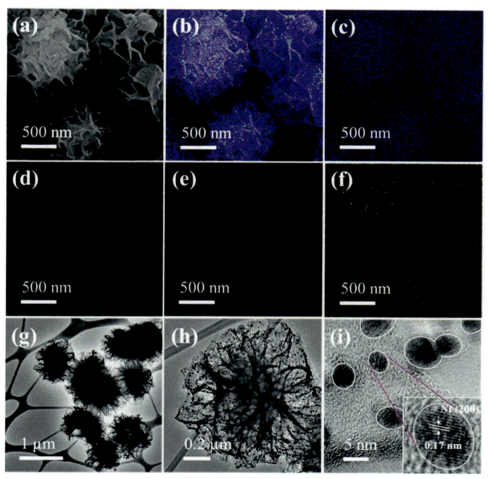

Figure 2.15 (A) FESEM image of Ni@NC–DC, elemental mapping of (B) mixture of C, N, O and Ni, (C) carbon, (D) nitrogen, (E) nickel, (F) oxygen, and (G–I) HRTEM images of Ni@NC–DC at different magnifications. *(Reprinted with permission from Ref. 27. Copyright 2020 Royal Chemical Society).*

1.4 Chemical methods of nanoparticle synthesis using various functional surface

Due to the improved control over the shape and size of NPs, chemical methods have gained much attention as compared to physical methods. Chemical methods mainly include microemulsion, sol-gel, solvothermal, hydrothermal, vapor deposition, ion exchange reduction, polyol, and so forth. The chemicals used in the synthesis of NPs require toxic reagents and are not decomposable, which results in limiting the production scale. Furthermore, some toxic substances may also infect the surface of NPs and make them inappropriate for certain biomedical applications.[28–31] The physical and chemical techniques employed for the synthesis of NPs are widely available. However, these methods require high temperature for the preparation of NPs and make the overall process energy-intensive. Hence, to overcome all the limitations of physical and chemical approaches, scientists and researchers have now focused on identifying a benign alternative approach for the synthesis of NPs.

In terms of biological applications, the surface properties of the nanoparticles play a vital role. These nanoparticles should be biocompatible and uniformally dispersed in the medium, depending on the type of targeted application. However, if the NPs are not designed with good dispersion, agglomeration may occur, making the system incompatible for biological applications. Thus, the functionalization of these nanoparticles by using an environmentally benign approach helps to overcome the aforementioned issues. The covalent bonding between the nanoparticles and the selected organic or inorganic molecules of interest may serve to functionalize the nanoparticles giving them the desirable dispersion in the medium along with stability. Such bonding may benefit from the oxidized CNT surface, and different functional groups like alkoxy, alkyl, amine, amino, and so forth, are shown in Fig. 2.16. Generally, these materials can have different physical

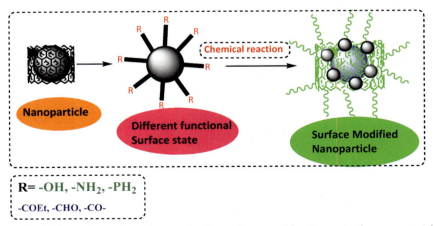

Figure 2.16 General reaction schemes for the surface modification onto the nanoparticles.

and chemical properties to their bulk-form equivalents. The enormous use and development efforts in nanoscience are to design new materials and technologies for practical use.

Recently, the surface modifications of the nanoparticles received much attention owing to their versatility not only by providing expected stability but also by increasing the reactivity profile. The change in the surface of the NPs via the functionalization can alter the properties like crystal structure, wettability, electronic property, surface adsorption, and reaction characteristics of the nanoparticles.[32–36]

Two major modification methods are identified for surface functionalization. The first method involves the adsorption/integration of small molecules like silanes with nanoparticles. In contrast, the second method involves the chemical grafting of the surface hydroxyl groups with different molecules.[37] The latter involves the chemical treatment of the low molecular agents with the nanoparticle surface, providing the nanoparticles with better dispersity and stability. The reaction involving the nanoparticles and the modifier results in the change in the surface structure and state of the nanoparticles. This chemical modification allows ease of functionalization with an outstanding switch over the monolayer structure, which causes such modified materials to be used in several applications. Some recent examples include the formation of C—C, Si—C, P—C, C—O, Si—O, O—N, C—N, and P—O bond functionalized NPs on oxide-free silicon (Fig. 2.17).

Different strategies for the surface modifications of the NPs are described below.
(1) Surface physical modification methods of the nanoparticles
(2) Surface chemical modification methods of nanoparticles

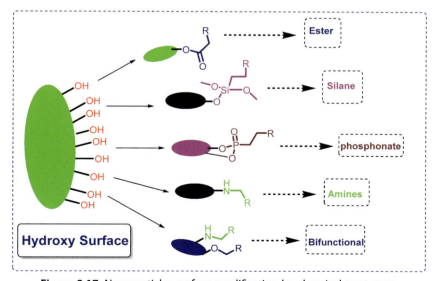

Figure 2.17 Nanoparticles surface modification by chemical treatment.

These methods are generally used, as per requirement, for the modification of nanoparticle surfaces to provide different functionalities, which in turn offer different properties to the nanomaterials that can be used for specific applications.

2. Esterification reaction method

The formation of ester via a chemical reaction between the acid and alcohols is termed esterification.[38] This can be further extended to functionalize different materials having −OH functionality. For instance, the metal oxides often possess −OH groups on the surface. These metal oxides can be further functionalized by treating with −COOH group containing molecules to form an ester bond, giving a functionalized metal oxide NPs as shown in Fig. 2.18.[39]

This functionalization is rather simple and can transform the hydrophilic metal oxide surface into a hydrophobic surface. This is the most effective method being practiced for the modification of weekly acidic or neutral nanoparticles surfaces. Thus, polycondensation reaction is highly useful for surface modification.

For the modification of silica, the above mentioned two different strategies are being practiced. The first consists of the surface adsorption with silanes, and the second involves the covalent attachment of different molecules to the hydroxyl groups present on the nanoparticle surface. Such a surface modification can lead to the generation of nanoparticles with expectable properties, depending on the attached molecule and process conditions. The condensation of silica-containing hydroxyl groups and the metal nanoparticles (MNPs) to produce active catalysts has been widely used due to intrinsic and size-dependent catalytic properties. In monomers, the NPs penetrate the carbon matrix, but in macrocycles, the NPs are partially filled. The aggregated polymer matrix can be easily separated (see Fig. 2.19). MNPs can easily catalyze various chemical reactions provided the possible interaction with reacting substrates is highly desirable.

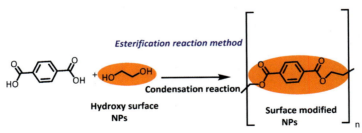

Figure 2.18 General esterification process for nanoparticle synthesis.

Fundamental concepts on surface chemistry for nanoparticle modifications 47

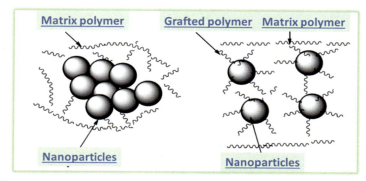

Figure 2.19 Nanoparticles surface silica modification methods. *(Reprinted with permission from Ref. 10. Copyright 2001 Elsevier Ltd).*

Generally, the chemical reaction of surface hydroxyl interacts with the metal surface to cause MNPs. Therefore, the silica microsphere plays an important role in surface modification. In SiO$_2$@Pd-PAMAM, the addition of amine-terminated G4 poly(amido-amine) helps the electrostatic interaction of Pd(II) ions and SiO$_2$. At the same time, it helps to reduce the Pd(II) into Pd (0) NPS. The synthesis of Pd-PAMAM, firstly SiO$_2$ microspheres were functionalized with vinyl groups by grafting their surfaces with vinyl-triethoxysilane is shown in Fig. 2.20.

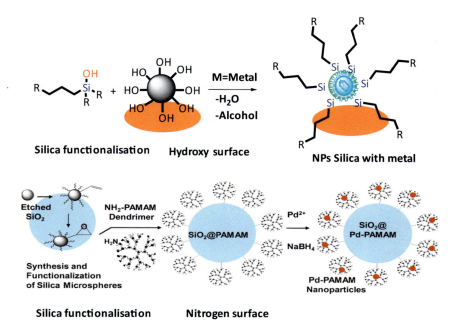

Figure 2.20 Reaction schemes for the surface modification onto the nanoparticles. *(Reprinted with permission from Ref. 10. Copyright 2011 American Chemical Society).*

3. Phosphate ester method

Similar to the esterification reaction, phosphate ester grafting aids in the functionalization of the nanoparticles resulting in switching the property of material from hydrophilic to hydrophobic.[39,40] The organic phosphate contains a phosphate acid and an alkyl group. This organic phosphate reacts with the nanoparticle surface via the acidic group while the alkyl chain remains unaltered. This reaction generates a phosphate ester with the metal oxide with an elimination of a small water/alcohol molecule as shown in Fig. 2.21. The presence of a high amount of the surface hydroxyl groups facilitates such functionalization inducing hydrophobicity.

Also, couling agent method uses a linker with two different functionalities, where one of the functionality reacts with the surface of the nanoparticles. In contrast, the other reacts with the polymer or organic molecular films.[39–43] This gives a closely linked functionalization and improved catalytic performance. The current coupling agents deployed for this purpose are silane-based agent, titanate-based agent, and aluminum acid-based coupling agent are mainly adopted. The coupling of the polymer or organic molecular films results in the high dispersion properties of the nanoparticles.

4. In situ modification method

Among the various surface modification or functionalization methodologies, efforts have been recently focused on the in situ modifications of the nanoparticles. This method allows the NP dispersion and polymerization to occur simultaneously.[39] The in situ modification involves the synthesis of nanoparticles, followed by the modification of the surface in the same medium. The nucleation of the nanoparticles gives it high surface energy, making it thermodynamically unstable. The polar groups on this nanoparticle are immediately functionalized to reduce this surface energy. This kind of functionalization

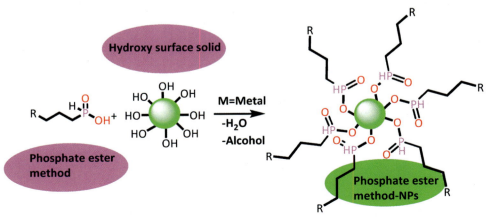

Figure 2.21 Reaction schemes for the surface modification onto the nanoparticles.

results in highly dispersed and stable nanoparticles. The reduction in the cost of production, owing to the production of nanoscale particles and the removal of postmodification steps, makes it a promising and ideal method of synthesizing and functionalizing the nanoparticles. Various materials like metal plates, RO-membrane, Si—Al metal, and plasmatic material take advantage of such modification methodology are shown in Fig. 2.22.

For instance, the extraordinary tendency of the Au NPs to confine electromagnetic field, ease of surface functionalization makes it an interesting material for various other potential applications like in photonic and optoelectronic technologies, chemical sensing and bio-sensing applications, energy harvesting, imaging, data storage, and optical tweezer devices for nanomaterial manipulation.[34,44–50] Due to their unique physical properties, the polymeric nanoparticles (PNPs) has been extensively used as a base for the distribution of therapeutic agents. The nature of these surface-functionalized PNPs can interface with the biological environment, aid in fine-tuning, and thus increasing the circulation half-life. Another role for these PNPs is to target the minimization of off-target drug exposure. The major surface functionalization strategies can be categorized based on the nature of the conjugating ligand:

(1) The use of macromolecules like lipids, antibodies, peptides, polysaccharides, and nucleic acids (aptamers)
(2) small ligands such as mono or oligosaccharides, steroids, and vitamins
(3) hydrophilic polymers

Similarly, as discussed in the earlier sections, various methods involve the in situ synthesis of nanoparticles along with the formation of different carbon and silica supports.[34,47–50] The in situ formation of the metal nanoparticles on carbon supports via carbonization/pyrolysis strategies has proven to provide highly dispersed and stable nanoparticles over carbonaceous support. At the same time, the silica materials have shown the ability to protect the nanoparticles from sintering at high temperature. Moreover, the high surface area of the support and the presence of different dopants (like sulfur or nitrogen, in the case of carbon supports), aids in the high dispersion of the nanoparticles and stabilization. Moreover, different functionalities on the carbon/silica supports may provide bifunctionality (metal nanoparticles and acidity/basicity depending on the support surface functionality) to the final catalysts.

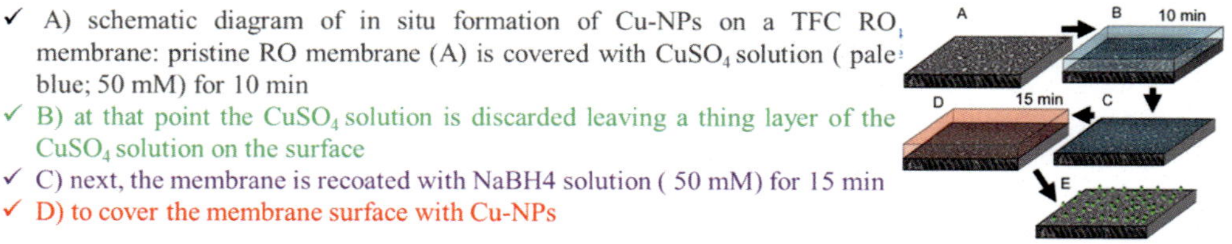

✓ A) schematic diagram of in situ formation of Cu-NPs on a TFC RO membrane: pristine RO membrane (A) is covered with CuSO₄ solution (pale blue; 50 mM) for 10 min
✓ B) at that point the CuSO₄ solution is discarded leaving a thing layer of the CuSO₄ solution on the surface
✓ C) next, the membrane is recoated with NaBH4 solution (50 mM) for 15 min
✓ D) to cover the membrane surface with Cu-NPs

Figure 2.22 In situ surface modification. *(Reprinted with permission from Ref. 44. Copyright 2016 Elsevier Ltd).*

5. Conclusions

With the growing socioeconomic developments, especially in developing and underdeveloped countries, nanoscience has proved to positively impact human society and life, both in terms of sustainability and economics. Nanoparticles play a considerable role in promoting the development of this field of science. Diverse physiochemical strategies are used to preserve the stability and integrity of nanoparticles. The most common approaches in practice for nanoparticle synthesis and stabilization have been briefly discussed. The surface modifications of the nanoparticles can alter the surface structure of nanoparticles, giving new functions and improved physical and chemical performance to the nanostructures. Although groundbreaking studies have been reported for the synthesis and stabilization of nanoparticles, more understanding of mechanical aspects may pave the path for sustainable and cost-effective methods. Socioeconomic growth calls for the development of different modern materials and technologies. Thus, the need for in-depth studies of the physical and chemical properties of the nanoparticles is highly required to promote the development of new materials and upgrade the traditional industries.

References

[1] Teleanu DM, Chircov C, Grumezescu AM, Teleanu RI. Neurotoxicity of nanomaterials: an up-to-date overview. *Nanomaterials* 2019;**9**:1−14.
[2] Khan I, Saeed K, Khan I. Nanoparticles: properties, applications and toxicities. *Arab J Chem* 2019;**12**: 908−31.
[3] Jamkhande PG, Ghule NW, Bamer AH, Kalaskar MG. Metal nanoparticles synthesis: an overview on methods of preparation, advantages and disadvantages, and applications. *J Drug Deliv Sci Technol* 2019; **53**:101174.
[4] Hasan S. A review on nanoparticles: their synthesis and types. *Res J Recent Sci* 2014;**4**:1−3.
[5] Al-Kayiem H, Lin S, Lukmon A. Review on nanomaterials for thermal energy storage technologies. *Nanosci Nanotechnol - Asia* 2013;**3**:60−71.
[6] Singh J, Dutta T, Kim KH, Rawat M, Samddar P, Kumar P. Green synthesis of metals and their oxide nanoparticles: applications for environmental remediation. *J Nanobiotechnol* 2018;**16**:1−24.
[7] Khandel P, Kumar R, Deepak Y, Soni K, Kanwar L, Kumar S, Silver A. *Biogenesis of metal nanoparticles and their pharmacological applications: present status and application prospects*. Springer Berlin Heidelberg; 2018.
[8] Khandel P, Shahi SK. Microbes mediated synthesis of metal nanoparticles: current status and future prospects. *Int J Nanomater Biostructures* 2016;**6**:1−24.
[9] Kamran U, Bhatti HN, Iqbal M, Nazir A. Green synthesis of metal nanoparticles and their applications in different fields: a review. *Z Phys Chem* 2019;**233**:1325−49.
[10] Biradar AV, Asefa T. Nanosized gold-catalyzed selective oxidation of alkyl-substituted benzenes and n-alkanes. *Appl Catal Gen* 2012;**435−436**:19−26.
[11] Pasricha R, Bala T, V Biradar A, Umbarkar S, Sastry M. Synthesis of catalytically active porous platinum nanoparticles by transmetallation reaction and proposition of the mechanism. *Small* 2009: 1467−73.
[12] Wang Y, V Biradar A, Asefa T. Assembling nanostructures for effective catalysis: supported palladium nanoparticle multicores coated by a hollow and nanoporous zirconia shell. *ChemSusChem* 2012;**5**: 132−9.

[13] Chandra P, Doke DS, Umbarkar B, Biradar AV. One-pot synthesis of ultrasmall MoO$_3$ nanoparticles supported on SiO$_2$, TiO$_2$, and ZrO$_2$ nanospheres: an efficient epoxidation catalyst. *J Mater Chem A* 2014;**2**:19060−6.

[14] Nandi S, Patel P. Nitrogen-rich graphitic-carbon stabilized cobalt nanoparticles for chemoselective hydrogenation of nitroarenes at milder conditions. *Inorg Chem Frontiers* 2018;**5**:806−13.

[15] Advani JH, Khan NH, Bajaj HC, Biradar AV. Stabilization of palladium nanoparticles on chitosan derived N-doped carbon for hydrogenation of various functional groups. *Appl Surf Sci* 2019;**487**:1307−15.

[16] Ott LS, Hornstein BJ, Finke RG. A test of the transition-metal nanocluster formation and stabilization ability of the most common polymeric stabilizer, poly(vinylpyrrolidone), as well as four other polymeric protectants. *Langmuir* 2006;**22**:9357−67.

[17] Rothenberg G, Dur L. Transition-metal nanoparticles: synthesis, stability and the leaching issue. *Appl Organomet Chem* 2008;**22**:288−99.

[18] Sultana S, Alzahrani N, Alzahrani R, Alshamrani W. Stability issues and approaches to stabilized nanoparticles based drug delivery system. *J Drug Target* 2020;**28**:468−86.

[19] Sadeghpour A, Szilagyi I, Vaccaro A, Borkovec M. Electrostatic stabilization of charged colloidal particles with adsorbed polyelectrolytes of opposite charge. *Langmuir* 2010;**26**:15109−11.

[20] Umbarkar SB, V Kotbagi T, Biradar AV, Pasricha R, Chanale J, Dongare MK, Mamede A, Lancelot C, Payen E. Acetalization of glycerol using mesoporous MoO$_3$/SiO$_2$ solid acid catalyst. *J Mol Catal A* 2009;**310**:150−8.

[21] Zanganeh N, Guda VK, Toghiani H, Keith JM. Sinter-resistant and highly active sub-5 nm bimetallic Au-Cu nanoparticle catalysts encapsulated in silica for high-temperature carbon monoxide oxidation. *ACS Appl Mater Interfaces* 2018;**10**:4776−85.

[22] Zhu J, Wei S, Haldolaarachchige N, Young DP, Guo Z. Electromagnetic field shielding polyurethane nanocomposites reinforced with core−shell Fe−silica nanoparticles. *J Phys Chem C* 2011;**115**:15304−10.

[23] Yang T, Ling H, Lamonier JF, Jaroniec M, Huang J, Monteiro MJ, Liu J. A synthetic strategy for carbon nanospheres impregnated with highly monodispersed metal nanoparticles. *NPG Asia Mater* 2016;**8**.

[24] Jo WK, Moru S, Tonda S. Cobalt-coordinated sulfur-doped graphitic carbon nitride on reduced graphene oxide: an efficient metal-(N,S)-C-class bifunctional electrocatalyst for overall water splitting in alkaline media. *ACS Sustain Chem Eng* 2019;**7**:15373−84.

[25] Oh JG, Oh HS, Lee WH, Kim H. Preparation of carbon-supported nanosegregated Pt alloy catalysts for the oxygen reduction reaction using a silica encapsulation process to inhibit the sintering effect during heat treatment. *J Mater Chem* 2012;**22**:15215−20.

[26] Advani JH, Ravi K, Naikwadi DR, Bajaj HC, Gawande MB, Biradar AV. Bio-waste chitosan-derived N-doped CNT-supported Ni nanoparticles for selective hydrogenation of nitroarenes. *Dalton Trans* 2020;**49**:10431−40.

[27] Ravi K, Advani JH, Bankar BD, Singh AS, Biradar AV. Sustainable route for the synthesis of flowers-like Ni@N-doped carbon nanosheets from bagasse and its catalytic activity towards reductive amination of nitroarenes with bio-derived aldehydes. *New J Chem* 2020;**44**:18714−23.

[28] Markovsky E, Baabur-Cohen H, Eldar-Boock A, Omer L, Tiram G, Ferber S. Administration, distribution, metabolism and elimination of polymer therapeutics. *J Contr Release* 2012;**161**:446−60.

[29] Sleep D, Cameron J, Evans LR. Albumin as a versatile platform for drug half-life extension. *Biochim Biophys Acta* 2013;**1830**:5526−34.

[30] Jeong B, Kim SW, Bae YH. Thermosensitive sol−gel reversible hydrogels. *Adv Drug Deliv Rev* 2012;**64**:154−62.

[31] Mahapatro A, Singh DK. Biodegradable nanoparticles are excellent vehicle for site directed. *J Nanobiotechnol* 2011;**9**:55−65.

[32] Fei L, Xihai H, Zhenzhong W, Huimin L. Progress on Surface Modification of Nano-powder. *Guangdong Chemical Industry* 2010:05.

[33] Zhijun X, Ruiqing C. *Nano materials and nano technology*. Beijing: Chemical Industry Press; 2010.

[34] Lin G, Shang M, Wang Y, Xue M. Research on surface modification methods of nanoparticles. In: *International Conference on Mechatronics, Electronic, Industrial and Control Engineering (MEIC-15)*; 2015. p. 1140—3.
[35] Chen J, Zhang W, Zhang J, Chen M, Hai T, Wei L. Progress in research on chitosan/semiconductor nanocomposites. *J Xi'an Univ Arts Sci* 2009:2.
[36] Jinling L. Research of nano-silica surface modification. *Mater Dev Appl* 2011;**2**:18—21.
[37] Pujari SP, Scheres L, M Marcelis AT, Zuilhof H. Covalent surface modification of oxide surfaces. *Angew Chem Int Ed* 2014;**53**:6322—56.
[38] Park JW, Park YJ, Jun CH. Post-grafting of silica surfaces with pre-functionalized organosilanes: new synthetic equivalents of conventional trialkoxysilanes. *Chem Commun* 2011;**47**:4860—71.
[39] Peng L. Latest developments of nanoparticles surface modification. In: *4th international conference on mechatronics, materials, chemistry and computer engineering*; 2015.
[40] Liang Y, Wu C, Zhao Q, Wu Q, Jiang B, Weng Y. Gold nanoparticles immobilized hydrophilic monoliths with variable functional modification for highly selective enrichment and on-line deglycosylation of glycopeptides. *Anal Chim Acta* 2015;**900**:83—9.
[41] Binshi XU. *Nano surface engineering*. Chemical Industry Press; 2003. p. 4—5.
[42] Libo W, Ming L. Lubrication nanometer additive surface chemical modification research progress. *Lubr Sealing* 2008;**9**:95.
[43] Zhidong L. *Nano material basis and application*. Beijing University press; 2010. p. 121—6.
[44] Ben-Sasson M, Lu X, Nejati S, Jaramillo H, Elimelech M. In situ surface functionalization of reverse osmosis membranes with biocidal copper nanoparticles. *Desalination* 2016;**388**:1—8.
[45] Ben-Sasson M, Zodrow KR, Genggeng Q, Kang Y, Giannelis EP, Elimelech M. Surface functionalization of thin-film composite membranes with copper nanoparticles for antimicrobial surface properties. *Environ Sci Technol* 2014;**48**:384—93.
[46] Ronen A, Semiat R, Dosoretz CG. Impact of ZnO embedded feed spacer on biofilm development in membrane systems. *Water Res* 2013;**47**:6628—38.
[47] Patra JK, Bae KH. Green nanobiotechnology: factors affecting synthesis and characterization techniques. *J Nanomater* 2014;**219**.
[48] Yegorov AS, Ivanov VS, Antipov AV, Wozniak AI, Tcarkova KV. Chemical modification methods of nanoparticles of silicon carbide surface. *Orient J Chem* 2015;**31**:1269—75.
[49] Ahmed SR, Kim J, Suzuki T, Neethirajan S, Lee J, Park EY. In situ self-assembly of gold nanoparticles on hydrophilic and hydrophobic substrates for influenza virus-sensing platform. *Sci Rep* 2017;**7**:1—11.
[50] Biradar AV, Biradar AA, Asefa T. Silica—dendrimer core-shell microspheres with encapsulated ultrasmall palladium nanoparticles: efficient and easily recyclable heterogeneous nanocatalysts. *Langmuir* 2011;**27**:14408—18.

CHAPTER 3

Synthesis of surface-modified nanomaterials

Gianvito Vilé

Department of Chemistry, Materials, and Chemical Engineering "Giulio Natta", Politecnico di Milano, Milan, Italy

1. Introduction

Nanomaterials and nanocomposites are widely used for agricultural, pharmaceutical, and personal care applications. In catalysis, they are employed to conduct specific transformations with the goal of preventing pollution, reduce energy consumption, and synthesize chemical species. The properties of a nanomaterial change significantly depending on its size and, in general, nanoscale materials have different properties compared to the bulk samples. In fact, as the nanomaterial size decreases, the amount of surface atoms increases, and this makes the sample useful for applications where a large fraction of surface atoms is needed. This is often the case in catalysis,[1] the science of speeding up a chemical reaction by using tiny amounts of nanomaterials. The catalyst reactivity can be correlated with the availability of these surface sites (though other factors, such as their exact location and orientation may play a role as well), which depends on the dispersion of the catalytic agent, being this the ratio between the number of surface atoms and the total number of atoms in the catalyst volume.

Nanoparticles (NPs) are a class of nanomaterials with distinctive mechanical, electronic, and optical properties that explicitly occur from the presence of an higher fraction of surface atoms. As a result, they are greatly important in catalysis.[2,3] NPs can be synthesized following various top-down and bottom-up routes (Fig. 3.1).[6] The top-down methods include vapor deposition techniques and are based on the principle of dividing the bulk precursor into smaller components. The bottom-up approaches involve reducing the precursor species in the presence of a template, followed by the removal of the additives and the controlled aggregation of atoms.

When preparing an NP, the synthesis method depends strongly on the type of properties that are needed in the sample. However, bottom-up strategies have proved to be more convenient and effective than top-down methods, enabling the rapid construction of well-defined NPs with different composition, size, shape, and induced chemical or physical properties. Table 3.1 summarizes the most applied synthesis methods to make nanomaterials encountered in catalysis.

It is important to recognize that nanomaterials used in catalysis are often composites. Metal NPs are, for examples, deposited on inorganic or organic carriers, which are named

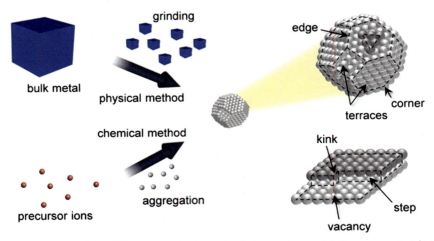

Figure 3.1 Overview of the different synthesis methods to produce catalytic nanomaterials and surface structure of a nanoparticle. *(The figure is adapted from Refs. 4 and 5).*

Table 3.1 Synthesis route for selected nanoparticles used in catalysis.

Nanoparticle	Synthesis routes
TiO₂	Hydrothermal, sonochemical, sol-gel, or flame spray pyrolysis
ZnO	Sol-gel, precipitation, organometallic synthesis, thermal evaporation, microwave, flame spray pyrolysis, mechanical milling, or mechanochemical synthesis
Al₂O₃	Flame spray pyrolysis, sol-gel, precipitation, or freeze drying
SiO₂	Sol-gel, flame spray pyrolysis, or water-in-oil microemulsion processes
Ag	Microwave, ultrasonic, colloidal routes followed by chemical reduction by inorganic and organic reducing agents, laser ablation, gamma irradiation, photochemical method, thermal decomposition of silver oxalate in water, or electrochemical synthesis
Au	Photochemical reduction, solvent evaporation techniques, colloidal routes followed by chemical reduction by inorganic and organic reducing agents, or microwave irradiation

Adapted from Ref. 7.

"support" because of their function of supporting the metal and giving the catalysts texture and mechanical resistance. The support provides an area for the dispersion of the catalytically active agent, which is particularly important when expensive metals, such as platinum, gold, iridium, or palladium are used as catalytic agents. The final structure of a supported nanocatalyst is schematically shown in Fig. 3.2 and some examples of these catalysts, with their respective applications, are summarized in Table 3.2. In most cases, the active component comprises a metal (*e.g.*, Pt, Au, Ir, Pd), a metal oxide

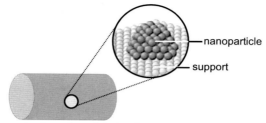

Figure 3.2 Structure at different length scales of a catalyst pellet and surface view on the presence of supported nanoparticles.

Table 3.2 Selected nanomaterials and their main catalytic applications.

Catalyst	Application
Pd/γ-Al$_2$O$_3$	Alkyne hydrogenation
Ag/α-Al$_2$O$_3$	Epoxidation
Pt/C	Fuel cells and hydrogenations
CrO$_x$/SiO$_2$	Olefin polymerization
V$_2$O$_5$/TiO$_2$	NO$_x$ abatement

(*e.g.*, CrO$_x$, V$_2$O$_5$), or a metal sulfide (*e.g.*, FeS), dispersed on a high-surface-area support (*e.g.*, TiO$_2$, Al$_2$O$_3$, SiO$_2$, various types of carbon). Typically, the active component is in the form of an NP, with a weight loading of 1–20 wt.%, a size between 1 and 10 nm; the support has a surface area is in the range of 1–1000 m^2/g. Support may be inert or interact with the active component. In the latter case, the interaction may affect the metal activity and selectivity.

2. Nanomaterials surface chemistry and Zeta potential

It is at the nanomaterials surface that a reaction takes place. Thus, before describing methods to prepare or modify a nanomaterials, this section introduces key concepts on surface chemistry.

The catalytic performance of a nanomaterial can be affected by several parameters (Fig. 3.3), beyond the size and shape described earlier. The specific surface area provides an estimate of the space available for catalysis and this value is typically determined by measuring the amount of physically adsorbed nitrogen according to the Brunauer–Emmett–Teller (BET) method.[8] In particular, prior to measuring the physisorption of nitrogen, the samples are degassed to ensure that water molecules adsorbed on the sample are properly removed. The BET surface area provides an estimate of the total surface area,

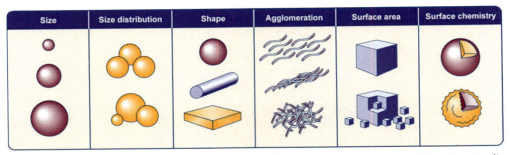

Figure 3.3 Characteristics of a nanomaterial that could affect catalysis include particle size, size distribution, particle shape, particle agglomeration, surface area, and surface chemistry. *(Source: https://bit.ly/3sm5R3l).*

assuming that all surface atoms behave similarly in adsorption and catalysis. This is not often the case: from a geometric viewpoint, particles are enclosed by several terraces, whose intersection creates edges, and the intersection of edges generates corners (Fig. 3.1). Because of their low coordination number, atoms on edges and corners usually show a higher catalytic reactivity compared to atoms on inner terraces. Tuning the morphology of a nanomaterial corresponds to the tuning of the terraces, edges, and corners present in there, which is often an important aspect not determined by simple BET analysis. For this reason, surface characterization involving more in-depth investigations into the surface structure, via a combination of methods detailed in Table 3.3, are often needed.

Table 3.3 Overview of important techniques used to characterize surface-modified nanomaterials.

Method	Insights gained
Powder X-ray diffraction	Presence and structure of crystalline phases
Inductively coupled plasma optical emission spectroscopy, X-ray fluorescence, CHNO analysis	Composition
Transmission electron microscopy, high-angle annular dark-field imaging scanning transmission electron microscopy, scanning electron microscopy	Structure and morphology of the catalyst; shape, size, and uniformity of the nanomaterials
Nuclear magnetic resonance spectroscopy	Atomic structure
Thermogravimetry	Thermal stability of the materials
Infrared spectroscopy	Surface structure and surface composition
Adsorption of gas and probe molecules	Surface areas, surface properties, chemisorption abilities
X-ray photoelectron spectroscopy	Chemical state of the material and structural integrity
X-ray absorption spectroscopy	Chemical environment of the atoms

An additional parameter directly influencing the materials properties is the surface charge. Metal and metal oxide particles are typically uncharged, but the surface may become negatively or positively charged depending on the pH of the solution surrounding the material. With an acidic solution, i.e., at low pH, the equilibrium of a surface is typically shifted toward the generation of positive charges. As the pH increases, the surfaces become less positively charged and can switch to negative charges. For example, metal oxide particles, which are often used as support materials, are commonly terminated by hydroxyl groups. Depending on the metal cations, lattice structure, and bonding chemistry of the −OH groups, their nature can be considered as acidic, neutral, or basic. For instance, infrared spectroscopy shows that the surface of γ-Al_2O_3 has a variety of −OH groups which are basic (terminal −OH, bonded to Al atoms), neutral (bridge −OH, bonded between two Al atoms), and acidic (bridge −OH, bonded between three Al atoms). At low pH, these hydroxyl groups will pick up a proton from the solution and become positively charged, while at elevated pH, deprotonation of the −OH group gives negative charges (Fig. 3.4). The pH at which the net charge of the surface is zero is referred to as the point of zero charge (PZC). This exact PZC is a critical parameter for controlling the synthesis and surface modification of a catalytic nanomaterial, since synthetic protocols are designed to introduce partial (positive or negative) charges before structure, ligand, or atom incorporation. For example, during impregnation of an oxide (e.g., Al_2O_3) with a transition metal (e.g., Pd), the introduction of partial (positive or negative) charges is practically done by choosing the most appropriate metal precursor (e.g., $PdCl_2$, $Pd(CH_3COO)_2$, or $Pd(NO_3)_2$) depending on the pH that it is desired for an optimal impregnation.

It is important to recognize that the effective charge of a surface can be partially neutralized by counterions present in the solution during nanomaterials synthesis and modification. Theses counterions can generate a charge space, part of which is held strongly to be carried along as the particles move (Fig. 3.5). The result is a Zeta potential layer. Both the original charge and the neutralizing counterions respond to pH changes. If the charge is high, particles efficiently repel one another and avoid interaction. If it is

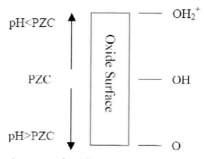

Figure 3.4 Surface charging of oxides in aqueous systems, depending on pH.

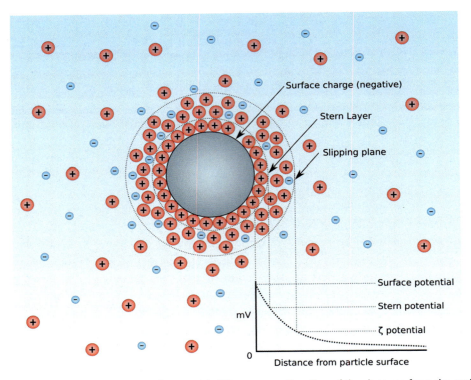

Figure 3.5 Ionic concentration and potential difference as a function of the distance from the surface of a particle suspended in a liquid medium.

low, thermal motion leads to collision and coalescence. These rates are the highest at the isoelectric point where the Zeta potential is zero, and this phenomenon can be exploited to induce a controlled aggregation of catalytic nanoparticles, effectively tuning the particle size.

3. Surface modification of catalytic nanoparticles by thermal and plasma treatment

Thermal treatment is one of the methods to induce surface modification of a nanomaterial. During this process, multiple changes can occur, including (1) generation of the active phase, where the hydroxide form is converted to the oxide form; (2) stabilization of the mechanical properties, to ensure stability of textural and structural properties during catalytic use; (3) loss of chemically adsorbed water; (4) changes in surface area and pore size distribution due to sintering; and (5) change in phase type and distribution at high temperature.

Thermal treatments have been widely employed in catalysis. For example, highly dispersed NiO nanoparticles on mesoporous silica SBA-15 can be obtained via incipient wetness impregnation of nickel nitrate, followed by washing, drying, and calcination.[9] In particular, as shown in Fig. 3.6, calcination in flowing air or stagnant air gives extensive surface redistribution and sintering of the NiO phase, leading to 10–100 nm NiO particles. On the contrary, changing the calcination medium from air to inert He atmosphere leads to less extensive redistribution of the nickel phase, with a NiO particle size between 5 and 10 nm. Finally, calcination in 1 vol.% NO in He gives a monomodal particle size distribution with well-defined NiO particles around 3–4 nm. Similar studies have corroborated the importance of thermal treatment in the preparation of supported catalysts.[10,11]

Plasma technologies can also be a valuable alternative to conventional thermal treatments for the nanomaterial surface amendment. A plasma is a quasi-neutral ionized gas comprising ions and free electrons. In a laboratory, plasma is made by exposing a gas to strong electromagnetic fields, accelerating the free electrons available in the gas. The kinetic energy reached by those electrons is transferred to other species and converted into internal energy through inelastic collisions. In particular, the reactive plasma species can interact with the NPs surfaces, forming a coating layer that alters the surface chemistry of a nanomaterial. Plasma-modified nanoparticles are more resistant to a temperature rise, and the approach can be utilized to avoid particle size reconstructing at high temperature and better control the NPs coalescence and sintering.

4. Silane chemical treatment

Surface modification of a nanoparticle by silane chemical treatments is an alternative method to control the materials properties and ultimately tune the catalytic performance. This can be accomplished by adsorption of (amino)silane coupling agents on the NPs external layers. This type of surface modification has been applied to a broad variety of catalytic NPs with the scope of improving their dispersibility in aqueous media and their catalytic activity. In fact, it has been reported that silane-functionalized nanoparticles show better dispersibility in organic solvents or polymer matrixes, and this is particularly advantageous when the surface-modified nanoparticles are used as such for organic synthesis.[12]

The notion of silane coupling was originally reported by Plueddemann and colleagues in the 1970s.[13] As mentioned earlier and also shown in Fig. 3.7, the surface of a pristine nanoparticle (i.e., the oxide) is typically covered with −OH groups. By addition of the 3-methacryloxypropyl trimethoxysilane, high-temperature treatment at *ca.* 80–100°C, and ultrasonication, condensation and bond formation between the surface −OH groups and the silane take place, resulting in NPs shielded with 3-methacryloxypropyl trimethoxysilanes.

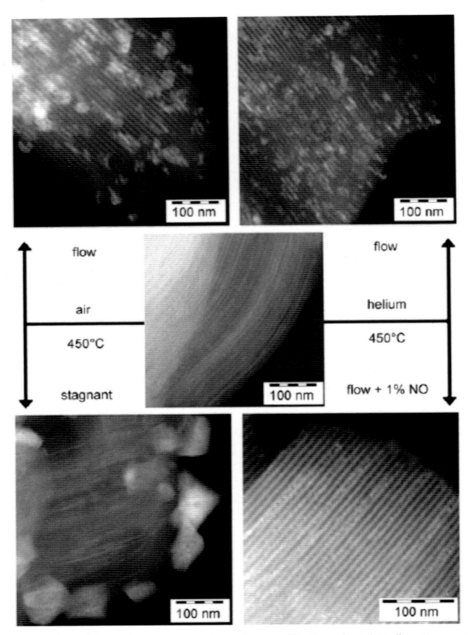

Figure 3.6 Effect of thermal treatment on the surface modification of catalytically active nanoparticles. *(Taken from Ref. 9).*

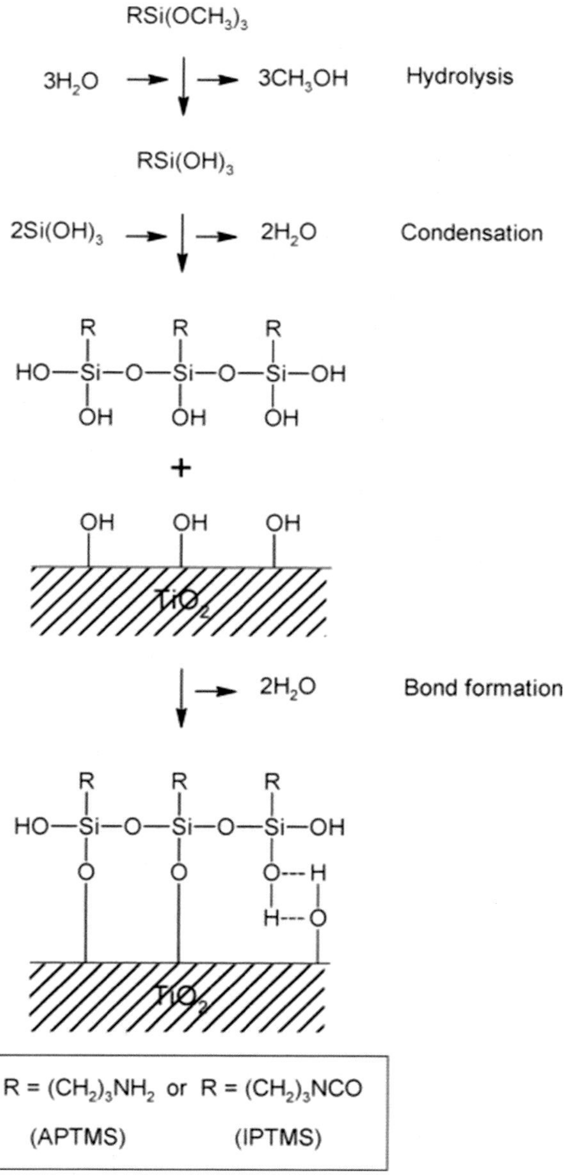

Figure 3.7 Schematic representation of the modification of TiO$_2$ with 3-methacryloxypropyl trimethoxysilane groups. *(Reprinted from Ref. 14).*

The nanocatalyst surface may be modified as well using alkyl/aryl isocyanates, epoxides, and metal alkoxides. For instance, Ukaji et al.[15] surface-modified TiO₂ using both 3-aminopropyltrimethoxysilane and N-propyltriethoxysilane silane coupling agents. The authors examined the influence of such modification on the photocatalytic activity and UV-shielding ability of titania, proving that the modified particles showed enhanced properties in comparison with the pristine particles. Similarly, Guo et al.[16] altered γ-Al₂O₃ nanoparticles with a single, bifunctional coupling agent, (3-methacryloxypropyl)trimethoxysilane. The authors studied the photocatalytic antibacterial properties of surface modification against *E. Coli* and Salmonella, showing excellent antimicrobial performance.

The concept of silane chemical treatment has broader applications and is at the basis for the fabrication of NP@SiO₂ core@shell nanomaterials, which are obtained by introduction of silanols, followed by their condensation and formation of siloxane bonds according to the general reaction: $Si(OH)_4 + 2H_2O \rightarrow SiO_2 + 4ROH$. A schematic illustration of this process is depicted in Fig. 3.8.

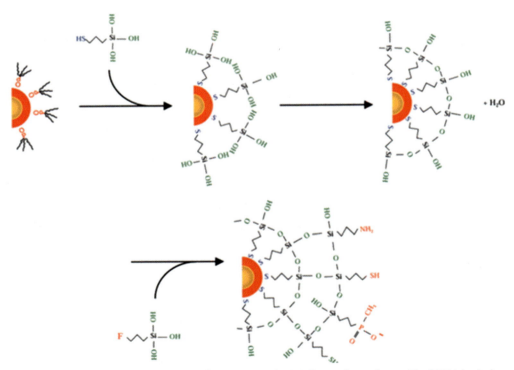

Figure 3.8 Sketch for the silanization of a nanoparticle. Briefly, methoxysilanes (Si—OCH₃) hydrolyze into silanol groups (Si—OH) and produce a polymer layer. The condensation of this layer leads to the development of siloxane bonds while water is released, leaving a SiO₂ layer surrounding the NPs. In some cases, silane precursors may be added into the shell to further functionalize the NPs. *(Taken from Ref. 17)*.

These type of heterostructures show enhanced catalytic properties, leading to high-temperature-stable catalytic systems that could be impossible to prepare otherwise. For example, Somorjai and coworkers have prepared Pt metal NPs coated with a mesoporous silica shell (Pt@mSiO$_2$).[18] In particular, the material was obtained in three steps, consisting of the (1) synthesis of Pt nanoparticles using tetradecyltrimethylammonium bromide (TTAB) as capping agent, (2) high-temperature silica polymerization around the Pt cores, generating Pt@SiO$_2$ structures, and (3) removal of the TTAB by calcination to produce the Pt@mSiO$_2$ core-shell species. The obtained system was catalytically active, selective, and stable up to 750°C for C$_2$H$_4$ hydrogenation and CO oxidation. The high thermal stability of the Pt@mSiO$_2$ NPs made possible to conduct high-temperature CO oxidation studies, including ignition behavior analyses which are not possible over bare Pt nanoparticles because of their tendency to deform and sinter.

It is important to highlight that silane chemical treatment may lead to adjustment of the surface adsorption properties (an important aspect in catalysis). For instance, surface hydroxyl groups on specific catalyst supports can be replaced by silanols, and this may provide mechanistic hints on the function of the —OH groups of the support in metal catalysis, particularly when those are expected to play a catalytic or promoting role.[19]

5. Ligand immobilization techniques to modify catalytic nanoparticles

Different from the use of inorganic (i.e., silanols) agents, another approach to modify the surface of a catalytic nanomaterials is based on grafting organic or polymer moieties, enhancing the chemical functionality of the pristine nanomaterial and altering its surface topology. Such nanoparticles grafted with organic elements are considered to be hybrid (organic-inorganic) nanocomposites and the grafting species is called "ligand."

The ligand can play a variety of functions, going from regulating the nanoparticle stability to the postsynthetic minimization of the surface free energy of a nanoparticle. The ligands can be a small organic compound (e.g., trisodium citrate), a large polymer (e.g., polyethylene), or even a biomolecule (e.g., a peptide, a protein, and an oligonucleotide). The use of ionic liquids as ligands has been established recently, with advantages when electron-transfer properties resulting from the N-donor complexes of these species are desired in the final materials. However, the catalytic application of ionic liquid-modified NPs is still limited.

The appropriate choice of a ligand and the type of protocol to be followed have a major effect on the final materials performance in catalysis. To select the most suitable ligands, two important aspects have to be considered: (1) *the chemical composition of the NP surface*, that is, the strength and nature of the bonding between the ligand and the NP. For example, thiols have strong affinity to gold, while hydroxyl groups bind better to iron oxide NPs (Table 3.4); (2) *the environment*, that is, the appropriate ligand to coat an

Table 3.4 Typical anchoring groups for a variety of nanoparticles (NPs).

NPs type	Anchoring group
Noble metal	−SH
	−NH$_2$
	−COOH
	−PR$_3$
Semiconducting quantum dot	−SH
	−COOH
	−PO(OR)$_2$
	−O=PR$_3$
Metal oxide	−NH$_2$
	−COOH
	−OH
	−PO(OR)$_2$

Taken from Ref. 17.

NP to be chosen based on the environment (e.g., pH, solvent media, etc.) in which the NP will be used; and (3) *the desired NP morphology* because ligands have diverse binding energies and packing characteristics based on the NP morphology (e.g., cubic, octahedra, etc.).

Synthetically, these hybrid nanocatalysts are prepared through wet chemical routes. An organic reducing agent (e.g., hydrazine, borohydride, or superhydrides such as LiB(C$_2$H$_4$)$_3$H) is dissolved in a solvent and mixed with a metal precursor and a stabilizing agent. Upon stirring, colloidal metal nanoparticles are generated, which are then deposited on a specific support. In some cases, specific surface reactions between the ligands may be induced to further develop the materials at the nanoscale. For example, Tsubokawa and colleagues[20] reported the synthesis of titania nanoparticle surfaces with branched polymers having azo groups on it. These organic moieties underwent then a radical polymerization of vinyl monomers from the azo groups, to give polymer-grafted NPs of titania. Ungrafted polymer coming from the leaching of the ligand was also partially formed during reaction (Fig. 3.9). Similarly, Bach et al. deposited

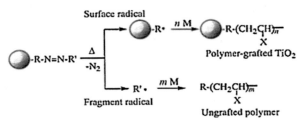

Figure 3.9 Example of surface modification via radical polymerization of vinyl monomers initiated by azo groups.

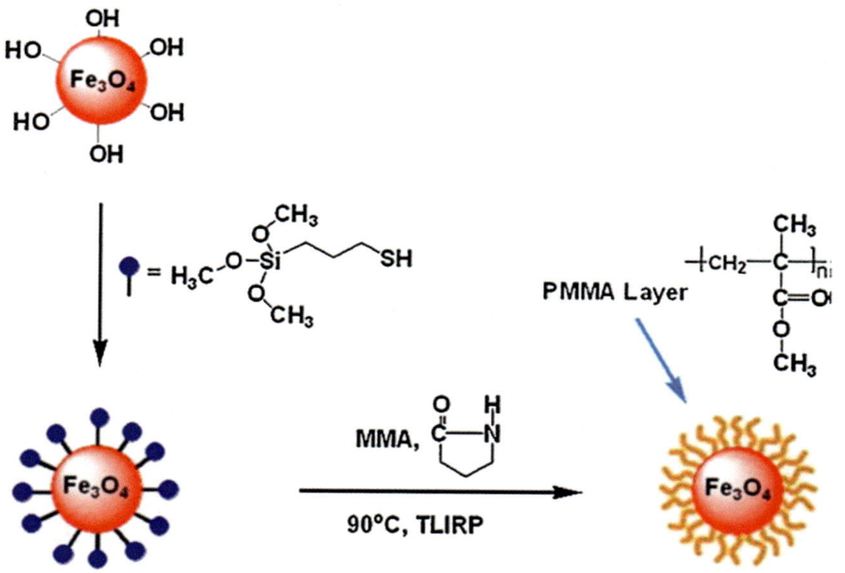

Figure 3.10 Synthesis of PMMA-modified NPs via thiol-lactam polymerization grafting approach.

poly(methyl methacrylate) on Fe$_3$O$_4$ magnetic nanoparticles using a similar approach based on thiolactam-based radical polymerization (Fig. 3.10).[21]

From a catalytic viewpoint, Marshall and Medlin have demonstrated that the modification of a metal nanoparticle with organic moieties can improve the product selectivity in a reaction.[22] This enhanced selectivity is independent of the tail type, and can only be ascribed to electronic effects produced by the presence of the ligand head group and by geometric effects due to the capping layers on the surface that induce precise adsorption conformations.

Notwithstanding the versatility of these surface-modified materials, the industrial scale up and utilization has been long hampered by some disadvantages, such as the necessity for organic solvents and expensive or noxious reducing agents during synthesis. The discovery of hexadecyl-2-hydroxyethyl-dimethyl ammonium dihydrogen phosphate (HHDMA) as a ligand that combines reducing and stabilizing functions in a single, water-soluble molecule has permitted the first industrial-scale preparation of colloidally prepared Pd, Pt, and mixed Pd—Pt nanoparticles supported on carbon and silica carriers. Vilé and colleagues[23] have proven the versatility of this new generation of materials to catalyze the hydrogenation of alkynes and nitroaromatics. To understand the experimental results, dedicated theoretical calculations have revealed that the ligand not only isolate and tailor the accessibility to the active site and the adsorption modes of reactants and intermediates (geometric effects), but alter the energy landscape of the catalytic reaction as well (electronic effects; Fig. 3.11). In fact, the head group of the ligand

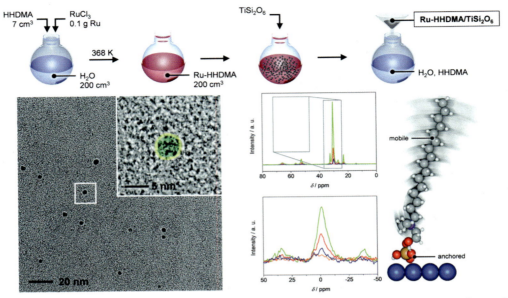

Figure 3.11 Synthesis and structure of HHDMA-modified Ru-based catalytic nanoparticles and surface characterization methods to visualize the ligand and observe the ligand motion via Nuclear Magnetic Resonance spectroscopy. *(Taken from Refs. 23 and 24)*.

may partake the reaction, absorbing some of the reaction intermediates (in the case of hydrogenation reactions, these are H species coming from the H_2 activation), thus acting as an effective cocatalyst.

6. Surface modification via single-atom anchoring

Heterogeneous catalysts modified with isolated metal species dispersed on solid supports have recently emerged as a new frontier in catalyst design and are attracting a considerable attention. These materials feature individual metals stabilized on appropriate carriers and represent a paradigm of site isolation, hailed as the limit of heterogeneous catalysis for their ability to "economize" the metal content. Owing to this maximum efficiency of atom-utilization and their unique structures and electronic properties, single-atom catalysts have shown potential to enhance the segregation of the active phases and maximize the product selectivity.[25] Being heterogeneous in nature, this merges the benefits of heterogeneous and homogeneous catalysis (Fig. 3.12). Typically, single-atom catalysts (SACs) are distinguished from single-site catalysts, a term more broadly applicable to materials with immobilized enzymes and organometallic complexes, as well as heteropolyatoms like polyoxometalates.

Synthesis of surface-modified nanomaterials 67

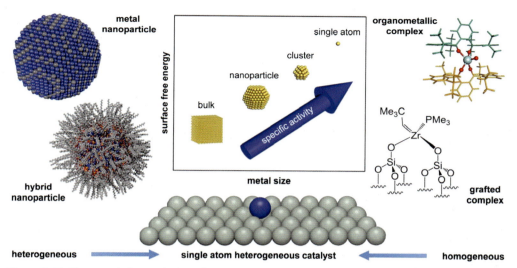

Figure 3.12 The search for catalysts with maximal activity and selectivity that can be easily recovered and reused has led to an increasing convergence between the two main pillars of heterogeneous and homogeneous catalysis. This has driven toward the increasingly precise engineering of (hybrid) metal nanoparticles, supported homogeneous catalysts, and ultimately single-atom heterogeneous catalysts.

One of the earliest SACs was reported in 1999 and featured atomically dispersed Pt species on MgO.[26] Evaluated in propane oxidation, the material was as active as the reference catalyst made of MgO-stabilized Pt nanoparticles. The catalyst, however, was not stable and the difficulty to precisely characterize the metal atoms, together with the active site restructuring under reaction conditions, left questions on the nature of the "active site." In 2003 Flytzani-Stephanopoulos and colleagues[27] used a model catalyst to provided evidence that cationic Au^{3+} or Pt^{2+} species on CeO_2 were responsible for the activity observed in the water-gas shift reaction. A few years later, Wilson and coworkers[28] reported single-atom Pd on mesoporous Al_2O_3 for the selective oxidation of allylic alcohols. Proof for the presence of single-atom came from X-ray absorption spectroscopy (XAS), which gave bond lengths consistent with Pd—O and not with Pd—Pd. In 2011, Zhang reported that single Pt atoms deposited on FeO_x were active for CO oxidation.[29] The group also used high-angle annular dark-field scanning transmission electron microscopy (HAADF-STEM), XAS, and computational techniques to elucidate both the bonding between the Pt atoms and the FeO_x carrier, as well as the catalytic mechanism. A strong metal support interaction was thought to be critical to preventing aggregation of the single atoms, but how to exploit this property to stabilize these active species was unclear. In this context, Vilé et al. demonstrated that by tuning the support structure, and choosing porous carbon-based materials, it is possible to tenaciously entrap isolated single metals into inert carriers.[25] This enabled the first application of stable single-atom catalysts in a relevant industrial process (i.e., the hydrogenation of alkynes to make vitamins). It also demonstrated the very first class of SACs based on carbon materials as support.

After these early contributions, the topic of SACs has become very hot. Table 3.5 overviews some of the SACs reported in the last few years and the targeted applications. A prerequisite is the attainment of an atomic dispersion of the desired metal species. However, fabrication of such SACs is a major challenge because of the tendency of aggregation of single-metal atoms, which can occur during both the synthesis and application of these materials.

Synthetically, SACs are prepared by incorporation of single-metal species on carbon nitride or by the creation of a joint electronic network. In the first case, the metal precursor is added after synthesis of the carrier, and the impregnation is done under microwave irradiation. In the second case, the metal precursor is added during synthesis of the carrier, and copolymerizes with it. Typically, SACs feature very low loadings of transition metal deposited on high-surface-area supports. In terms of applications, SACs have proven to be powerful catalysts in a variety of oxidation, hydrogenation, and electron-driven reactions. For the SACs identified to date, the majority are based on mesoporous graphitic carbon nitride (mpg-C_3N_4) carrier, a material offering very attractive features: it is inexpensive, easily and reproducibly prepared, and has a high thermal and chemical stability. Moreover, the support can be prepared with high surface area and in an exfoliated form, with widely separated layers; this enhances the accessibility of the metal centers. Finally, the specific interaction of the metal with the support, isolated in sixfold cavities and strongly coordinated with neighboring nitrogen atoms, has been found to enable a unique reaction mechanism. The possible active participation of the support in

Table 3.5 Overview of selected SACs reported in the last years.

Catalyst	Metal content	Synthesis	Application
Pt/FeO$_x$	0.17 wt%	Coprecipitation	CO oxidation
Pt/Cu(111)	0.02 ML	UHV deposition	Selective hydrogenations
Pt/graphene	1.52 wt%	Atomic layer deposition	CH_3OH oxidation
Ir/FeO$_x$	0.01 wt%	Coprecipitation	Water-gas shift
Pd/Cu(111)	0.01 ML	UHV deposition	Selective hydrogenations
Pd/graphene	0.25 wt%	Atomic layer deposition	Selective hydrogenations
Pd/g-C_3N_4	0.5 wt%	Deposition-reduction	Selective hydrogenations
Pd/γ-alumina	0.2 wt%	Wet impregnation/sol-gel	Selective hydrogenations
Pd–La/γ-alumina	0.5 wt%	Dry impregnation	CO oxidation
Ag/g-C_3N_4	1 wt%	Reactive copolymerization	Selective hydrogenations
Au/FeO$_x$	0.09 wt%	Wet impregnation	CO oxidation

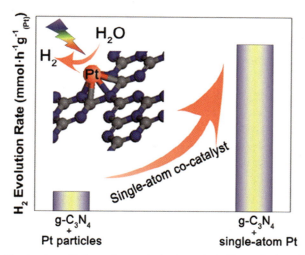

Figure 3.13 Single-atom Pt for enhanced photocatalytic H$_2$ evolution. *(From Ref. 30)*.

the catalytic reaction has been confirmed with density functional theory calculations, which have shown that the carrier can dissociate hydrogen atoms and coordinate intermediates, generating a positive charge that pushes the Pd slightly out of the support.

From a catalytic viewpoint, surface-modified nanomaterials with SAC often show outstanding performance compared to the nanoparticle-based catalysts. For example, Li et al.[30] have shown that single-atom Pt can significantly enhance the photocatalytic H$_2$ evolution (Fig. 3.13), reaching a 8.6 times higher performance than that of Pt nanoparticles and up to 50 times that for bare g-C$_3$N$_4$.

7. Industrial-scale utilization of synthetic methods to prepare surface-modified nanomaterials

As we have seen in this chapter, surface-modified nanomaterials are designed and made to enhance the properties of the pristine material and are now having a rapidly expanding range of applications in our everyday life. Most efforts have focused on the use of these surface-modified nanomaterials (mainly based on metal oxides and carbon-based carriers) in catalysis. These nanomaterials are used not only in research laboratories, but also in industrial practice. As illustrated by Vance et al.[31] more than 1800 commercial surface-modified nanocatalysts based on surface-modified nanomaterials are used at industrial scale (from more than 600 companies around the world).

8. Challenges, future perspective, and conclusions

Nanomaterials are widely used as catalysts because of their unique properties coming from the highly tuned surface structure. In this chapter, key synthesis methods to engineer surface-modified nanomaterials have been presented. These methods apply to a variety of materials classes, such as oxides, carbon, graphene, and so on. Surface-functionalized nanomaterials were found to enhance the adsorption capacity and catalytic property of the nanocatalysts due to change in surface chemistry.

Despite promising results on the use of those nanomaterials in industry, there are still some bottlenecks that need to be overcome for the practical applications. Toxicity and reusability are some key challenges to be looked for in the years to come. Generally, nanomaterials tend to form agglomerates in aqueous solution through various surface forces and the use in liquid-phase applications is often challenging. Another aspect of key interest is the potential leaching of the adsorbed modifier (i.e., silanol nests, ligands, or surface atoms); systematic plant scale studies need to be undertaken to evaluate this aspect. At industrial scale, materials cost is the most important factor, which is higher in case of carbon-based nanomaterials that can be bypassed by converting agricultural waste to carbon-based products.

References

1. Somorjai GA. Modern concepts in surface science and heterogeneous catalysis. *J Phys Chem* 1990;**94**: 1013−23.
2. Muzzio M, Li J, Yin Z, Delahunty IM, Xie J, Sun S. Monodisperse nanoparticles for catalysis and nanomedicine. *Nanoscale* 2019;**11**:18946−67.
3. Xia Y, Yang H, Campbell CT. Nanoparticles for catalysis. *Acc Chem Res* 2013;**46**:1671−2.
4. Jeyaraj M, Gurunathan S, Qasim M, Kang M-H, Kim J-H. A comprehensive review on the synthesis, characterization, and biomedical application of platinum nanoparticles. *Nanomaterials* 2019;**9**:1719.
5. Vilé G, Albani D, Almora-Barrios N, López N, Pérez-Ramírez J. Advances in the design of nanostructured catalysts for selective hydrogenation. *ChemCatChem* 2016;**1**:21−33.
6. Dhand C, Dwivedi N, Loh XJ, Ying ANJ, Verma NK, Beuerman RW, et al. Methods and strategies for the synthesis of diverse nanoparticles and their applications: a comprehensive overview. *RSC Adv* 2015;**5**:105003−37.
7. Franzel L, Bertino MF, Hub ZJ, Carpenter EE. Synthesis of magnetic nanoparticles by pulsed laser ablation. *Appl Surf Sci* 2012;**261**:332−6.
8. Brunauer S, Emmett PH, Teller E. Adsorption of gases in multimolecular layers. *J Am Chem Soc* 1938; **60**:309−19.
9. Sietsma J, Meeldijk J, den Breejen J, Versluijs-Helder M, van Dillen A, de Jongh P, et al. The preparation of supported NiO and Co_3O_4 nanoparticles by the nitric oxide controlled thermal decomposition of nitrates. *Angew Chem Int Ed* 2007;**46**:4547−9.
10. Sietsma J, Friedrich H, Broersma A, Versluijs-Helder M, van Dillen AJ, de Jongh P, et al. How nitric oxide affects the decomposition of supported nickel nitrate to arrive at highly dispersed catalysts. *J Catal* 2008;**260**:227−35.
11. Vilé G, Baudouin D, Remediakis IN, Copéret C, López N, Pérez-Ramírez J. Silver nanoparticles for olefin production: new insights into the mechanistic description of propyne hydrogenation. *ChemCatChem* 2013;**5**:3750−9.

12. Wang C, Mao H, Wang C, Fu S. Dispersibility and hydrophobicity analysis of titanium dioxide nanoparticles grafted with silane coupling agent. *Ind Eng Chem Res* 2011;**50**:11930—4.
13. Plueddemann EP. Adhesion through silane coupling agents. *J Adhes* 1970;**2**:184—201.
14. Zhao J, Milanova M, Warmoeskerken M, Dutschk V. Surface modification of TiO_2 nanoparticles with silane coupling agents. *Colloids Surf A* 2012;**413**:273—9.
15. Ukaji E, Furusawa T, Sato M, Suzuki N. The effect of surface modification with silane coupling agent on suppressing the photo-catalytic activity of fine TiO_2 particles as inorganic UV filter. *Appl Surf Sci* 2007;**254**:563—9.
16. Guo Z, Pereira T, Choi O, Wang Y, Hahn HT. Surface functionalized alumina nanoparticle filled polymeric nanocomposites with enhanced mechanical properties. *J Mater Chem* 2006;**16**:2800—8.
17. Heuer-Jungemann A, Feliu N, Bakaimi I, Hamaly M, Alkilany A, Chakraborty I, et al. The role of ligands in the chemical synthesis and applications of inorganic nanoparticles. *Chem Rev* 2019;**119**:4819—80.
18. Joo SH, Park JY, Tsung C-K, Yamada Y, Yang P, Somorjai G. Thermally stable Pt/mesoporous silica core-shell nanocatalysts for high-temperature reactions. *Nat Mater* 2009;**8**:126—31.
19. Conte M, Miyamura H, Kobayashi S, Chechik V. Spin trapping of Au-H intermediate in the alcohol oxidation by supported and unsupported gold catalysts. *J Am Chem Soc* 2009;**131**:7189—96.
20. Tsubokawa N, Hayashi S, Nishimura J. Grafting of hyperbranched polymers onto ultrafine silica: postgraft polymerization of vinyl monomers initiated by pendant azo groups of grafted polymer chains on the surface. *Prog Org Coating* 2002;**44**:69—74.
21. Bach LG, Islam MR, Kim JT, Seo SY, Lim KT. Encapsulation of Fe_3O_4 magnetic nanoparticles with poly(methyl methacrylate) *via* surface functionalized thiol-lactam initiated radical polymerization. *Appl Surf Sci* 2012;**258**:2959—66.
22. Marshall ST, O'Brien M, Oetter B, Corpuz A, Richards RM, Schwartz DK, et al. Controlled selectivity for palladium catalysts using self-assembled monolayers. *Nat Mater* 2010;**9**:853.
23. Albani D, Li Q, Vilé G, Mitchell S, Almora-Barrios N, Witte PT, et al. Interfacial acidity in ligand-modified ruthenium nanoparticles boosts the hydrogenation of levulinic acid to gamma-valerolactone. *Green Chem* 2017;**19**:2361—70.
24. Albani D, Vilé G, Mitchell S, Witte PT, Almora-Barrios N, Verel R, et al. Ligand ordering determines the catalytic response of hybrid palladium nanoparticles in hydrogenation. *Catal Sci Technol* 2016;**6**:1621—31.
25. Vilé G, Albani D, Nachtegaal M, Chen Z, Dontsova D, Antonietti M, et al. A stable single-site palladium catalyst for hydrogenations. *Angew Chem Int Ed* 2015;**54**:11265—9.
26. Asakura K, Nagahiro H, Ichikuni N, Iwasawa Y. Structure and catalytic combustion activity of atomically dispersed Pt species at MgO surface. *Appl Catal* 1999;**188**:313—24.
27. Fu Q, Saltsburg H, Flytzani-Stephanopoulos M. Active nonmetallic Au and Pt species on ceria-based water-gas shift catalysts. *Science* 2003;**301**:935—8.
28. Hackett SFJ, Brydson RM, Gass MH, Harvey I, Newman AD, Wilson K, et al. High-activity, single-site mesoporous Pd/Al_2O_3 catalysts for selective aerobic oxidation of allylic alcohols. *Angew Chem Int Ed* 2007;**46**:8593—6.
29. Qiao B, Wang A, Yang X, Allard LF, Jiang Z, Cui Y, et al. Single-atom catalysis of CO oxidation using Pt_1/FeO_x. *Nat Chem* 2011;**3**:634—41.
30. Li X, Bi W, Zhang L, Tao S, Chu W, Zhang Q, et al. Single-atom Pt as co-catalyst for enhanced photocatalytic H_2 evolution. *Adv Mater* 2016;**28**:2427—31.
31. Vance ME, Kuiken T, Vejerano EP, McGinnis SP, Hochella MF, Rejeski D, et al. Nanotechnology in the real world: redeveloping the nanomaterial consumer products inventory. *Beilstein J Nanotechnol* 2015;**6**:1769.

CHAPTER 4

Surface modification of nano-based catalytic materials for enhanced water treatment applications

Eleni Petala, Amaresh C. Pradhan and Jan Filip
Regional Centre of Advanced Technologies and Materials, Czech Advanced Technology and Research Institute, Palacký University Olomouc, Olomouc, Czech Republic

1. Nanomaterials for water treatment

During the last two decades, water and wastewater treatment procedures gain new perspective due the exploitation of nanosized materials. Nanoscale materials perform adsorptive, degrading, and antimicrobial properties toward a large range of pollutants, for example, organic compounds, heavy metals, and waterborne microbes. The extraordinary features of the nanomaterials such as high surface area, high chemical reactivity, and functionalization potential can contribute to an efficient enhanced environmental remediation scenarios. Many different nanoscale materials have been studied and exploited toward environmental applications, such as nanoscale zeolites, metal and metal oxides, and carbon structures, for example, nanotubes and fibers.[1] The nanomaterials used for water treatment could be generally classified according to their function as shown in Table 4.1.

2. Advancement on modification of nanomaterials

For the environmental technologies sector to achieve optimum remediation but also economic results, it must reduce operational costs and increase the toxics removal capacity. Several nanomaterials can meet these needs, performing high remediation capacity and applicability. However, because of the ever-expanding complexity and toxicity of the polluted sites, the required water treatment processes became more complex and expensive, and thus advanced approaches in terms of technological/chemical engineering and cost/benefit perspectives are needed. All the aforementioned nanomaterials exhibit advantages as well as disadvantages from the technological and chemical perspectives, while they perform various degrees of toxic compounds removal efficiencies under different conditions. For instance, TiO_2-based materials do not absorb visible light, a fact that limits their applicability; thus a modification that extends the application range in terms

Table 4.1 Examples of nanomaterials for water treatment.

No	Nanomaterial	Examples
1	1. Nanoadsorbents	Carbon structures as carbon nanotubes and graphene,[2] manganese oxide,[3] zinc oxide,[4] titanium oxide,[5] magnesium oxide,[6] and ferric oxides[7]
2	2. Nanocatalysts, i.e., *photocatalysts, electrocatalysts, chemical oxidants*	TiO_2, ZnO, CdS, WO_3[8,9], graphene-based materials, noble metal, Au, Ag, Pt, and Pd, nanoparticles, zero-valent metal nanoparticles and metal oxide nanoparticles[10]
3	Nanomembranes	Two-dimensional (2D) material-based membranes, for example, graphene, metal-organic framework (MOFs), graphitic carbon nitride (g-C_3N_4), zeolites,[11] polymeric membrane doped with nanoparticles[12]
4	Biomaterials	Nanoscale cellulosic materials,[13] metal oxide-biochar nanocomposites loaded with bacteria[14]

of light conditions of these materials is necessary. Moreover, magnetic nanoparticles possess attractive magnetic interactions that leads to agglomeration and aggregation, affecting their mobility, applicability, and the overall remediation capacity. Furthermore, a series of parameters, such as water solubility, chemical reactivity, and presence of other substances that can bind onto the surface of the particles can also affect the mobility of the nanoparticles in aquifer media.[15]

By various modification routes of the nanoparticles, properties such as hydrophilicity/phobicity, porosity-surface area, mechanical strength, mobility, and dispersibility can be optimized and tuned, targeting specific water treatment conditions.[16] Polymer-coated nanoparticles consist of a wide part of research toward that approach, since polymers can add to the final material values such as dispersibility and increase of the binding sites for adsorption of toxic compounds.[17] Most common polymers that are used are the biopolymers, such as polysaccharides, that by coating the surface of the particles can offer stability and homogeneous size distribution (Fig. 4.1).[18]

Furthermore, a big trend has been prevailed toward the preparation of advanced composites, that possess multifunctional properties and synergistic effects toward water treatment, combining different removal mechanisms, that is, absorption and reduction. The obtained composites can be easily modified and functionalized further to increase their removal ability. For instance, magnetic photocatalytic materials composed of zero-valent iron nanoparticles immobilized on the surface of porous TiO_2 matrix have shown synergistic photocatalytic and reduction properties toward Cr(VI) removal,[20] while graphene-based nanocomposites are promising candidates for removal of various pollutants in water such as antibiotics,[21] oils, and dyes.[22]

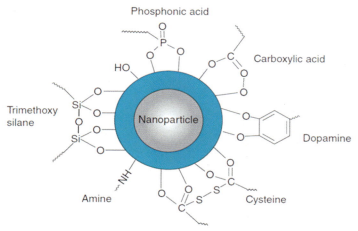

Figure 4.1 Anchored polymers and functional groups onto iron-oxide nanoparticles surface.[19]

3. Examples of nanoscale modified catalytic materials

3.1 TiO$_2$-based materials

The drawbacks of TiO$_2$ photocatalyst are the fast recombination of electrons and holes and the fact that it does not absorb visible light. The photocatalytic efficiency of TiO$_2$ depends strongly on the degree of recombination of electrons and holes which occurs within a few nanoseconds in the absence of any suitable "electron or hole scavenger" that could limit this recombination.[23] Metal ions and metal nanoparticles, such as Pt, Nd, and Fe, dispersed on the TiO$_2$ surface,[24] have been used as electron scavengers to trap the photo-generated electrons.[25,26] This inhibits their recombination with positive holes and, hence, improves the photocatalytic efficiency of the TiO$_2$ surface. Moreover, substitution of various transition elements to the Ti site, and various anions to the oxygen site (e.g., N, C, B, and S) can shift the light absorption to visible-light region.[27] In Fig. 4.2 it is shown that a doped TiO$_2$-based material can exhibit different optical properties in comparison with simple TiO$_2$. Coupling between diverse semiconductors with different energy levels, that is, coupling of TiO$_2$ with other narrow bandgap semiconductors (e.g., CdS, Ag$_3$PO$_4$), is of great interest while can extend the absorption of TiO$_2$ in the visible light range.[28,29]

Furthermore, composite materials based on TiO$_2$ (e.g., incorporation of carbon nanomaterials such as carbon nanotubes and graphene) can merge the benefits for every component and enhance the properties of each other through a synergetic effect. For instance, decorating the particle surface with a noble metal and coating with a lower-bandgap material have been mentioned to enhance the efficiency of TiO$_2$ as a catalyst.

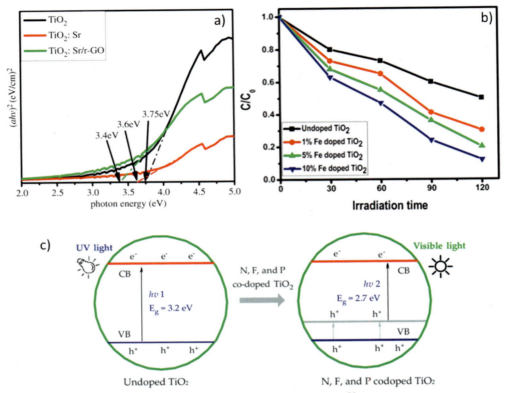

Figure 4.2 (A) Band gap estimation of TiO$_2$, TiO$_2$: Sr, TiO$_2$: Sr/r-GO.[30] (B) Compared photocatalytic degradation of methylene blue by undoped and Fe doped TiO$_2$.[31] (C) Illustration of the band gap energy levels of doped TiO$_2$ material with enhanced visible-light response.[32]

Doping with other metal oxides can extend the light absorption to visible-light region by providing trapping sites and changing the bandgap thus also the requisite energy to activate the TiO$_2$.[33,34]

3.2 Carbon-based materials for water treatment

Carbon materials have been widely used for environmental applications due to their significant sorption properties. The surface sites or free valences at the edges of a carbon material are crucial for any adsorption process. Thus, various pollutants can be removed from aqueous solution by such materials via different adsorption mechanism such as electrostatic and nonelectrostatic interactions, hydrophobic, or donor–acceptor interactions where the electron donors are the surface carbonyl groups of the material. In some cases, a carbon material can incorporate more than one of the possible interactions, thus, increasing the selectivity. One of these examples are the carbon nanotubes, CNTs, where noncovalent forces, such as hydrogen bonding, π-π stacking, electrostatic forces, van der Waals forces, and hydrophobic interactions can be combined boosting the efficiency of these materials.[35]

Carbon materials can be easily modified and functionalized to increase their adsorption ability, for example, by treatment with oxidizing agents that increases the hydrophilicity, inducing a large number of oxygen-containing groups that enhance the chelation ability with metal species, and by treatment with basic agents that enhances uptake of organics.[36] The type of contaminant that can be adsorbed and the efficiency of carbon materials are strongly dependent on their surface chemical characteristics. Thus, the modification of carbon materials can lead to materials focused on specific applications. Table 4.2 shows numerous carbon materials and their derivatives in combination with iron species toward various pollutants removal.

3.3 NZVI-based materials for water treatment

Iron-based nanoparticles are among the most promising materials for the removal of various toxic compounds. In general, three iron phases have been widely used, namely, magnetic nanoparticles of zero-valent iron (NZVI), and two types of iron oxides, that is, nanoparticles of magnetite (Fe_3O_4) and maghemite (γ-Fe_2O_3). All of these phases show excellent efficiency in removing heavy metals, for example, Cr(VI), As(V), Ni(II), and organic compounds; azo dyes, pesticides, pharmaceuticals, and inorganic ions; and nitrate (NO_3^-) and perchlorate (ClO_4^-).[53]

Among these materials, NZVI is widely considered to be an attractive remediation medium with significantly high efficiency in removing toxic compounds,[54] low-cost, and with low environmental impact associated to iron handling[55] due to their versatility[56] and multimodal action.[57] NZVI particles have been used both at laboratory scale and in

Table 4.2 Various carbon materials that have been tested toward different pollutants.

Nanomaterial/nanocomposite	Chemical and biological removal
MWNT/NZVI	Heavy metals, e.g., Cr,[37] organics, e.g., dyes[38]
GO/NZVI or RGO/NZVI	Heavy metals, e.g., U,[39] Pb,[40] organics, e.g., *chlorinated organic compounds (trichloroethylene)*[41]
GO/Fe_3O_4 or rGO/Fe_3O_4	Heavy metals, e.g., Co,[42] Pb,[43] Se,[44] organics, e.g., *1-naphthol, 1-naphthylamine, dyes (malachite green and rhodamine)*[45]
GO/α-FeOOH	Oils, e.g., *gasoline, paraffin*, heavy metals, e.g., Cr, Pb[46]
AC/Fe_2O_3 and AC/Fe_2O_3	VOCs, e.g., *phenol, chloroform, and chlorobenzene*,[37] heavy metals, e.g., As, Hg and Pb,[47] herbicides, e.g., *atrazine*[48]
AC/NZVI	Heavy metals, e.g., Cr,[49] organics, e.g., *chlorinated organic compounds*,[50] oxyanions, e.g., *bromate*,[51] *phosphate*[52]

the field[58] where they proved to be very effective in various environmental remediation processes. Particularly in water treatment, NZVI's ability to remove a wide range of toxic compounds has been thoroughly investigated.

NZVI has been shown to have a much greater capacity, particularly in terms of metal ions removal, than conventional technologies,[59] that is, ion exchange, filtration, electrochemical precipitation, adsorption, and bioremediation.[60] Moreover, the reaction of NZVI with pollutants, and their subsequent remediation, are generally rapid, as proved by numerous studies. The great performance of NZVI relies on the combination of reducing properties with high sorption capabilities.

The main routes of pollutants sequestration by NZVI include electrochemical reduction (i.e., conversion of metals to less toxic and less soluble forms by lowering their oxidation state), reductive mineral precipitation and coprecipitation (i.e., due to a significant reduction in oxidation–reduction potential (ORP), from oxidative to neutral or anoxic conditions), the formation of insoluble hydroxides, and sorption mechanisms onto the surface of nanocrystalline iron oxyhydroxides that are formed due to the occurrence of NZVI oxidation.[61] As it can be seen in Fig. 4.3, the application of NZVI has the potential to trap several metals at the same time. Additionally, the metals trapped on the NZVI surface could potentially be collected and recycled.

NZVI is considered an excellent electron donor that can initiate redox reactions. The relative number of electrons transferred from NZVI to any competing substance can be defined as its overall efficiency. NZVI can react with a variety of natural oxidants that coexist in a solvent, including water, contaminants, and dissolved oxygen. He et al.[10] defined efficiencies for each individual process, as it is depicted in Fig. 4.4, providing significant insights in the quantification of selectivity and efficiency of NZVI. Generally, a model system regarding prediction of efficiency and selectivity of NZVI is a key tool that can determine its overall performance according to the environmental conditions each time (Fig. 4.5).

3.3.1 Modification needs and routes of modification of ZVI

Despite the known removal activity of NZVI particles toward organic and inorganic pollutants, their large-scale use is usually limited by their relatively low flux rate in aquifers, their limited mobility and longevity, and concerns about cost-benefit. The application of most as-synthesized NZVI particles face specific obstacles, which eliminate the overall performance of NZVI. Bare NZVI nanoparticles typically show some disadvantages including the strong tendency to rapid aggregation, sedimentation, and oxidation phenomena.[64]

Due to its unstable colloidal nature and the particles' extended attractive magnetic interactions, NZVI shows a strong tendency to agglomerate and aggregate, which affects its specific surface area. Gravitational sedimentation of the particles is a consequence of the aggregation phenomena that occur. NZVI tends to aggregate[65] and adhere onto various

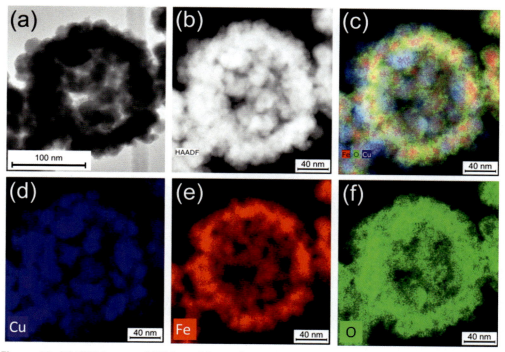

Figure 4.3 (A) TEM image of NZVI particles with 3D arrangement recorded after 90 min of copper entrapment; (B) STEM-HAADF of NZVI with 3D arrangement and Cu adsorbed; (C) STEM-XEDS mapping with overlay of elemental Fe (*red*), Cu (*blue*), and O (*green*); and panel (D) to (E) show individual STEM-XEDS images of Cu, Fe, and O, respectively. *(Reprinted with permission from Ref. 62.)*

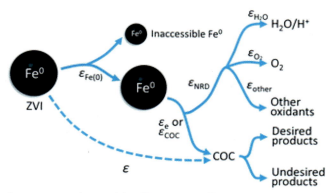

Figure 4.4 Dependence among the particle efficiency (ε), efficiency of Fe(0) utilization ($\varepsilon_{Fe(0)}$), electron efficiency (ε_e or ε_{coc}) of contaminant reduction (where COC is: contaminants of concern), and the electron efficiency of natural reductant demand, NRD, (e.g., H_2O, O_2, and other oxidants) reduction (ε_{NRD}). *(Reprinted with permission from Ref. 63.)*

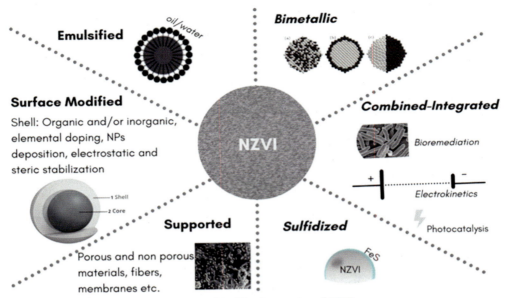

Figure 4.5 Modification routes of NZVI.

mineral grains (such as quartz and clay minerals in sedimentary rocks), and their migration is not allowed for more than a few meters from the site of a subsurface application.[66] Consequently, properties and capacities related to the overall performance and applicability, such as delivery, mobility, and reactivity, are strongly affected.

Moreover, NZVI particles often display unpredictably fast oxidation, both in air and aqueous environments, leading to undesirable oxidation-dissolution or passivation.[67] Thus, the reactivity of NZVI can be potentially affected by this oxidation, causing a gradual loss of efficiency in interacting with pollutants. Thus, in the case of complex real-world scenarios, where the degree of reaction-selectivity is challenged, NZVI's efficiency in target contaminants removal decreases.

These phenomena are highly detrimental since they cause sedimentation and inactivation of nanoparticles. Therefore, specific strategies should be employed to modify NZVI particles in a such way to successfully improve reactivity, in terms of reduced aggregation potential, reduced interparticle interactions, and limited adhesion to the surface of mineral grains.[68] The morphological factor of nanoparticles is considered another key feature that determines the activity and stability of the nanoparticles. Therefore, the chemical ability to control and devise morphologically homogeneous NZVI particle architectures is extensively investigated.

These reasons have moved the scientific community to the field of modification and stabilization of the NZVI particles, leading to a new perspective related to their remediation ability, mobility, and stability. It has to be mentioned that according to the targeted application and properties of the nanomaterial, it is possible and feasible to combine more than one modification pathway in one material, for example, support material and surfactant, which remarkably improves the remediation performance of the materials. This section presents the most efficient and well-studied methods that successfully contribute to overcoming different drawbacks of this technology and to expanding the range of nanoparticle applications. Thus, surface-modification routes such as the use of polyelectrolytes or nonionic surfactants as coaters, emulsified dispersions, hybrid composites using a support material, bimetallic particles, and sulfidation are described. Their contribution to overcome the particles magnetic attraction and change the surface and interfacial features as well as the stability, mobility, aggregation, and reactivity is further explained.

3.3.2 Surface modifiers

Surface modification of NZVI by various surfactants can lead to electrostatic and steric stabilization and the combination of them, named as electrosteric stabilization (Fig. 4.6). Electrostatic stabilization interferes in the surface charge that allows the repulsive forces to prevail over aggregation while steric stabilization involves long-chain hydrophilic polymers onto the surface of the particles to succeed the prevention of aggregation (Fig. 4.7). In that way, it is possible to tune the properties of NZVI and provide customized features targeting specific enhanced application. Generally, the surface modifiers facilitate to change the magnetic interactions and the surface and interfacial

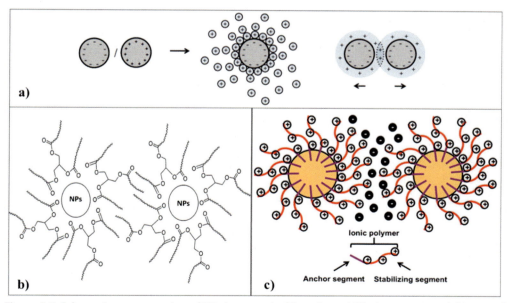

Figure 4.6 Schematic representation of (A) electrostatic, (B) steric, and (C) electrosteric stabilization of nanoparticles. *(Reprinted with permission from (A) Ref. 69, (B) Ref. 70.)*

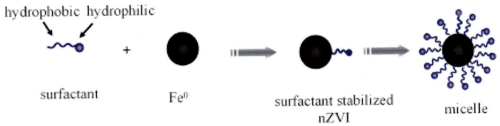

Figure 4.7 Example of surface modification of NZVI-Schematic diagram of the mechanism of NZVI stabilized by sodium dodecyl sulfate (SDS). *(Reprinted with permission from Ref. 71.)*

properties. Consequently, they can enhance the colloidal stability, mobility, well dispersion, and high efficacy. Electrosteric stabilization with ionic polymeric molecules is the most promising approach, which can lead to decreased agglomeration of NZVI particles and higher surface contaminant concentration and sorption. Features such as mobility in porous media, the hydrodynamic diameter, the activation energy and oxidation can be also tuned. Table 4.3 presents the most used surface modifiers.

3.3.3 Bimetallic particles

Deposition of a transition or a noble metal (e.g., Pd, Ni, Cu, Pt, and Ag) onto the NZVI surface, forming iron-based bimetallic nanoparticles, has shown a sufficiently enhanced degradation capacity.[92] The additive metal causes the increase of the reaction rate by lowering its activation energy. For instance, bimetallic nanoparticles, especially the ones involving Fe–Pd combination, have been reported to degrade organic halides to halogen-free products more efficiently, with reaction rates several orders of magnitude higher than bare NZVI.[93] This improvement is mainly determined by the formation of galvanic couple of the transition or noble metal with NZVI, which accelerates the

Table 4.3 Examples of different surface modifiers employed for NZVI stabilization.

Xanthan gum (XG)[72]	Polyethylenimine (PEI)[73]
3-Aminopropyltriethoxysilane (APS)[74]	Pluronic F-127[75]
Carboxymethyl cellulose (CMC)[76]	Polydopamine (PDA)[77]
Sodium dodecyl sulfate (SDS)[78]	Ethylene glycol (EG)[79]
Polyaspartate (PAP)[80]	Polyvinyl alcohol (PVA)[81]
Polystyrene sulfonate (PSS)[82]	Sodium laurate (SL)[83]
Polyvinylpyrrolidone (PVP)[84]	Sodium oleate (NaOA)[85]
Poly(acrylic acid) (PAA)[86]	Cetrimonium bromide (CTAB)[87]
Polyacrylamide (PAM)[88]	Sodium dodecylbenzene sulfonate (DBS)[89]
Methoxyethoxyethoxyacetic acid (MEEA)[90]	Hexadecylpyridinium chloride surfactant (HDPCl)[91]

electron transfer from the metallic core (Fe^0) to the NZVI surface where the metal additive is deposited; therefore, the metal additive promotes hydrogenation and hydrodechlorination reactions.[94] In these bimetallic particles incorporating NZVI, the deposited transition or noble metals have much higher redox potential (E_o), leading to a galvanic cell, where iron acts as the anode and the electron release reactions are significantly accelerated.[95] Bimetallic nanoparticles exhibited Fenton-like properties under aerobic conditions, forming hydroxyl radicals, while the dechlorination mechanism is prevailing among the reduction and sorption processes, enhancing the pollutants removal selectivity.[96] Moreover, it has been shown that embedded noble metals could protect NZVI against self-passivation.[97] In this context, NZVI-based bimetallic particles have been used in cases where a high reaction rate is needed compared to bare NZVI, for example, aromatics and polychlorinated biphenyls (PCBs), or where selectivity toward favorable end products is required, for example, catalytic effect of Pd/Fe for nitrate reduction[98] and Ni/Fe high dechlorination efficiency toward 1,1,1-TCA[99] (Figs. 4.8 and 4.9).

3.3.4 Sulfidation

Even though sulfidation of NZVI is a relatively new field in comparison to other modification routes, it has proven to be a very promising chemical NZVI "upgrade" approach. Chemical modification of NZVI by reduced sulfur compounds can significantly increase the reactivity of NZVI while greatly enhancing its selectivity, thus expanding its range of applications. Sulfidation of NZVI is considered technologically simple and cheap, as well as environmentally acceptable. It can be achieved by both one-step and two-step (post-synthesis treatment) synthesis processes. During this surface modification, Fe^{2+} ions, generated from the NZVI metallic core, precipitate with S^{2-} ions, forming an FeS layer

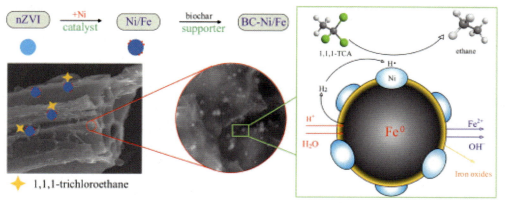

Figure 4.8 SEM images of biochar (BC)-supported Ni/Fe bimetallic nanoparticles and schematic representation of 1,1,1-TCA degradation in the BC-Ni/Fe bimetallic system. *(Reprinted with permission from Ref. 100.)*

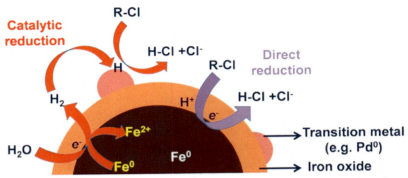

Figure 4.9 Mechanisms of reductive dechlorination by NZVI with or without Pd catalyst. *(Reprinted with permission from Ref. 101.)*

on the surface of the particles (Fig. 4.10). It has been shown that sulfidized NZVI has enhanced the remediation capacity, compared to bare NZVI, since it can inhibit the H_2 evolution reaction (HER, reduction of water by Fe^0 to form hydrogen), increase the electron efficiency and selectivity for dehalogenation, and reach greater long-term performance.[103] Cao et al. have indicated that sulfur doped onto the NZVI, forms a highly chemically reactive surface that alters the dechlorination and defluorination pathways. They have shown that S sites are the main factor for the direct electron transfer, Fe sites for the atomic H-mediated reaction, and β-elimination for the primary defluorination pathway.[104]

3.3.5 NZVI supported on various materials

The combination of a support material and NZVI in one single hybrid material or system, have led to the development of innovative methods that have been thoroughly investigated, exhibiting remarkable benefits, such as high sorption capacity, large surface area, and formation of well-dispersed nonagglomerated iron particles. The immobilization and stabilization of NZVI onto a support material can tremendously decrease the aggregation phenomena occurring in NZVI (Fig. 4.11), transforming the nanoparticles into materials with higher specific area, colloidal stability, homogeneous dispersion, narrower size distribution, and higher activity than nonsupported NZVI.[106]

Moreover, such a combination of a matrix and nanoparticles brings about synergetic properties combining absorption, reduction, and precipitation removal mechanisms, that is,[107] contributing to the maximum degree of remediation capacity and performing multifunctional altered features, for example, when it is combined with a photocatalytic material. The use of a support material is of great interest because it can extend the applicability, sustainability, and the range of NZVI's properties.

Since NZVI shows compatibility with most porous and nonporous support materials, its stability on numerous support materials has been investigated, proving that this

Surface modification of nano-based catalytic materials for enhanced water treatment applications 85

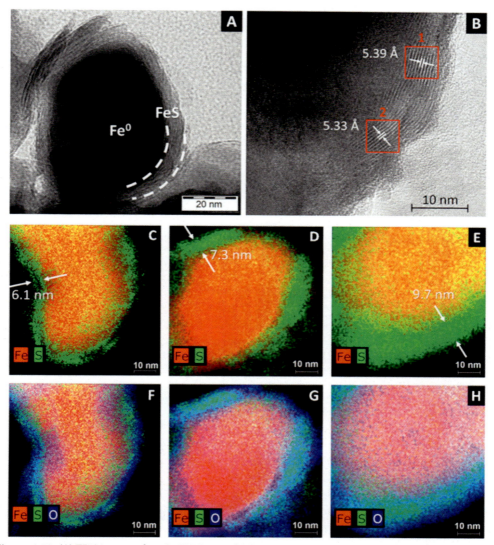

Figure 4.10 (A) TEM image of an S—NZVI particle; (B) HRTEM image on the sulfide shell structure of an S-nZVI particle; (C—H) overlay of Fe—S and Fe—S—O high-resolution EDS mapping of NZVI particles sulfidated with the following S/Fe ratios: 0.0094 (C and F), 0.0195 (D and G), and 0.0629 (E and H). *(Reprinted with permission from Ref. 102.)*

modification route is more than efficient and desirable. There is not usually a restriction in the candidate support materials and both inorganic and organic structures of advanced as well as conventional materials can be involved. It has to be noted that chelating agents are not commonly required as NZVI particles can be attached onto the support surface

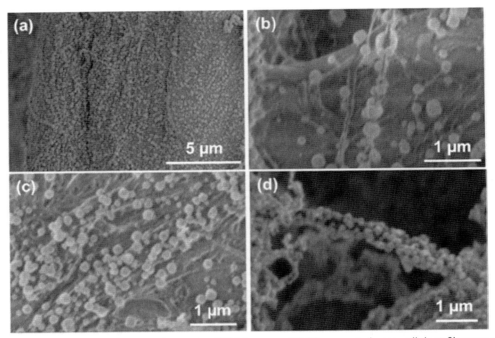

Figure 4.11 Scanning electron microscopy (SEM) images of NZVI supported onto cellulose filter paper loaded with 5% in weight of NZVI. NZVI particles are well-dispersed onto the cellulose paper, possess narrow size distribution and nonagglomerated formation. *(Reprinted with permission from Ref. 105.)*

due to the formed Fe—O— bridges,[108] which facilitates the synthesis process even more and eliminates the additional use of chemicals. Those that have been successfully combined with NZVI are clays,[109] zeolites,[110] polymer resins,[111] amorphous silica structures,[112] ordered mesoporous silica structures,[113] membranes,[114] activated carbon[115,116] and biochars,[117] graphite oxide[118] and graphene structures,[119] and ordered mesoporous carbon[120] (Fig. 4.12). Below are described two common hybrid materials categories, carbon–NZVI, and clay–NZVI structures. Porous carbon-supported NZVI particles are excellent candidates for pollutants reduction, enhancing their mobility and capability. The high surface area of the porous materials enables the increase in the reactive sites through successful dispersion of NZVI particles. Moreover, the porous carbon helps to increase their surface area and to avoid the formation of passivation layer and agglomeration of NZVI particles. Various porous carbon materials, such as carbon microspheres,[121] mesoporous carbon,[122] ordered mesoporous carbon,[123] multiwalled carbon nanotubes (MWCNTs)[124] and carbon fibers,[125] exhibit great interest due to their great potential for supporting NZVI particles, which inhibits the aggregation and passivation

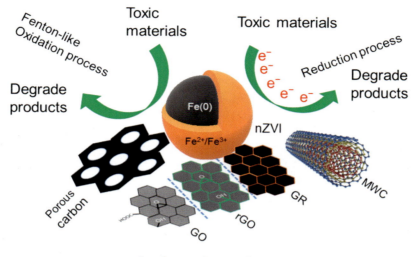

Figure 4.12 Schematic diagram of NZVI supported with carbon-based materials.

while maintaining their magnetic separation ability avoiding the environmental risk of nanoparticles. Moreover, these porous materials provide extra features while they contribute to the removal of pollutants by adsorption processes.[126]

Another category of support materials that is chosen because of its simplicity and cost-efficiency is the inorganic clays[109] such as kaolinite[127] and bentonite,[128] which have been employed as porous support for NZVI particle deposition (Fig. 4.13). These supports aid

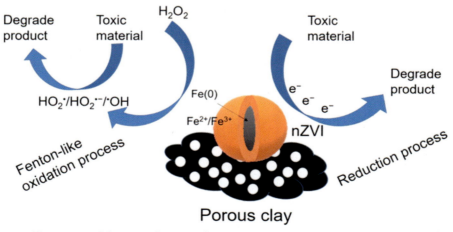

Figure 4.13 Schematic diagram of NZVI supported with clay-based material.

in preventing of aggregation, increasing dispersion of NZVI particles and performing overall stability. The inorganic porous clays mineral decorated with NZVI particles have received increasing attention as effective adsorbents for heavy metals removal. For instance, Lv et al. investigated a new strategy for hybridization of NZVI with layered double hydroxide (LDH) and reduced graphene oxide (rGO). The as-prepared Fe@LDH/rGO material possess features such as improved dispersibility, hydrophilicity, and increased positive surfaces that led to Cr(VI) higher removal capacity.[129]

3.3.6 Emulsification

Emulsified nanoscale zero-valent iron (ENZVI) is as a surface modified approach to NZVI, targeting to hinder the passivation of NZVI, thus facilitating the delivery of reactive NZVI particles in dense nonaqueous phase liquid (DNALP) source zones and ensuring the contact of pollutants with NZVI.[130] Emulsified NZVI (ENZVI) is a biodegradable emulsion, that consists of a surfactant, biodegradable vegetable oil, water, and NZVI particles that forms a biodegradable oil-water membrane cladded on NZVI.[131] This membrane serves as a protective shell to preserve the properties and reactivity of NZVI and to enhance the corrosion resistance while electrostatic and steric repulsion serves as a barrier to prevent particles aggregation (Fig. 4.14). Additionally, emulsified oil provides an effective and long-lasting medium that enhances anaerobic biodegradation by slowly releasing electrons. ENZVI contains an encapsulation solution that precludes interaction with other water constituents which could exhaust the reducing capacity of NZVI, resulting in DNALP contaminants, for example, ethenes, diffusion through the liquid membrane, and exclusively reacting with NZVI until the oil membrane is degraded by biological activity. This modification increases the reactivity of the particles, leading to reduced required amount of EZVI compared to bare ZVI. For instance, EZVI has exhibited enhanced removal efficiency from subsurface aqueous systems compared with bare NZVI of volatile organic compounds (VOCs), for example, trichloroethylene (TCE)[134] and tetrachroroethylene (PCE),[135] pesticides and inorganic species, for example, nitrate and phosphorus[136] lead, copper, nickel, and cadmium.[137]

3.3.7 Combined technologies

Because of the expanding complexity and toxicity of polluted sites, the required water treatment processes have become more complex and expensive, which needs advanced approaches in terms of technological/chemical engineering and cost/benefit perspectives. To ensure that the environmental technologies sector achieves optimum remediation but also economic results, the integration of more than one remediation approaches is deemed necessary. Specifically, the combination of nanoremediation with other remediation technologies was found to improve the biological aspect of the soil, such as bioremediation of phytoremediation.[138] NZVI is compatible with several chemical/physical

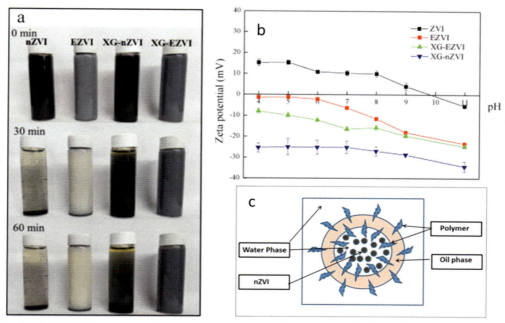

Figure 4.14 (A) Photographs of settlement processes; (B) zeta potential of modified and unmodified NZVI nanoparticles (NZVI, EZVI, xanthan gum (XG)-NZVI and XG-EZVI) at selected time interval or at different pH, respectively, proving that both emulsion and xanthan gum modification enhanced the NZVI stabilization, that is, hinder aggregation, while it is suggested that the reactivity, higher stability and migratory ability of NZVI can be simultaneously achieved by the coexistence of EZVI and XG; and (C) schematic illustration of EZVI. *(Reprinted with permission from (B) Ref. 132, (C) Ref. 133.)*

methods, for example, (1) chemical precipitation processes, (2) novel ion exchange materials and membrane filtration, (3) electrochemical methods, (4) bioremediation using aerobic and anaerobic microorganisms, (5) chemical reduction/adsorption using various materials, for example, magnetic nanoparticles or environmentally green and inert matrices, (6) semiconductor-based photocatalysis, and (7) electrokinetics.

For instance, the synergistic use of NZVI and bioremediation is a beneficial approach for contaminated sites uptake[139] because the integration of electrokinetics and NZVI[140] can couple the electrolysis reaction advantages with reduction reactions, thus increasing the benefits regarding water remediation by electromigration phenomena that extend the mobility, advection, and the operational life of NZVI.[141] Moreover, synergetic photocatalytic and reduction properties in one system enhance the capacity of NZVI by increased stability, reactivity, and possibility of reusing the photocatalytic material in consecutive cycles.[20]

3.3.8 Examples of water-treatment enhancement of NZVI-based materials

All the modification routes of NZVI can drastically change structural, morphological, chemical, and colloidal properties of the particles. A decrease in aggregation and sedimentation, an increase in the surface area, homogeneous dispersion, and tuning of the surface or interfacial properties are crucial ways that can modify the properties of NZVI particles. The modification of the properties is reflected in NZVI's performance related to water treatment application. Thus, modified particles exhibit significantly enhanced reactivity in various aspects such as increased delivery and mobility, higher remediation capacity, and long-term reactivity. Such alteration of properties can overcome any obstacles hindering NZVI large-scale application, that was previously suppressed due to their relatively low flow rate, limited mobility and lifespan, and cost-benefit risks and limitations. Herein, some representative results found in literature related to the reactivity enhancement of modified NZVI are presented.

Table 4.4 comparatively presents some results related to NZVI supported on carbon-based materials. The use of carbon materials as an ideal support for iron species (e.g., iron-oxide, iron carbide, and zero-valent iron particles) has been thoroughly investigated and showed remarkable advantages such as high sorption capacity, large surface area and well-dispersed nonagglomerated supported iron particles. In addition, such a combination brings about synergetic properties, combining absorption, reduction, and precipitation removal mechanisms, contributing to the maximum degree of remediation capacity.[147]

The enhancement of NZVI's stability and mobility by various modifications routes is important for appropriate flow rates, which leads to enhanced transport and effective application of NZVI particles in soil and groundwater remediation activities. As it is depicted in Fig. 4.15, Ibrahim et al. have shown that the relative concentration of the NZVI effluent and the flow rate were increased while the aggregation and sedimentation of particles was decreased when the modification of NZVI conducted by sodium carboxy-methyl-cellulose, CMC. This fact can definitely increase the remediation efficiency and applicability of the particles.

It is important to note the different removal mechanism pathways that occur when NZVI modification is achieved. It can lead to a different enhanced perspective of sequestration of toxic compounds from contaminated water. Singh et al.[149] synthesized sulfide-modified NZVI, S—NZVI, that possessed the characteristic core-shell structure, but with a shell being a mixture of iron-oxide and FeS. S—NZVI performed significantly enhanced As(III) and As(V) removal, compared to bare NZVI. The highest As removal occurred when the S/Fe ratio was 0.1 under acidic conditions. The main findings indicated the difference in the removal mechanism between bare NZVI and S—NZVI. In the first case, the main removal mechanism involves the reduction of the adsorbed As(III) and As(V), while in the case of S—NZVI, the removal of As(III) and As(V) combines adsorption as As(III) and As(V) oxyanion with subsequent precipitation of As_2S_3.

Table 4.4 Supported NZVI catalysts onto carbon structures for Cr (VI) degradation.

Carbon-supported NZVI	Remediation mechanism	Conditions	Time	Reactivity	Mechanism
NZVI-g-C$_3$N$_4$ nanohybrid[142]	Photoreduction/Oxidation	Catalyst dose = 30 mg, Mixed Cr (VI): RhB = 20 mg/L, visible light	150 min	92.9% Cr (VI), 99% RhB	NZVI as e$^-$ trapper, lowest e$^-$-h$^+$ recombination
Activated carbon fiber-nZVI[143]	Reduction	Acidic medium, Cr(VI) = 10 mg/L	60 min	100% Cr (VI) reduction	ACF as an electron transfer mediator
NZVI@Fe$_2$O$_3$ NPs-graphene[144]	Reduction and adsorption	Cr(VI) = 100 mg/L, catalyst = 0.41 g/500 mL, pH = 3, Temp. = 30°C	99.8% Cr (VI) reduction	120 min	NZVI as an e$^-$ donor
NZVI@modified biochar[145]	Reduction	Catalyst = 200 mg/L, Cr(VI) = 10 mg/L, 120 min, lower pH	35.29% Cr(VI) reduction	30 min	Synergy between NZVI and oxygen functional group of biochar
PEG-NZVI@BC)[146]	Reduction and adsorption	Catalyst dose = 0.05 g/40 mL, Cr(VI) = 100 mg/L, pH = 5.6	100% Cr(VI) reduction	24 h	Electron transfer between Fe and BC

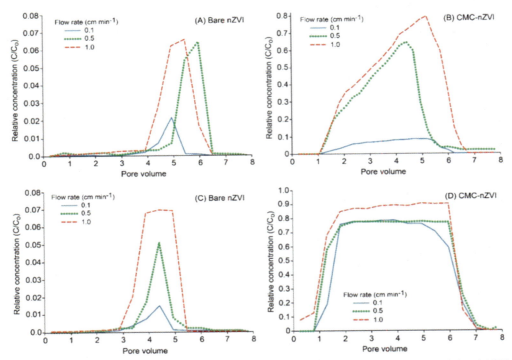

Figure 4.15 Effect of the flow rate on the breakthrough curves of CMC polymer stabilized NZVI particles at pH 9, and particle concentration of 0.1 (A and B) and 1.0 g/L (C and D). *(Reprinted with permission from Ref. 148.)*

Moreover, Du et al. have shown that sulfidation of NZVI promotes the electron transfer from the metallic core iron, Fe(0), to surface-bound iron, Fe(III), leading to single-electron transfer for the generation of H_2O_2 and hydroxyl radicals. Sulfidation of NZVI enhances both adsorption and oxidation capacities compared to bare NZVI, with the possibility of controlling the number of adsorption sites and mechanism according to the sulfidation rate (Fig. 4.16).

Lin at al. provided important findings to understand the relationship between the iron aggregate size and the transport process, showing that the transport of iron suspension was strongly affected by the main aggregate sizes.[151] Moreover, to obtain lower agglomeration of NZVI and therefore sufficient stability and mobility, an alkaline and low ionic strength environment is needed.[152] In the field of applying NZVI, especially in groundwater remediation, these parameters are difficult to be controlled. Thus, modification and stabilization methods are valuable with regard to increasing the transport ability of NZVI. Vinod et al.,[153] using a tree polysaccharide, gum karaya (GK), were able to stabilize

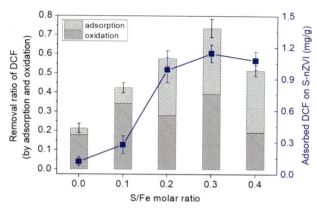

Figure 4.16 Diclofenac, DCF, removal performance of different S—NZVI with different S/Fe molar ratio. *(Reprinted with permission from Ref. 150.)*

NZVI particles for 3 months, avoiding any aggregation, and sedimentation phenomena. NZVI—GK composite exhibited much higher reduction efficiency on hexavalent chromium, Cr(VI) reduction, and volatile organic compounds, VOCs, degradation compared with bare NZVI particles (Fig. 4.17).

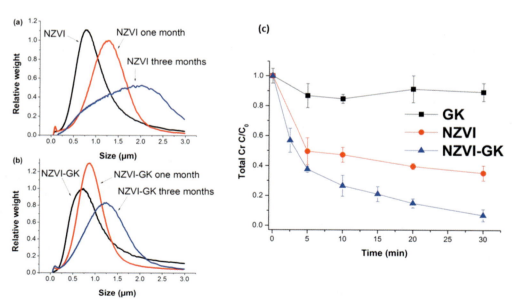

Figure 4.17 Particle size distributions of (A) bare NZVI and (B) gum karaya-modified NZVI, NZVI—GK, particles after one week, one month, and three months storage under room temperature; (C) total chromium decontamination by GK, NZVI, and NZVI—GK under the same conditions. *(Reprinted with permission from Ref. 104.)*

4. Conclusions

Nanobased materials rank among the best candidates for water treatment applications, demonstrating great water remediation activity. Their modification has received a great deal of attention because it can bring tremendous improvements, offering homogeneous small size, nonagglomeration, uniform dispersion of the particles, and higher reactivity. This chapter presented the most common and effective modification routes of nanomaterials, especially focused on NZVI, that can improve various properties of the particles in a multifunctional way. The modification of nanobased materials can eliminate drawbacks and obstacles to real-field applications of this technology, thus bringing a new technological respective to water remediation methods.

Acknowledgment

The authors gratefully acknowledge the support by the Operational Program Research, Development and Education - European Regional Development Fund, (Project No. CZ.02.1.01/0.0/0.0/16_019/0000754) of the Ministry of Education, Youth and Sports of the Czech Republic. We also thank Viktorie Víchová and Monika Klimparová for technical assistance.

References

1. Anjum M, Miandad R, Waqas M, Gehany F, Barakat MA. Remediation of wastewater using various nano-materials. *Arab J Chem* 2019;**12**(8):4897−919.
2. Nazal M. An overview of carbon-based materials for the removal of pharmaceutical active compounds. In: Bartoli M, editor. *Carbon-based material for environmental protection and remediation*. IntechOpen; 2020.
3. Husnain SM, Asim U, Yaqub A, Shahzad F, Abbas N. Recent trends of MnO_2-derived adsorbents for water treatment: a review. *New J Chem* 2020;**44**:6096−120.
4. Spoiala A, Ilie C-I, Trusca R-D, Oprea OC, Surdu V-A, Vasile BS, et al. Zinc oxide nanoparticles for water purification. *Materials* 2021;**14**:4747.
5. Lee SY, Park SJ. TiO_2 photocatalyst for water treatment applications. *J Ind Eng Chem* 2013;**19**(6):1761−9.
6. Borgohain X, Boruah A, Sarma GK, Rashid MDH. Rapid and extremely high adsorption performance of porous MgO nanostructures for fluoride removal from water. *J Mol Liq* 2020;**305**:112799.
7. Ahmad SZN, Al-Gheethi A, Hamdan R, Othman N. Efficiencies and mechanisms of steel slag with ferric oxides for removing phosphate from wastewater using a column filter system. *Environ Sci Pollut Res* 2020;**27**:35184−94.
8. Loeb SK, Alvarez PJJ, Brame JA, Cates EL, Choi W, Crittenden J, et al. The technology horizon for photocatalytic water treatment: sunrise or sunset? *Environ Sci Technol* 2019;**53**(6):2937−47.
9. Ren G, Han H, Wang Y, Liu S, Zhao J, Meng X, et al. Recent advances of photocatalytic application in water treatment: a review. *Nanomaterials* 2021;**11**(7):1804.
10. Lu F, Astruc D. Nanocatalysts and other nanomaterials for water remediation from organic pollutants. *Coord Chem Rev* 2020;**408**:213180.
11. Rehman F, Thebo KH, Aamir M, Akhtar J. Chapter 8 - nanomembranes for water treatment. In: *Micro and nano technologies, nanotechnology in the beverage industry*. Elsevier; 2020. p. 207−40.
12. Madhura L, Singh S. A review on the advancements of nanomembranes for water treatment. In: *Nanotechnology in environmental science*; 2018.

13. Ray SS, Ofondu Chinomso Iroegbu A. Nanocellulosics: benign, sustainable, and ubiquitous biomaterials for water remediation. *ACS Omega* 2021;**6**(7):4511−26.
14. He S, Zhong L, Duan J, Feng Y, Yang B, Yang L. Bioremediation of wastewater by iron oxide-biochar nanocomposites loaded with photosynthetic bacteria. *Front Microbiol* 2017;**8**:823.
15. Gehrke I, Geiser A, Somborn-Schulz A. Innovations in nanotechnology for water treatment. *Nanotechnol Sci Appl* 2015;**8**:1−17.
16. Pandey N, Shukla SK, Singh NB. Water purification by polymer nanocomposites: an overview. *Nanocomposites* 2017;**3**(2):47−66.
17. Berber M. Current advances of polymer composites for water treatment and desalination. *J Chem* 2020:1−19.
18. Dias AM, Hussain A, Marcos AS, Roque AC. A biotechnological perspective on the application of iron oxide magnetic colloids modified with polysaccharides. *Biotechnol Adv* 2011;**29**(1):142−55.
19. Nassar NN. The application of nanoparticles for wastewater remediation. *Future Sci Book Series Appl Nanomater Water Qual* 2013:52−65.
20. Petala E, Baikousi M, Karakassides MA, Zoppellaro G, Filip J, Tuček J, et al. Synthesis, physical properties and application of the zero-valent iron/titanium dioxide heterocomposite having high activity for the sustainable photocatalytic removal of hexavalent chromium in water. *Phys Chem Chem Phys* 2016;**18**:10637−46.
21. Li MF, Liu YG, Zeng GM, Liu N, Liu SB. Graphene and graphene-based nanocomposites used for antibiotics removal in water treatment: a review. *Chemosphere* 2019;**226**:360−80.
22. Li Y, Zhang R, Tian X, Yang C, Zhou Z. Facile synthesis of Fe$_3$O$_4$ nanoparticles decorated on 3D graphene aerogels as broad-spectrum sorbents for water treatment. *Appl Surf Sci* 2016;**369**:11−8.
23. Prairie MR, Evans LR, Stange BM, Martinez SL. An investigation of titanium dioxide photocatalysis for the treatment of water contaminated with metals and organic chemicals. *Environ Sci Technol* 1993;**27**:1776−82.
24. Fan JW, Liu XH, Zhang J. The synthesis of TiO$_2$ and TiO$_2$-Pt and their application in the removal of Cr (VI). *Environ Technol* 2011;**32**:427−37.
25. Rengaraj S, Venkataraj S, Yeon JW, Kim Y, Li XZ, Pang GKH. Preparation, characterization and application of Nd−TiO$_2$ photocatalyst for the reduction of Cr(VI) under UV light illumination. *Appl Catal B Environ* 2007;**77**:157−65.
26. Kamat PV, Meisel D. Nanoscience opportunities in environmental remediation. *Compt Rendus Chem* 2003;**6**:999−1007.
27. Rawal SB, Bera S, Lee D, Jang DJ, Lee WI. Design of visible-light photocatalysts by coupling of narrow bandgap semiconductors and TiO$_2$: effect of their relative energy band positions on the photocatalytic efficiency. *Catal Sci Technol* 2013;**3**:1822−30.
28. Tang L, Guo X, Yang Y, Zha Z, Wang Z. Gold nanoparticles supported on titanium dioxide: an efficient catalyst for highly selective synthesis of benzoxazoles and benzimidazoles. *Chem Commun* 2014;**50**:6145−8.
29. Navarro RM, Alvarez-Galvan MC, Villoria de la Mano JA, Al-Zahrani SM, Fierro JLG. A framework for visible-light water splitting. *Energy Environ Sci* 2010;**3**:1865−82.
30. Mandal S, Jain N, Pandey MK, Sreejakumari SS, Shukla P, Chanda A, et al. Ultra-bright emission from Sr doped TiO$_2$ nanoparticles through r-GO conjugation. *R Soc Open Sci* 2019;**6**(3):190100.
31. Shyniya CR, Bhabu KA, Rajasekaran TR. Enhanced electrochemical behavior of novel acceptor doped titanium dioxide catalysts for photocatalytic applications. *J Mater Sci Mater Electron* 2017;**28**:6959−70.
32. Abdullah AM, Pinilla MÁ, Pillai S, O'Shea K. UV and visible light-driven production of hydroxyl radicals by reduced forms of N, F, and P codoped titanium dioxide. *Molecules* 2019;**24**(11):2147.
33. Wang JA, Limas-Ballesteros R, López T, Moreno A, Gómez R, Novaro O, et al. Quantitative determination of titanium lattice defects and solid-state reaction mechanism in iron-doped TiO$_2$ photocatalysts. *J Phys Chem B* 2001;**105**:9692−8.
34. Zhou M, Yu J, Cheng B. Effects of Fe-doping on the photocatalytic activity of mesoporous TiO$_2$ powders prepared by an ultrasonic method. *J Hazard Mater* 2006;**137**:1838−47.

35. Gupta VK, Saleh TA. Sorption of pollutants by porous carbon, carbon nanotubes and fullerene- an overview. *Environ Sci Pollut Res Int* 2013;**20**:2828—43.
36. Yin CY, Aroua MK, Daud W. Review of modifications of activated carbon for enhancing contaminant uptakes from aqueous solutions. *Separ Purif Technol* 2007;**52**:403—15.
37. Lv X, Xu J, Jiang G, Xu X. Removal of chromium(VI) from wastewater by nanoscale zero-valent iron particles supported on multiwalled carbon nanotubes. *Chemosphere* 2011;**85**:1204—9.
38. Xiao S, Shen M, Guo R, Huang Q, Wang S, Shi X. Fabrication of multiwalled carbon nanotube-reinforced electrospun polymer nanofibers containing zero-valent iron nanoparticles for environmental applications. *J Mater Chem* 2010;**20**:5700—8.
39. Sun Y, Ding C, Cheng W, Wang X. Simultaneous adsorption and reduction of U(VI) on reduced graphene oxide-supported nanoscale zerovalent iron. *J Hazard Mater* 2014;**280**:399—408.
40. Jabeen H, Kemp KC, Chandra V. Synthesis of nano zerovalent iron nanoparticles — graphene composite for the treatment of lead contaminated water. *J Environ Manag* 2013;**130**:429—35.
41. Ahmad A, Gu X, Li L, Lv S, Xu Y, Guo X. Efficient degradation of trichloroethylene in water using persulfate activated by reduced graphene oxide-iron nanocomposite. *Environ Sci Pollut Res* 2015;**1—10**.
42. Liu M, Chen C, Hu J, Wu X, Wang X. Synthesis of magnetite/graphene oxide composite and application for cobalt(II) removal. *J Phys Chem C* 2011;**115**:25234—40.
43. Yang X, Chen C, Li J, Zhao G, Ren X, Wang X. Graphene oxide-iron oxide and reduced graphene oxide-iron oxide hybrid materials for the removal of organic and inorganic pollutants. *RSC Adv* 2012;**2**:8821—6.
44. Fu Y, Wang J, Liu Q, Zeng H. Water-dispersible magnetic nanoparticle—graphene oxide composites for selenium removal. *Carbon N. Y.* 2014;**77**:710—21.
45. Sun H, Cao L, Lu L. Magnetite/reduced graphene oxide nanocomposites: one step solvothermal synthesis and use as a novel platform for removal of dye pollutants. *Nano Res* 2011;**4**:550—62.
46. Cong HP, Ren XC, Wang P, Yu SH. Macroscopic multifunctional graphene-based hydrogels and aerogels by a metal ion induced self-assembly process. *ACS Nano* 2012;**6**:2693—703.
47. Reed BE, Vaughan R, Jiang L. As(III), As(V), Hg, and Pb removal by Fe-oxide impregnated activated carbon. *J Environ Eng* 2000;**126**:869—73.
48. Castro CS, Guerreiro MC, Gonçalves M, Oliveira LCA, Anastácio AS. Activated carbon/iron oxide composites for the removal of atrazine from aqueous medium. *J Hazard Mater* 2009;**164**:609—14.
49. Wu L, Liao L, Lv G, Qin F, He Y, Wang X. Micro-electrolysis of Cr (VI) in the nanoscale zero-valent iron loaded activated carbon. *J Hazard Mater* 2013;**254—255**:277—83.
50. Tseng H-H, Su J-G, Liang C. Synthesis of granular activated carbon/zero valent iron composites for simultaneous adsorption/dechlorination of trichloroethylene. *J Hazard Mater* 2011;**192**:500—6.
51. Wu X, Yang Q, Xu D, Zhong Y, Luo K, Li X, et al. Simultaneous adsorption/reduction of bromate by nanoscale zerovalent iron supported on modified activated carbon. *Ind Eng Chem Res* 2013;**52**:12574—81.
52. Li X, Wang C. Phosphate removal from water by zero-valent iron/activated carbon fiber felt galvanic couples. In: *Second international conference on mechanic automation and control engineering*; 2011. p. 2367—70.
53. Tang SC, Lo IM. Magnetic nanoparticles: essential factors for sustainable environmental applications. *Water Res* 2013;**47**(8):2613—32.
54. Tuček J, Prucek R, Kolařík J, Zoppellaro G, Petr M, Filip J, et al. Zero-valent iron nanoparticles reduce arsenites and arsenates to As(0) firmly embedded in Core—Shell superstructure: challenging strategy of arsenic treatment under anoxic conditions. *ACS Sustain Chem Eng* 2017;**5**(4):3027—38.
55. Kanel SR, Manning B, Charlet L, Choi H. Removal of arsenic(III) from groundwater by nanoscale zero-valent iron. *Environ Sci Technol* 2005;**39**(5):1291—8.
56. Zhu H, Jia Y, Wu X, Wang H. Removal of arsenic from water by supported nano zero-valent iron on activated carbon. *J Hazard Mater* 2009;**172**(2—3):1591—6.
57. Kanel SR, Grenèche JM, Choi H. Arsenic(V) removal from groundwater using nano scale zero-valent iron as a colloidal reactive barrier material. *Environ Sci Technol* 2006;**40**(6):2045—50.

58. Li X, Elliot DW, Zhang W. Zero-valent iron nanoparticles for abatement of environmental pollutants: materials and engineering aspects. *Crit Rev Solid State Mater Sci* 2006;**31**(4):111—22.
59. Filip J, Kolařík J, Petala E, Petr M, Šráček O, Zbořil R. Nanoscale zerovalent iron particles for treatment of metalloids. In: Phenrat T, Lowry G, editors. *Nanoscale zerovalent iron particles for environmental restoration*. Cham: Springer; 2019. p. 157—79.
60. Li L, Hu J, Shi X, Fan M, Luo J, Wei X. Nanoscale zero-valent metals: a review of synthesis, characterization, and applications to environmental remediation. *Environ Sci Pollut Control Ser* 2016;**23**:17880—900.
61. Li S, Wang W, Liang F, Zhang W. Heavy metal removal using nanoscale zerovalent iron (NZVI): theory and application. *J Hazard Mater* 2017;**322**(Pt A):163—71.
62. Slovák P, Malina O, Kašlík J, Tomanec O, Tuček J, Petr M, et al. Zero-valent iron nanoparticles with unique spherical 3D architectures encode superior efficiency in copper entrapment. *ACS Sustain Chem Eng* 2016;**4**(5):2748—53.
63. He F, Gong L, Fan D, Tratnyek PG, Lowry GV. Quantifying the efficiency and selectivity of organohalide dechlorination by zerovalent iron. *Environ Sci Proc Impacts* 2020;**22**:528—42.
64. Lu HJ, Wang JK, Ferguson S, Wang T, Bao Y, Hao H. Mechanism, synthesis and modification of nano zerovalent iron in water treatment. *Nanoscale* 2016;**8**:9962—75.
65. Johnson RL, Johnson GO, Nurmi JT, Tratnyek PG. Natural organic matter enhanced mobility of nano zerovalent iron. *Environ Sci Technol* 2009;**43**(14):5455—60.
66. Kanel SR, Choi H. Transport characteristics of surface modified nanoscale zero-valent iron in porous media. *Water Sci Technol* 2007;**55**(1—2):157—62.
67. Wang Y, Yanga K, Lin D. Nanoparticulate zero valent iron interaction with dissolved organic matter impacts iron transformation and organic carbon. *Environ Sci Nano* 2020;**7**:1818—30.
68. Chen X, Ji D, Wang X, Zang L. Review on nano zerovalent iron (nZVI): from modification to environmental applications. *IOP Conf Ser Earth Environ Sci* 2017;**51**(012004).
69. Matter F, Luna AL, Niederberger M. From colloidal dispersions to aerogels: how to master nanoparticle gelation. *Nano Today* 2020;**30**(1748—0132):100827.
70. Mohamed MA, Nafady NA. Nanoparticle-mediated chaetomium, unique multifunctional bullets: what do we need for real applications in agriculture? In: Abdel-Azeem A, editor. *Recent developments on genus chaetomium. Fungal biology*. Cham: Springer; 2020. p. 267—300.
71. Huang DL, Chen GM, Zeng GM, Xu P, Yan M, Lai C, et al. Synthesis and application of modified zero-valent iron nanoparticles for removal of hexavalent chromium from wastewater. *Water Air Soil Pollut* 2015;**226**:375.
72. Ren L, Dong J, Chi Z, Li Y, Zhao Y, Jianan E. Rheology modification of reduced graphene oxide based nanoscale zero valent iron (nZVI/rGO) using xanthan gum (XG): stability and transport in saturated porous media. *Colloids Surf A Physicochem Eng Asp* 2019;**562**(0927—7757):34—41.
73. Lin KS, Mdlovu NV, Chen CY, Chiang CL, Dehvari K. Degradation of TCE, PCE, and 1,2—DCE DNAPLs in contaminated groundwater using polyethylenimine-modified zero-valent iron nanoparticles. *J Clean Prod* 2018;**175**(0959—6526):456—66.
74. Liu Q, Bei Y, Zhou F. Removal of lead(II) from aqueous solution with amino-functionalized nanoscale zero-valent iron. *Cent Eur J Chem* 2009;**7**(1):79—82.
75. Li Y, Zhang Y, Jing Q, Lin Y. The influence of pluronic F-127 modification on nano zero-valent iron (NZVI): sedimentation and reactivity with 2,4-dichlorophenol in water using response surface methodology. *Catalysts* 2020;**10**(4):412.
76. Ibrahim HM, Awad M, Al-Farraj AS, Al-Turki AM. Stability and dynamic aggregation of bare and stabilized zero-valent iron nanoparticles under variable solution chemistry. *Nanomaterials* 2020;**10**(2):192.
77. Wang X, Lian W, Sun X, Ma J, Ning P. Immobilization of NZVI in polydopamine surface-modified biochar for adsorption and degradation of tetracycline in aqueous solution. *Front Environ Sci Eng* 2018;**12**:9.

78. Su Y, Zhao YS, Li LL, Qin CY, Wu F, Geng NN, et al. Transport characteristics of nanoscale zero-valent iron carried by three different "vehicles" in porous media. *J Environ Sci Health A Tox Hazard Subst Environ Eng* 2014;**49**(14):1639−52.
79. Zhang J, Zhu Q, Xing Z. Preparation of new materials by ethylene glycol modification and Al(OH)$_3$ coating NZVI to remove sulfides in water. *J Hazard Mater* 2020;**390**(0304−3894):122049.
80. Singh R, Misra V. Stabilization of zero-valent iron nanoparticles: role of polymers and surfactants. In: Aliofkhazraei M, editor. *Handbook of nanoparticles*. Cham: Springer; 2015. p. 1−19.
81. Ren J, Yao M, Woo YC, Tijing LD, Kim JH, Shon HK. Recyclable nanoscale zerovalent iron (nZVI)-immobilized electrospun nanofiber composites with improved mechanical strength for groundwater remediation. *Compos B Eng* 2019;**171**(1359−8368):339−46.
82. Cirtiu CM, Raychoudhury T, Ghoshal S, Moores A. Systematic comparison of the size, surface characteristics and colloidal stability of zero valent iron nanoparticles pre- and post-grafted with common polymers. *Coll Surf A Physicochem Eng Asp* 2011;**390**(1−3):95−104.
83. Wang Z, Choi F, Acosta E. Effect of surfactants on zero-valent iron nanoparticles (NZVI) reactivity. *J Surfactants Deterg* 2017;**20**:577−88.
84. Tian H, Liang Y, Yang D, Sun Y. Characteristics of PVP−stabilised NZVI and application to dechlorination of soil−sorbed TCE with ionic surfactant. *Chemosphere* 2020;**239**:124807.
85. Li J, Fan M, Li M, Liu X. Cr(VI) removal from groundwater using double surfactant-modified nanoscale zero-valent iron (nZVI): effects of materials in different status. *Sci Total Environ* 2020;**717**:137112.
86. Silva L, Abdelraheem W, Nadagouda MN, Rocco AM, Dionysiou D, Fonseca F, et al. Novel microwave-driven synthesis of hydrophilic polyvinylidene fluoride/polyacrylic acid (PVDF/PAA) membranes and decoration with nano zero-valent-iron (nZVI) for water treatment applications. *J Membr Sci* 2021;**620**(0376−7388):118817.
87. Tian H, Liang Y, Zhu T, Zeng X, Sun Y. Surfactant-enhanced PEG-4000-NZVI for remediating trichloroethylene-contaminated soil. *Chemosphere* 2018;**195**:585−93.
88. Liu J, Liu A, Guo J, Zhou T, Zhang W. Enhanced aggregation and sedimentation of nanoscale zero-valent iron (nZVI) with polyacrylamide modification. *Chemosphere* 2021;**263**:127875.
89. Dong H, Ning Q, Li L, Wang Y, Wang B, Zhang L, et al. A comparative study on the activation of persulfate by bare and surface-stabilized nanoscale zero-valent iron for the removal of sulfamethazine. *Separ Purif Technol* 2020;**230**:115869.
90. Harm U, Schuster J, Mangold K-M. *Modification of iron nanoparticles for ground water remediation*. DECHEMA, Karl-Winnacker Institut; 2010 [dechema-dfi.de].
91. Abd El-Lateef HM, Khalaf Ali MM, Saleh MM. Adsorption and removal of cationic and anionic surfactants using zero-valent iron nanoparticles. *J Mol Liq* 2018;**268**:497−505.
92. Bayat M, Nasernejad B, Falamaki C. Preparation and characterization of nano-galvanic bimetallic Fe/Sn nanoparticles deposited on talc and its enhanced performance in Cr(VI) removal. *Sci Rep* 2021;**11**:7715.
93. He F, Li Z, Shi S, Xu W, Sheng H, Gu Y, et al. Dechlorination of excess trichloroethene by bimetallic and sulfidated nanoscale zero-valent iron. *Environ Sci Technol* 2018;**52**(15):8627−37.
94. Huang X, Zhang F, Peng K, Liu J, Lu L, Li S. Effect and mechanism of graphene structured palladized zero-valent iron nanocomposite (nZVI-Pd/NG) for water denitration. *Sci Rep* 2020;**10**(9931).
95. Elliott DW, Zhang W. Field assessment of nanoscale bimetallic particles for groundwater treatment. *Environ Sci Technol* 2001;**35**:4922−6.
96. Ulucan-Altuntas K, Debik E. Dechlorination of dichlorodiphenyltrichloroethane (DDT) by Fe/Pd bimetallic nanoparticles: comparison with nZVI, degradation mechanism, and pathways. *Front Environ Sci Eng* 2020;**14**:17.
97. Ma Y, Lv X, Yang QI, Wang Y, Chen X. Reduction of carbon tetrachloride by nanoscale palladized zero-valent iron@ graphene composites: kinetics, activation energy, effects of reaction conditions and degradation mechanism. *Appl Catal Gen* 2017;**542**(0926−860X):252−61.
98. Shi J, Long C, Li A. Selective reduction of nitrate into nitrogen using Fe−Pd bimetallic nanoparticle supported on chelating resin at near-neutral pH. *Chem Eng J* 2016;**286**(1385−8947):408−15.

99. Wang Y, Zhao H, Zhao G. Iron-copper bimetallic nanoparticles embedded within ordered mesoporous carbon as effective and stable heterogeneous Fenton catalyst for the degradation of organic contaminants. *Appl Catal B Environ* 2015;**164**(0926−3373):396−406.
100. Li H, Qiu Y, Wang X, Yang J, Yu YJ, Chen YQ, et al. Biochar supported Ni/Fe bimetallic nanoparticles to remove 1,1,1-trichloroethane under various reaction conditions. *Chemosphere* 2017; **169**(0045−6535):534−41.
101. Zhao X, Liu W, Cai Z, Han B, Qian T, Zhao D. An overview of preparation and applications of stabilized zero-valent iron nanoparticles for soil and groundwater remediation. *Water Res* 2016;**1**(100): 245−66.
102. Brumovský M, Filip J, Malina O, Oborná J, Sracek O, Reichenauer TG, et al. Core-shell Fe/FeS nanoparticles with controlled shell thickness for enhanced trichloroethylene removal. *ACS Appl Mater Interfaces* 2020;**12**(31):35424−34.
103. Wang H, Zhong Y, Zhu X, Li D, Deng Y, Huang P, et al. Enhanced tetrabromobisphenol A debromination by nanoscale zero valent iron particles sulfidated with S^0 dissolved in ethanol. *Environ Sci Proc Impacts* 2021;**23**:86−97.
104. Cao Z, Li H, Lowry GV, Shi X, Pan X, Xu X, et al. Unveiling the role of sulfur in rapid defluorination of florfenicol by sulfidized nanoscale zero-valent iron in water under ambient conditions. *Environ Sci Technol* 2021;**55**(4):2628−38.
105. Datta KKR, Petala E, Datta KJ, Perman JA, Tucek J, Bartak P, et al. NZVI modified magnetic filter paper with high redox and catalytic activities for advanced water treatment technologies. *Chem Commun* 2014;**50**:15673−6.
106. Zou Y, Wang X, Khan A, Wang P, Liu Y, Alsaedi A, et al. Environmental remediation and application of nanoscale zero-valent iron and its composites for the removal of heavy metal ions: a review. *Environ Sci Technol* 2016;**50**(14):7290−304.
107. Zhou Q, Li J, Wang M, Zhao D, Zhou Q. Iron-based magnetic nanomaterials and their environmental applications. *Crit Rev Environ Sci Technol* 2016;**46**:783−826.
108. Petala E, Dimos K, Douvalis A, Bakas T, Tucek J, Zbořil R, et al. Nanoscale zero-valent iron supported on mesoporous silica: characterization and reactivity for Cr(VI) removal from aqueous solution. *J Hazard Mater* 2013;**261**(0304−3894):295−306.
109. Ezzatahmadi N, Ayoko GA, Millar GJ, Speight R, Yan C, Li J, et al. Clay-supported nanoscale zero-valent iron composite materials for the remediation of contaminated aqueous solutions: a review. *Chem Eng J* 2017;**312**(1385−8947):336−50.
110. Li Z, Wang L, Meng J, Liu X, Xu J, Wang F, Brookes P. Zeolite-supported nanoscale zero-valent iron: new findings on simultaneous adsorption of Cd(II), Pb(II), and As(III) in aqueous solution and soil. *J Hazard Mater* 2018;**344**(0304−3894):1−11.
111. Eljamal R, Eljamal O, Maamoun I, Yilmaz G, Sugihara Y. Enhancing the characteristics and reactivity of nZVI: polymers effect and mechanisms. *J Mol Liq* 2020;**315**:113714.
112. Li S, Li S, Wen N, Wei D, Zhang Y. Highly effective removal of lead and cadmium ions from wastewater by bifunctional magnetic mesoporous silica. *Separ Purif Technol* 2021;**265**(1383−5866):118341.
113. Guo Y, Chen B, Zhao Y. Fabrication of the magnetic mesoporous silica Fe-MCM-41-A as efficient adsorbent: performance, kinetics and mechanism. *Sci Rep* 2021;**11**(2612).
114. Kazemi M, Peyravi M, Jahanshahi M. Multilayer UF membrane assisted by photocatalytic NZVI@ TiO_2 nanoparticle for removal and reduction of hexavalent chromium. *J Water Proc Eng* 2020; **37**(2214−7144):101183.
115. Sun M, Zhang Z, Liu G, Lv M, Feng Y. Enhancing methane production of synthetic brewery water with granular activated carbon modified with nanoscale zero-valent iron (NZVI) in anaerobic system. *Sci Total Environ* 2021;**760**(0048−9697):143933.
116. Qu J, Liu Y, Cheng L, Jiang Z, Zhang G, Deng F, et al. Green synthesis of hydrophilic activated carbon supported sulfide nZVI for enhanced Pb(II) scavenging from water: characterization, kinetics, isotherms and mechanisms. *J Hazard Mater* 2021;**403**(0304−3894):123607.

117. Wang S, Zhao M, Zhou M, Li YC, Wang J, Gao B, et al. Biochar-supported nZVI (nZVI/BC) for contaminant removal from soil and water: a critical review. *J Hazard Mater* 2019;**373**(0304−3894): 820−34.
118. Goswami A, Kadam RG, Tuček J, Sofer Z, Bouša D, Varma RS, et al. Fe(0)-embedded thermally reduced graphene oxide as efficient nanocatalyst for reduction of nitro compounds to amines. *Chem Eng J* 2020;**382**(122469):1385−8947.
119. Pu S, Deng D, Wang K, Wang M, Zhang Y, Shangguan L, et al. Optimizing the removal of nitrate from aqueous solutions via reduced graphite oxide−supported nZVI: synthesis, characterization, kinetics, and reduction mechanism. *Environ Sci Pollut Res* 2019;**26**:3932−45.
120. Shi J, Wang J, Wang W, Teng W, Zhang W. Stabilization of nanoscale zero-valent iron in water with mesoporous carbon (nZVI@MC). *J Environ Sci* 2019;**81**(1001−0742):28−33.
121. Sunkara B, Zhan J, He J, McPherson GL, Piringer G, John VT. Nanoscale zerovalent iron supported on uniform carbon microspheres for the in situ remediation of chlorinated hydrocarbons. *ACS Appl Mater Interfaces* 2010;**2**:2854−62.
122. Cheng Y, Zhou W, Zhu L. Enhanced reactivity and mechanisms of mesoporous carbon supported zero-valent iron composite for trichloroethylene removal in batch studies. *Sci Total Environ* 2020; **718**(137256):1−7.
123. Ling X, Li J, Zhu W, Zhu Y, Sun X, Shen J, et al. Synthesis of nanoscale zero-valent iron/ordered mesoporous carbon for adsorption and synergistic reduction of nitrobenzene. *Chemosphere* 2012;**87**: 655−60.
124. Xu J, Cao Z, Liu X, Zhao H, Xiao X, Wu J, et al. Preparation of functionalized Pd/Fe-Fe$_3$O$_4$@-MWCNTs nanomaterials for aqueous 2,4-dichlorophenol removal: interactions, influence factors, and kinetics. *J Hazard Mater* 2016;**317**:656−66.
125. Chen S, Li M, Wu Y, Wang Y. Activated carbon fiber-supported nano zero-valent iron on Cr(VI) removal. *IOP Conf Ser Earth Environ Sci* 2021;**675**:012170.
126. Xu J, Liu X, Lowry GV, Cao Z, Zhao H, Zhou J, et al. Dechlorination mechanism of 2,4-dichlorophenol by magnetic MWCNTs supported Pd/Fe nanohybrids: rapid adsorption, gradual dechlorination, and desorption of phenol. *ACS Appl Mater Interfaces* 2016;**23**:7333−42.
127. Kakavandi B, Takdastan A, Pourfadakari S, Ahmadmoazzam M, Jorfic S. Heterogeneous catalytic degradation of organic compounds using nanoscale zero-valent iron supported on kaolinite: mechanism, kinetic and feasibility studies. *J Taiwan Inst Chem Eng* 2019;**96**:329−40.
128. Zhang M, Yi K, Zhang X, Han P, Liu W, Tong M. Modification of zero valent iron nanoparticles by sodium alginate and bentonite: enhanced transport, effective hexavalent chromium removal and reduced bacterial toxicity. *J Hazard Mater* 2020;**388**(121822):1−10.
129. Lv X, Qin X, Wang K, Peng Y, Wang P, Jiang G. Nanoscale zero valent iron supported on MgAl-LDH-decorated reduced graphene oxide: enhanced performance in Cr(VI) removal, mechanism and regeneration. *J Hazard Mater* 2019;**373**:176−86.
130. Pasinszki T, Krebsz M. Synthesis and application of zero-valent iron nanoparticles in water treatment, environmental remediation, catalysis, and their biological effects. *Nanomaterials* 2020;**10**:917.
131. Su C, Puls RW, Krug TA, Watling MT, O'Hara SK, Quinn JW, et al. A two and half-year-performance evaluation of a field test on treatment of source zone tetrachloroethene and its chlorinated daughter products using emulsified zero valent iron nanoparticles. *Water Res* 2012;**46**:5071−84.
132. Zhang M, Dong Y, Gao S, Cai P, Dong J. Effective stabilization and distribution of emulsified nanoscale zero-valent iron by xanthan for enhanced nitrobenzene removal. *Chemosphere* 2019;**223**: 375−82.
133. Ken DS, Sinha A. Recent developments in surface modification of nano zero-valent iron (nZVI): remediation, toxicity and environmental impacts. *Environ Nanotechnol Monit Manag* 2020;**14**(100344).
134. O'Hara S, Krug T, Quinn J, Clausen C, Geiger C. Field and laboratory evaluation of the treatment of DNAPL source zones using emulsified zero-valent iron. *Remediation* 2006;**16**:35−56.
135. Lee YC, Kwon TS, Yang JS, Yang JW. Remediation of groundwater contaminated with DNAPLs by biodegradable oil emulsion. *J Hazard Mater* 2007;**140**:340−5.

136. Eljamal R, Maamoun I, Sugihara Y, Eljamal O. Comparative study of bare and emulsified nanoscale zero-valent iron for nitrate and phosphorus removal. In: *Proceedings of international exchange and innovation conference on engineering & sciences (IEICES)*, vol. 6; 2020. p. 192–7.
137. Brooks KB, Quinn JW, Clausen CA, Geiger CL. reportApplication of emulsified zero-valent iron to marine environments, NASA *technical reports server (NTRS) collection*, Document ID: 20120000060.
138. Galdames A, Ruiz-Rubio L, Orueta M, Sánchez-Arzalluz M, Vilas-Vilela JL. Zero-valent iron nanoparticles for soil and groundwater remediation. *Int J Environ Res Publ Health* 2020;**17**(16):5817.
139. Galdames A, Mendoza A, Orueta M, De Soto García IS, Sánchez M, Virto I, et al. Development of new remediation technologies for contaminated soils based on the application of zero-valent iron nanoparticles and bioremediation with compost. *Resour Eff Technol* 2017;**3**(2):166–76.
140. Chang JH, Cheng SF. The remediation performance of a specific electrokinetics integrated with zero-valent metals for perchloroethylene contaminated soils. *J Hazard Mater* 2006;**131**(1–3):153–62.
141. Černík M, Hrabal J, Nosek J. Combination of electrokinetics and nZVI remediation. In: Filip J, Cajthaml T, Najmanová P, Černík M, Zbořil R, editors. *Advanced nano-bio technologies for water and soil treatment. Applied environmental science and engineering for a sustainable future*. Cham: Springer; 2020.
142. Lianga Z, Wena Q, Wanga X, Zhanga F, Yua Y. Chemically stable and reusable nano zero-valent iron/graphite-like carbon nitride nanohybrid for efficient photocatalytic treatment of Cr(VI) and rhodamine B under visible light. *Appl Surf Sci* 2016;**386**:451–9.
143. Qu G, Kou L, Wang T, Liang D, Hu S. Evaluation of activated carbon fiber supported nanoscale zero-valent iron for chromium (VI) removal from groundwater in a permeable reactive column. *J Environ Manag* 2017;**201**:378–87.
144. Lv X, Xue X, Jiang G, Wu D, Sheng T, Zhou H, et al. Nanoscale zero-valent iron (nZVI) assembled on magnetic Fe$_3$O$_4$/graphene for chromium (VI) removal from aqueous solution. *J Colloid Interface Sci* 2014;**417**:51–9.
145. Donga H, Denga J, Xiea Y, Zhanga C, Jianga Z, Chenga Y, et al. Stabilization of nanoscale zero-valent iron (nZVI) with modified biochar for Cr(VI) removal from aqueous solution. *J Hazard Mater* 2017;**332**:79–86.
146. Wu H, Wei W, Xu C, Meng Y, Bai W, Yang W, et al. Polyethylene glycol-stabilized nano zero-valent iron supported by biochar for highly efficient removal of Cr(VI). *Ecotoxicol Environ Saf* 2020;**188**(109902):1–8.
147. Baikousi M, Bourlinos AB, Douvalis A, Bakas T, Anagnostopoulos DF, Tuček J, et al. Synthesis and characterization of gamma-Fe$_2$O$_3$/carbon hybrids and their application in removal of hexavalent chromium ions from aqueous solutions. *Langmuir* 2012;**28**(8):3918–30.
148. Ibrahim HM, Awad M, Al-Farraj AS, Al-Turki AM. Effect of flow rate and particle concentration on the transport and deposition of bare and stabilized zero-valent iron nanoparticles in sandy soil. *Sustainability* 2019;**11**:6608.
149. Singh P, Pal P, Mondal P, Saravanan G, Nagababu P, Majumdar S, et al. Kinetics and mechanism of arsenic removal using sulfide-modified nanoscale zerovalent iron. *Chem Eng J* 2021;**412**(128667):1385–8947.
150. Su Y, Jassby D, Song S, Zhou X, Zhao H, Filip J, et al. Enhanced oxidative and adsorptive removal of diclofenac in heterogeneous fenton-like reaction with sulfide modified nanoscale zerovalent iron. *Environ Sci Technol* 2018;**52**(11):6466–75.
151. Duan R, Dong Y, Zhang Q. Characteristics of aggregate size distribution of nanoscale zero-valent iron in aqueous suspensions and its effect on transport process in porous media. *Water* 2018;**10**(670).
152. Lin D, Hu L, Lo IMC, Yu Z. Size distribution and phosphate removal capacity of nano zero-valent iron (nZVI): influence of pH and ionic strength. *Water* 2020;**12**(2939).
153. Vinod VTP, Wacławek S, Senan C, Kupčík J, Pešková K, Černík M, et al. Gum karaya (Sterculia urens) stabilized zero-valent iron nanoparticles: characterization and applications for the removal of chromium and volatile organic pollutants from water. *RSC Adv* 2017;**7**:13997–4009.

CHAPTER 5

Surface-modified nanomaterials-based catalytic materials for water purification, hydrocarbon production, and pollutant remediation

Ragib Shakil[1,a], Md. Mahamudul Hasan Rumon[1,a], Yeasin Arafat Tarek[1], Chanchal Kumar Roy[1], Al-Nakib Chowdhury[1] and Rasel Das[2]

[1]Department of Chemistry, Bangladesh University of Engineering and Technology (BUET), Dhaka, Bangladesh;
[2]Department of Chemistry, Stony Brook University, Stony Brook, NY, United States

1. Introduction

Wastewater treatment and hydrocarbon production have become rising issues regarding mass population and government policies. Due to rapid population growth and industrialization, there have been increases in the utilization of pure water and hydrocarbon fuel. Because the population growth is an upward trend, the dependency will never be reduced; rather it will be a catastrophic issue for human beings in the coming future. One may ask: Why does freshwater become polluted, which might shrink pure water availability? In general, water becomes toxic because of the presence of organic compounds, metallic or other inorganic ions and pathogenic viruses, and bacteria in water; and this contaminated water is called wastewater.[1,2] In wastewater, pH, biological oxygen demand (BOD), chemical oxygen demand (COD), and other essential parameters fluctuate and change its odor and color. These contaminations are severely threatening to human health as well as the ecological system. Therefore, water purification has become an indispensable area of research in the scientific community. For remediation of the wastewater, nanocatalysis appears to be the best technique. In catalyst mediated wastewater treatment process, some organic pollutants and others undergo redox reactions. As a result, the pollutants tend to break down into small molecules that may further degrade to CO_2, H_2, and other less toxic compounds.[3] Even, it is possible to recycle CO_2 gas to produce low hydrocarbon fuels. It thus helps the elimination of typically used thermochemical and biological processes, which are costly for hydrocarbon production.[4,5]

To meet the undeniable need for undesirable pollutant-free water and hydrocarbon production, nanocatalyst technology is one of the most popular techniques adopted

---

[a] These authors are equally contributed.

by scientists. Catalysts are the materials present in a chemical reaction with low activation energy that accelerates the reaction rate. Compared with bulk or conventional catalyst, catalytic nanomaterials are superior due to their large surface area, lower bandgap energy, and other leapfrogging properties.[6] Many semiconductors remove pollutants through the photocatalytic process. A bandgap between the valence band (VB) and conduction band (CB) of a semiconductor plays an important role in catalytic reactions. Because of the large bandgap, the activation energy becomes large.[7] As a result, efficient photocatalytic reactions cannot occur. Therefore, reducing the bandgap has become one of the mainstay topics of research in the scientific community.[8] Several efforts have been imposed to reduce the band energy gap of nanocatalyst materials, such as surface modification, combining two or more atoms or molecules, or reinforcing materials to the parental material. As an example, the bandgap of crystalline TiO_2 (anatase) is 3.2 eV ($\lambda \leq 387.5$ nm) which means the excitation energy required for this semiconductor will be mostly UV-region.[9] Visible light can only contribute up to 5% to excite pristine TiO_2. However, scientists are looking toward using renewable solar energy, and that only can be achieved by shifting the excitation energy of the material to the visible region.[10] It could be possible by doping of metal or metal oxides or nonmetals into the semiconductor material. For instance, when TiO_2 nanocatalyst is blended with CuS nanomaterial, the bandgap energy reduces from 3.2 to 1.89 eV. That is why the TiO_2/CuS nanocomposite becomes 1.4 times more efficient than the neat CuS.[9,10] Fig. 5.1 shows the catalytic process of CuS modified TiO_2 for water pollutants degradation and conversion of biofuel from biomass.

In this chapter, several nanocatalytic materials with their modification strategies for wastewater treatment and their possible conversion into hydrocarbons are highlighted. A special focus is also made on the metal-based nanoparticles and carbon materials that are used for catalytic degradation of pollutants and hydrocarbon generation.

2. Nanocatalyst materials

Catalysis, mediated through nanomaterials, reveals new dimensions toward sustainable chemistry. Numerous nanocatalysts have been adopted for many purposes ranging from refining, environmental remediation, and food processing to emerging research areas such as biorefinery processes, reforming sensors, and energy conversion and storage. Owing to extraordinary physicochemistry, optoelectronics, high surface area, enhanced reactivity, and stability, nanocatalysts play a central role in wastewater treatment and hydrocarbon production.[11-16] There are several types of nanocatalysts for selective processes. However, transition metal oxides are the most superior due to their small bandgap and lower activation energy.[17] Indeed, several researchers have reported transition metal oxide-based nanocatalysts such as TiO_2,[18] V_2O_5,[19] ZnO,[20] SnO_2,[21] WO_3,[22] CuS,[23] CdS,[24] SiO_2,[25] Fe_2O_3,[26] and so forth. Although metal oxide-based nanocatalysts are

Surface-modified nanomaterials-based catalytic materials   105

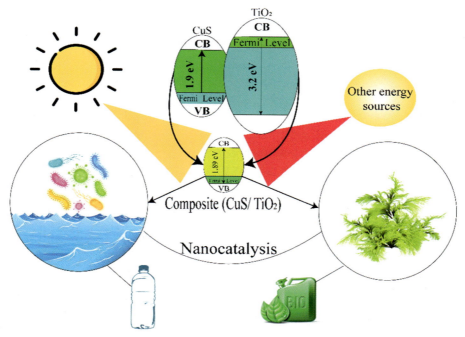

Figure 5.1 Illustration of the catalytic process of CuS/TiO$_2$ for removing contaminants from wastewater and converting biomass into hydrocarbon fuel.

moderately good in their activity due to their improper bandgap associate with the energy source, they cannot show their maximum efficacy. In this context, several research works have been performed to modify the neat catalysts by adding other nanomaterials or atoms with them. This modification results in a lower bandgap of the final material and hence the catalytic efficacy is enhanced several times compared to the neat one.[27,28]

3. Wastewater treatment

Nanocatalysts are very specific and efficient in their reaction. As a result, nanocatalyst technology for the treatment of wastewater should be an excellent selection. Several types of nanocatalysts have been reported for the removal of organic wastes and pathogens. Some of these nanocatalysts exhibited both waste degradation and antimicrobial properties. Though the photocatalytic and nonphotocatalytic nanocatalysts are being used for purification purposes, zero-valent metals, metal oxides, semiconductors, surface-modified polymers, and carbon materials are the most popular nanocatalysts for water treatment.[29]

3.1 Zero-valent metal nanocatalysts

Recently, various nanoscale zero-valent metal (nZVM) nanocatalysts, e.g., iron (Fe), zinc (Zn), aluminum (Al), and nickel (Ni), have been attracted broad research interests in the treatment of wastewater.[30] Ni shows lower negative standard reduction potential than Fe, which is a clear indication of lower reduction capability.[31] Despite a poorer reduction ability, Fe has many prominent advantages over Zn for pollutant remediation and wastewater treatment, including superior oxidation, adsorption, precipitation properties, and low cost.[32] Next, Al nanocatalysts are thermally unstable, compared to Fe nanocatalysts, and hence less studied for pollutants remediation. The hydroxyl generation and the redox reaction tend to be happened instantly due to the high reduction potential of these nZVM.[33] With a moderate standard reduction potential, nZVM (Fe or Zn) has a good potential to serve as a reducing agent as opposed to many redox-labile contaminants.

In the oxidation process, Fe^0 contributes to the generation of various iron oxides and classes of hydroxides, e.g., FeO, $FeOOH$, $Fe(OH)_3$, Fe_2O_3 and Fe_3O_4.[34] For degradation, adsorption, and precipitation of water contaminants, both the FeO core and the FeOOH shell are essential. Aqueous solution oxidation of Fe^0 contributes to the generation of H_2O_2. The produced H_2O_2 then reacts with Fe^0 to produce reactive surface hydroxyl radicals. Many organic water contaminants such as pathogens cell, p-nitrophenol, dye, and hexabromocyclododecane are then degraded by the radical.[35] The radical formation process during the pollutants degradation is given as

$$Fe^0 + 2H_2O \rightarrow Fe^{2+} + H_2 + 2OH^- \tag{5.1}$$

$$Fe^0 + 2H^+ \rightarrow Fe^{2+} + H_2 \tag{5.2}$$

$$Fe^0 + O_2 + 2H^+ \rightarrow Fe^{2+} + H_2O_2 \tag{5.3}$$

$$Fe^0 + H_2O_2 + 2H^+ \rightarrow Fe^{2+} + 2H_2O \tag{5.4}$$

$$Fe^{2+} + H_2O_2 \rightarrow Fe^{3+} + OH^\bullet + OH^- \tag{5.5}$$

3.2 Metal oxide nanocatalysts

3.2.1 TiO$_2$

TiO_2 is a semiconductor-based photoactive nanocatalyst material. Among all the metal and metal oxide, TiO_2 is the most efficient nanocatalyst due to several factors.[36] It is a noncorrosive nanoparticle that has excellent optoelectronic properties that makes it superior particularly in wastewater treatment and hydrocarbon production. Another most important and crucial factor for being an efficient photocatalytic material is the proper matching between the bandgap and the wavelength of the radiation. Researchers tend to prepare such a material that can utilize most of the portion of the electromagnetic radiation to get the efficient catalytic result. For instance, anatase is a crystalline form of TiO₂ (3.2 eV) which implies its inability to be fully excited by the visible light irradiation.[33] As the use of solar energy in the catalytic process is the prime goal in enhancing its maximum catalytic efficacy, visible light absorption is crucially important.[37] This goal can only be achieved by incorporating another metal or metal oxides or nonmetals into the TiO₂ semiconducting matrix.[31,32,34] TiO₂ and its composites-based water treatment process involve two processes: (a) degradation and (b) mineralization.[35,38,39] Degradation of pollutants means the splitting of pollutants into other products or unstable less toxic intermediate. On the contrary, mineralization means the total demolition of pollutants those result in H₂O, CO₂, or other inorganic components. The probable mechanism of these two processes in the presence of light is illustrated in Fig. 5.2.

When TiO₂ is excited with proper energy, a photogenerated hole (h⁺) and photogenerated electrons (e⁻) are generated in VB and CB, respectively. The generated (h⁺) can participate in the subsequent redox reactions while its surface comes in contact with the sorbed species. The (h⁺) travel through the whole surface of the catalyst and make it active toward the degradation or mineralization process. More specifically, the generated (h⁺) reacts with water instantly to produce OH• radicals and subsequently (e⁻) reacts with oxygen to generate superoxide radicals. The formation of radicals can be demonstrated by the following equations:

$$TiO_2 + h\nu \rightarrow e^-_{(CB)} + h^+_{(VB)} \quad (5.6)$$

$$O_2 + e^- \rightarrow {}^\bullet O_2^- \quad (5.7)$$

$$H_2O + h^+ \rightarrow {}^\bullet OH + H^+ \quad (5.8)$$

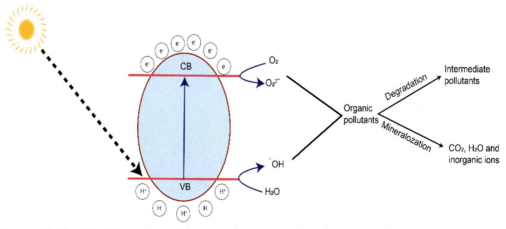

Figure 5.2 Simplified illustration of the degradation and mineralization mechanisms of the organic water pollutants by TiO₂ nanocatalyst. *(The figure is adapted from Ref. 3.)*

An example of the degradation and mineralization of typical lignocaine by TiO_2 photocatalyst is shown in Fig. 5.3.

3.2.2 Fe$_2$O$_3$

Compared to TiO_2, Fe_2O_3 nanoparticles are more active to the visible light due to their narrow bandgap (2.2 eV).[40] This interesting property makes it suitable for degrading wastes from water in the presence of visible light. Numerous oxides and oxyhydroxides form of Fe(III), e.g., α-Fe_2O_3, β-FeOOH, α-FeOOH, etc. have been reported in the literature which contributes to degrading the wastes and toxic pollutants from water.[41,42] A visual representation of the catalytic activity of Fe_2O_3 nanocatalysts toward organic dye and pathogens is shown in Fig. 5.4.

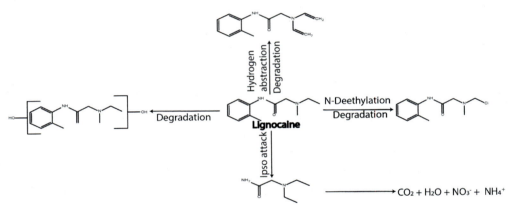

Figure 5.3 A typical degradation and mineralization process of Lignocaine by the TiO_2 nanocatalyst. *(The figure is adapted from Ref. 3.)*

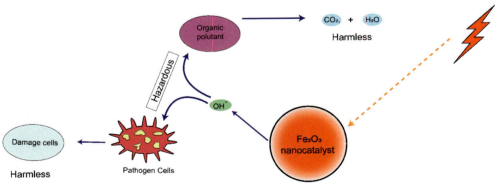

Figure 5.4 Illustration of the catalytic activities of Fe_2O_3 toward the removal of the organic pollutants and biological cell.

The total process can be represented by the following equations,

$$\text{Organic pollutants(OP)} + h\nu \rightarrow \text{OP}^{\bullet} \quad (5.9)$$

$$\text{OP}^{\bullet} + \alpha\text{-Fe}_2\text{O}_3 \rightarrow \text{OP}^{\bullet} + +\alpha\text{-Fe}_2\text{O}_3(e_{CB}^{-}) \quad (5.10)$$

$$\alpha\text{-Fe}_2\text{O}_3 + h\nu \rightarrow \alpha\text{-Fe}_2\text{O}_3(e_{CB}^{-} + h_{VB}^{+}) \quad (5.11)$$

$$H_2O \rightarrow H^{+} + OH^{-} \quad (5.12)$$

$$e_{CB}^{-} + O_2 \rightarrow O_2^{\bullet-} \quad (5.13)$$

$$O_2^{-} + H_{aq}^{+} \rightarrow HO_2^{\bullet} \quad (5.14)$$

$$HO_2^{\bullet} + HO_2^{\bullet} \leftrightarrow H_2O_2 + O_2 \quad (5.15)$$

$$h_{VB}^{+} + OH_{aq}^{-} \rightarrow OHo^{\bullet-} \quad (5.16)$$

$$\text{OP}^{\bullet}/\text{OP}^{\bullet+} + (O_2^{\bullet-}, HO_2^{\bullet}, H_2O_2, OH^{\bullet-}) \rightarrow \text{Degradation Products} \quad (5.17)$$

3.2.3 ZnO

ZnO has emerged as another useful nanocatalyst in wastewater treatment. This is because of their prominent characteristics, such as strong oxidation power and good photocatalytic activities. ZnO nanocatalysts are ecofriendly as they are compatible with bio-organisms,[43] making them ideal for wastewater treatment. Moreover, it has the advantage of low cost over TiO_2 nanocatalyst. Furthermore, the photocatalytic capability of ZnO is close to that of TiO_2 anatase, since their bandgap energies are almost the same (around 3.17 eV).[44] As a result, similar to the TiO_2, ZnO nanocatalysts cannot use the visible wavelength properly in the photocatalytic process. For this particular reason, scientists modify ZnO with other nanoparticles to enhance their catalytic activity. For example, when the Cu nanoparticle is coupled with ZnO nanoparticles, the bandgap of the composite (Cu–ZnO) appears to be 2.84 eV.[45] This bandgap is applicable for absorbing the visible light and hence the photocatalytic activity toward the removal of organic pollutants could be enhanced. As an antibacterial agent, ZnO and its composites are good choices for their superior antibacterial properties. In a typical removal process of pollutants or pathogen (bacteria, viruses, fungi, etc.), ZnO nanocatalysts generate photo-induced holes. This produces surface (OH•) radicals. Due to these (OH•), the oxidation process takes place that demolishes the pathogens cell walls that cause the pathogens to die quickly. At the surface of the ZnO, the generated electrons react with surface oxygen to form superoxide.[46] The superoxide is the source of the (OH•) radicals that are responsible for the whole microbes killing process. The removal process of pollutants and pathogens using ZnO is described in the following Fig. 5.5.

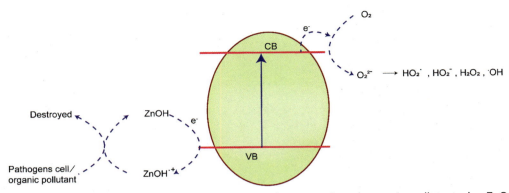

Figure 5.5 A typical mechanism of destroying pathogens cell and organic pollutants by ZnO nanocatalysts.

3.3 Carbon-based nanocatalysts

In comparison with metal-based nanomaterials, carbon-based nanomaterials have unique physicochemical properties such as large surface area, highly porous structure, enhance reactivity, good thermal stability, and variable hybridization possibility.[26,47,48] Variable hybridization states can result from different structural configurations like fullerene (C_{60} and C_{540}), single-walled nanotubes (SWNT), multiwalled nanotubes (MWNT), and graphene.[49] There are numbers of studies depict that surface treatment, activation, or functionalization of pristine carbon material are the first conditions to determine the suitability of carbon-based materials for applications in environmental remediation.[50,51] These materials are especially useful for the degradation of organic and inorganic pollutants due to their high adsorption properties.[52] Interestingly, the huge adsorption properties do not act directly toward catalysis, rather with the modification of the carbon-based materials by semiconductor-based metal oxides or any other nonmetal-based catalysts.[53] This happens because of the lower bandgaps of most of the carbon-based nanomaterials in between VB and CB.

However, surface-modified carbon nanocatalysts having metal oxides have well-known catalytic activities due to the synergistic effect of carbon and metal nanocatalysts. Carbon surfaces are naturally conductive and active, thus undergo adsorption while the metal oxides are good catalysts in nature. As a result, the two combined properties enhance the catalytic properties of neat metal oxides.[54] In this process, the recombination of (e^-) and (h^+) in the modified carbon-based nanocatalysts becomes slower due to the availability of electrons on the surface. Further functionalization in the carbon surface introduces new active sites for the incoming metal oxides in the carbon surface.[55] The term functionalization means the introduction of (−COOH, −OH, −C−O, −C=O, −C−N, −NH$_2$, −C=S), etc. in the surface of the carbon moiety.[56] The processes of carbon nanomaterials functionalization are oxidation, nitrogenation, and sulfuration as shown in Fig. 5.6.

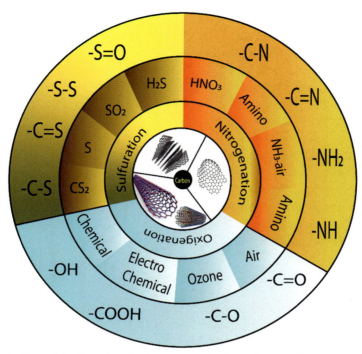

Figure 5.6 Illustration of the functionalization process in the carbon surface by different functional groups.

3.3.1 Functionalized graphene-based nanocatalysts

Pristine graphene and reduced graphene oxide (rGO) are important in the process of catalysis due to their less defective surface and high electrical conductivity. In general, pristine graphene and rGO show more improved catalytic activities in the oxidative and reductive process than GO. Alternatively, due to the presence of various oxygen-containing functional groups in the GO surface, it makes a complex with metals that would act as heterogeneous catalysts in the reduction reactions.[57,58] When graphene-based materials are modified with metal-based nanocatalysts, the photocatalytic effects of modified material enhance multiple times compared to neat metal oxide or graphene.[59] For example, when GO surface is modified with TiO_2 or any other metal oxides, the tendency to render a redshift in the visible light region. That is how the surface-modified material can use light more efficiently in the catalytic process.

3.3.2 Functionalized CNT-based nanocatalysts

Generally, CNT is of two types such as SWNT and MWNT. In SWNT, a hexagonal nanotube is covered by six others that create porous heterogeneous structures aligned in tube bundles.[60] Adsorption can take place at four different sites available for a typical open-ended CNT bundle, which are of two types: those with lower adsorption energy, positioned on the external surfaces of the external CNTs comprising the bundle; and those with higher adsorption energy, located either between two adjacent tubes or within an individual tube.[48] Compared to SWNT, MWNTs do not usually exist as bundles, even when unique preparation methods are used to generate such configurations.

Functionalized CNTs have $-OH$, $-C=O$, and $-COOH$ that positively influences adsorption capacities and catalytic activity, depending on the particular synthetic procedures and the related catalytic processes.[61] The adsorption mechanism and catalytic properties of CNTs are almost similar to graphene materials, but the photocatalytic properties of CNTs are different from GO.[62] It is because the CNTs have a very small band-energy gap that induces the photocatalytic performance multiple times. Graphene and other carbon-based materials do not possess this extraordinary photocatalytic property like CNT.[63] As well, the modification of CNTs with metal oxides enhances multiple times due to the synergistic effect. CNT-based nanocatalysts increase the catalytic rates as demonstrated in Fig. 5.7.

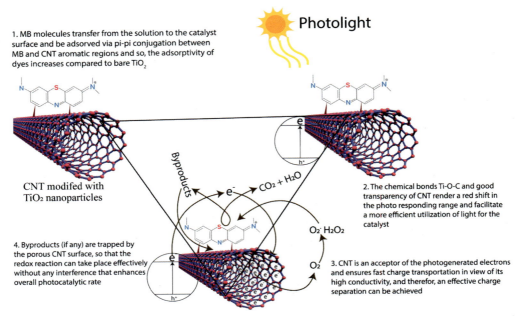

Figure 5.7 Illustration of surface modification of CNT with TiO_2 nanoparticles for the removal of methylene blue. *(The figure is adapted with permission from Ref. 26.)*

3.4 Polymer-based nanocatalysts

Generally, high porosity and permanent pore deformation can be observed in the catalytic process of polymeric materials.[64] A uniform distribution of homogenous nanocatalyst in a polymer matrix is very effective in the degradation of organic pollutants from wastewater.[65,66] Polymers are used to bind metal ions as chelating agents, thereby promoting the purification of water. The main feature of dendritic nanopolymers is that they have a lower trend to move through the pores of ultrafiltration membranes. Generally, linear polymers of comparable molecular weight have lower polydispersity and globular form and hence they cannot perform water purification.[67] However, upon modification with metal or metal oxides or any other polymers, polymer-based nanocatalysts show multiple times better catalytic activity toward the removal of various toxic metal ions, dyes, and microorganisms from water/wastewater streams.[68] For example, chitosan-based carbon nanofibers (CNFs) are incorporated with Fe_3O_4 nanoparticles along with polyvinyl alcohol nanocomposite films.[69] This unusual combination of materials shows a successful ability to adsorb Cr^{6+} from water and demonstrates a high metal uptake potential (~80 mg/g of chitosan/iron-CNF composite). Several other polymers are incorporated with some metal oxides to enhance the catalytic performances. Of them, metal oxide/polysulfonate, metal oxide/poly (ether) sulfone, metal oxide/starch, metal oxide/polyurethane, metal oxide/polyamide polymer composites are noteworthy.

3.5 Miscellaneous nanocatalysts

Besides nZVI, TiO_2, ZnO, Fe_3O_4, carbon, and polymer-based catalysts: perovskites, SiO_2, V_2O_5, WO_3, and metal-organic framework (MOF) are prospective nanoparticles toward the wastewater treatment.[70–73] Due to the intrinsic properties and modifications in the surface, the preceding materials have shown good catalytic properties. Calcium titanate ($CaTiO_3$) is an inorganic mineral which is called perovskite. It is named after Russian mineralogist, L. A. Perovski. Perovskite is generally expressed as ABX_3, where A refers to the rare earth metal, B refers to the ions of transition metals, and X represents the oxygen ions.[74] Perovskite is a very emerging material as it offers extremely magnetic, superconductive, and ferroelectric properties.[75–77] Besides other properties, perovskite shows superior catalytic reactions like hydrogenation and hydrolysis of hydrocarbons,[78] reduction in SO_2 by CO,[79] selective oxidation of C_3H_8,[80] and combustion of methane.[80] In the environmental pollutants removal, these materials act as adsorbents for the adsorption of several dyes, toxic pesticides, volatile organic compounds, and few gases.[81]

A typical photocatalytic reaction of perovskite material is represented in Fig. 5.8. As can be seen in Fig. 5.8, it produces oxidative free radicals necessary for pollutants degradation. This free radical generation is quite successive in the perovskite materials. As a result, they have mentionable catalytic activity toward organic pollutants and

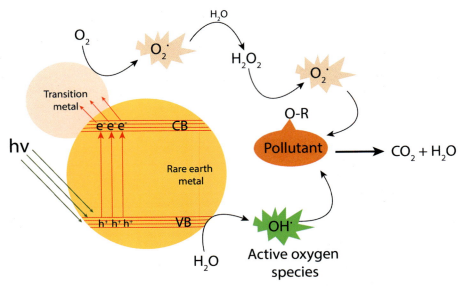

Figure 5.8 Photocatalytic degradation mechanism of water pollutants by perovskite. *(The figure is adapted with permission from Ref. 82.)*

biological cells. The collaborative contribution of rare earth metal and transition metal oxides makes it a perfect candidate for the adsorption and photochemical degradation of pollutants from air and water. Secondly, the MOF can act as a host matrix for the metal or metal oxide nanoparticles distribution (Fig. 5.9).[84] Additionally, the modification in the surface or the composition can be brought out in two ways. The first way is the selection of proper organic ligands from a wide range of ligands as a host matrix and the second one is obviously the guest materials for the host matrix. The fabrication process is truly scalable to generate active sites and enhance the catalytic activities of the modified materials from MOF.[85] This kind of MOF can be used as a template to form nanoarchitectured metal-based nanocatalysts due to the available expanded internal volume. The metal located at the nodes of the MOF structure such as Cr, Mn, or Cu possesses free coordination sites following thermal removal of water ligands without great structural change.

4. Nanocatalysts in hydrocarbon production

4.1 Biomass to hydrocarbon

Biomass is an abundant and diverse source toward the solution of hydrocarbon generation. To produce hydrocarbon fuel such as bioethanol, biodiesel, bio-oil, biosyngas, and biohydrogen, thermochemical conversion is a major path.[86] In enhancing product quality and attaining optimum operating conditions, both homogenous and heterogeneous

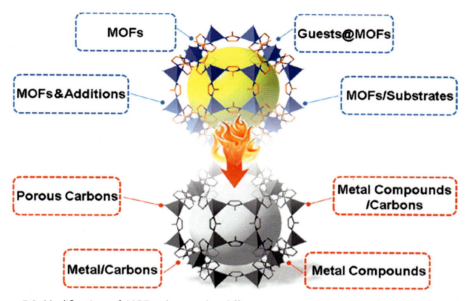

Figure 5.9 Modification of MOF substrate by different guest species. *(The figure is adapted with permission from Ref. 83.)*

nanocatalysts play an important role. The most common problems of bulk heterogeneous catalysts, such as mass transfer resistance, time consumption, rapid deactivation, and inefficiency may be solved by nanocatalysts modification with a high specific surface area and high catalytic activity. In this aspect, new types of nanocatalysts have been developed. The thermochemical involves quick pyrolysis, liquefaction, combustion, and gasification as seen in Fig. 5.10. In hydrocarbon fuel production, nanocatalysts emerged as an important catalyst to convert biomass to hydrocarbon or biofuels.

4.1.1 Pyrolysis

In a typical fast or rapid pyrolysis process, the heat exposure time becomes very short, and that results in the phase transition with enhanced heat and mass transfer during the reaction process.[88] Long residence times and decreased temperature range (200–350°C) favor the formation of charcoal. In terms of their effect on the quality and quantity of pyrolyzed materials, metal oxide catalysts have been extensively studied using diverse biomass species.[89,90] It is known that metal oxides have impacts on the temperature of decomposition. It is, therefore, noted that in a thermogravimetric study the transition metal-supported alumina has a strong effect on the decrease in devolatilization temperature.[91] The addition of catalysts to the pyrolytic process greatly reduces the bio-oil yield, but when the oxygenated groups present in the bio-oil are removed, the quality of the fuel is improved. In terms of biomass conversion, Al_2O_3 is found to be more efficient

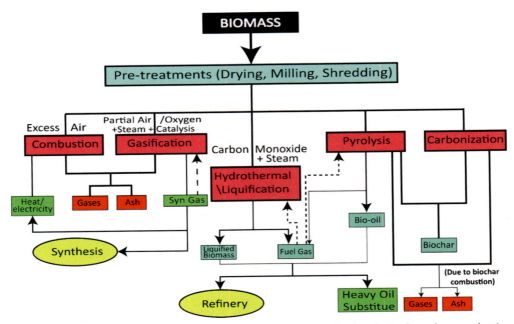

Figure 5.10 Illustration of various catalytic processes that are involved in the hydrocarbon production from biomass. *(The figure is adapted from Ref. 87.)*

than ZnO with a value of 79.94%. The effect of the catalyst on the bio-oil yield, however, differed from one another. For instance, when the amount of catalyst is high compared to the biomass, the conversion of bio-oil yield increases in the case of ZnO. Nevertheless, Al_2O_3 promoted gas formation in the pyrolysis process thus decrease the bio-oil yield with the increase in catalyst percentage.[92]

4.1.2 Liquefaction

Liquefaction of biomass to bio-oils needs higher temperatures and pressures. Water is used as a liquefaction reactant and solvent, avoiding any polymerization and preventing the creation of solid products such as coke and charcoal.[93] However, catalytic liquefaction is somewhat similar to hydrothermal liquefaction, but a catalyst is used to decrease the residence time, operating temperature, and pressure, thereby improving the consistency of liquid products. The crude bio-oils produced from catalytic liquefaction are found to flow easily and to be much less viscous than the biocrude of noncatalyzed liquefaction.[94] Moreover, in liquefaction processes, alkaline salts, Na_2CO_3, KOH, and so on are widely used as homogenous catalysts. The effects of several other catalysts on biomass liquefaction, such as $NaHCO_3$, $Ca(OH)_2$, $Ba(OH)_2$, $FeSO_4$, and so forth have also been investigated.[95,96] The heterogeneous catalysts such as Pd/C, Pt/C,

Ru/C, Ni/SiO$_2$–Al$_2$O$_3$, CoMo/γ-Al$_2$O$_3$, and so forth are also used in biomass catalytic conversion.[97] Liquid catalysts have the advantage of mono dispersion in reaction mixtures in the catalytic hydroconversion of biomass. In other words, solid catalysts have the superiority of greater catalytic activity. The promising materials for use in catalytic hydro biomass conversion are acid-functionalized paramagnetic nanoparticles. In the catalytic hydroconversion process, these functionalized nanoparticles can be easily separated and recycled.

4.1.3 Gasification

The gasification process involves the partial oxidation of biomass to fuel gases at high temperatures (>800°C). Typically, mixtures of CO, H$_2$, CO$_2$, and some light hydrocarbons are produced by air or steam. The combustion of biomass at high temperatures (>800°C) in the presence of air transforms the chemical energy into hot gases. There is a very significant impact of nanocatalysts on the removal of tar contents during the gasification process.[98] Nanocatalysts can minimize not only the tar content but also increase the consistency of gas products and the efficiency of conversion.[99] During the final phase of the gasification process, the presence of a nanocatalyst decreases the char yield while it increases the formation of char during the volatilization stage.[100] There are some requirements for successful gasification and catalysts, such as being efficient at removing tars, being resistant to deactivation due to carbon fouling or sintering, can be quickly regenerated, and are inexpensive. In general, catalysts are used to decrease the tar production, in the conversion of biomass to hydrocarbon is classified into mineral or synthetic.[101,102] Mineral catalysts include calcined rocks, olivine, clay minerals, and ferrous metal oxides. On the other hand, transition metals, activated alumina, alkali-metal carbonates, and chars are the main synthetic catalysts.[103] Dolomite catalysts are effective in tar cracking but they have some drawbacks, such as elevated pressure sensitivity and thermal instability, resulting in surface area loss due to sintering.[104] For catalytic hot gas cleanup, Ni-based catalysts are very effective during the gasification of biomass. Elimination of tar is achieved by Ni-based catalysts with a high rate. In addition, Ni catalysts are used for the manufacture of hydrogen gas-rich products. It is claimed that Ni catalysts are the most effective of all catalysts for converting tar into fuel gas.[105,106]

4.2 CO$_2$ to methanol and other hydrocarbons

With the increasing global warming effect, the conversion of atmospheric CO$_2$ to various hydrocarbons has become an indispensable need in the fields of environmental science and catalysis.[107,108] The conversion process of CO$_2$ requires hydrogenation to turn CO$_2$ into hydrocarbons depending on the conversion process. There are different possible routes to produce hydrocarbons from CO$_2$ hydrogenation (Fig. 5.11). Basically,

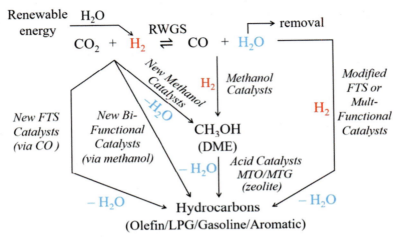

Figure 5.11 Illustration of possible routes of hydrogenation process to produce hydrocarbons from CO_2. *(The figure is adapted with permission from Ref. 109.)*

there are two common routes to convert CO_2 into biofuel and other hydrocarbons: (a) Direct process termed as Fischer−Tropsch synthesis (FTS),[110] (b) Indirect process involving methanol production as an intermediate in the hydrocarbon production.[109] Each route starts with syngas ($CO + H_2$) production.

The direct process has advantages over the indirect process as it is economical and ecofriendly. The direct process involves either the reduction of CO_2 to CO via the catalysis of modified FTS or multi-functional catalysts or new FTS catalysts or bi-functional catalytic process.[111−113] Literature studies show that metal oxides or carbon-based nanocatalysts have been reported toward the conversion of CO_2 into hydrocarbons. As an instance, Al_2O_3, TiO_2, and SiO_2 modified with Ni, Co, Ru, and Fe metals are effective in the conversion process.[114,115] CO_2 conversion can be carried out by photochemical catalysis using semiconductor-based metal oxides such as, TiO_2 nanocatalysts which facilitate hydrogenation reaction in the production of hydrocarbons from CO_2.[116] In a typical reaction, the e^-, provided by the TiO_2, gather on Ti^{2+} sites (Eq. 5.18), reacts with adsorbed CO_2 where they promote proton generation (Eqs. 5.19−5.21) by water oxidation. This can eventually reduce the CO_2 into CO (Eq. 5.22), CH_3OH (Eq. 5.23), and CH_4 (Eq. 5.24). A simplified catalytic process for the conversion of CO_2 into CH_4 and other higher hydrocarbons is illustrated in Fig. 5.12.

The total reaction processes that are involved in the conversion of CO_2 into methanol and other hydrocarbons are shown in the following equations:

$$TiO_2 \xrightarrow{h\nu} e_{cb}^- + h_{vb}^+ \quad (5.18)$$

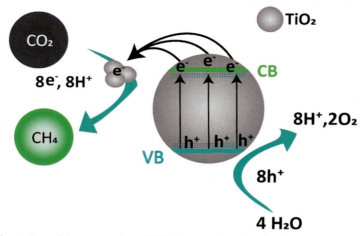

Figure 5.12 Illustration of the conversion of CO₂ into methanol and subsequent higher hydrocarbon.

$$H_2O + h^+ \rightarrow OH^{\bullet} + H^+ \qquad (5.19)$$

$$H_2O + OH + 3h^+ \rightarrow 3H^+ + O_2^- \qquad (5.20)$$

$$H_2 + 2h^+ \rightarrow 2H^+ \qquad (5.21)$$

$$CO_2 + 2H^+ + 2e^- \xrightarrow{h\nu,\ catalyst} CO + H_2O \qquad (5.22)$$

$$CO_2 + 6H^+ + 6e^- \xrightarrow{h\nu,\ catalyst} CH_3OH + H_2O \qquad (5.23)$$

$$CO_2 + 8H^+ + 8e^- \xrightarrow{h\nu,\ catalyst} CH_4 + 2H_2O \qquad (5.24)$$

5. Recent advancement and real-time utilization of nanocatalysts

Nanotechnology for wastewater treatment and pollutant remediation has gained attention due to its good efficiency and low cost. Photocatalysis is identified as the major revolutionary technology for water treatment and purification. In the elimination of trace compounds, photocatalytic devices may be used to supplement conventional methods. Such devices are commercially viable, for example, pool disinfection. To ensure successful isolation and recovery, Panasonic has developed a technology that attaches a photocatalyst (TiO₂) to an industrial adsorbent and a catalyst called zeolite.[117] TiO₂ can break down many pathogens, organic pollutants, pesticides, dyes, and various inorganic toxic materials like nitrate. In Table 5.1, a summary of surface-modified nanocatalysts reported in the last five years has been presented.

Table 5.1 Surface-modified nanocatalysts for wastewater treatment and pollutant remediation.

Photocatalysts	Band gap (eV)	Pollutants	References
CuS/TiO$_2$	1.89	Methyl orange	[9]
WO$_3$–TiO$_2$	2.63	Paracetamol	[118]
MnO$_2$-RGO	—	Neutral red	[119]
CaBi$_6$O$_{10}$	2.3	Ciprofloxacin	[120]
TiNPs-Fe$_2$O$_3$	2.0	Rhodamine B	[121]
AgNPs-Fe$_2$O$_3$	1.8		
CuO Nanosheets	1.92	Organic pollutant/Food dye, Allura red	[122]
Poly(thiophene-1,3,4-oxadiazole)	—	Organic dyes	[123]
Ag$_2$O/AgBr—CeO$_2$	—	Tetracycline	[124]
Zn(TPP) (IQNO), Zn(bba)$_2$(phen)	3.25	Crystal Violet, Fuchsin Basic, Methyl Orang	[125]
BiVO$_4$-GO-TiO$_2$-polyaniline (PANI)	—	Methylene blue, Phenol	[126]
TiO$_2$/Polypyrrole	—	Red 45	[127]
Poly-3-hexylthiophene-AgNPs	2.3	Rhodamine 6G	[128]
CdS/TiO$_2$	—	1,2,4-Trichlorobenzene	[129]
TiO$_2$@NH$_2$-UiO-66	2.82	Styrene	[130]
UiO-66/g-C$_3$N$_4$	2.92	Methylene blue	[131]
MIL-101(Fe)/TiO$_2$	—	Tetracycline	[132]
[Cu(4,4′-bipy)Cl]$_n$	1.78–2.14	Methylene blue	[133]
AgI/UiO-66	—	Sulfamethoxazole	[134]

The overall market of water and wastewater treatment applications amounted to 1.6 billion dollars in 2007, up by 6.6 billion dollars in 2015. In 2015, the United States, Germany, Japan, and China are proposed as the leading countries in water treatment using nanotechnological approaches.[135] Disinfection is one of the fastest-growing segments of the industry with large appliances and advantages; nanomaterial-based photocatalysis is a promising approach for disinfection. Furthermore, combined with filters, photocatalysts can greatly minimize membrane fouling and improve the performance of water cleaning. Small-scale UV-structure photocatalytic systems (http://www.ube.es/index.html) have been on the market for a number of years, while the demonstration phase is in solar photocatalytic water processing systems (http://www.raywox.com) and the pilot projects for drinking water purification have only begun in the developed world (http://www.rcsi.ie/sodis/). Wastewater with implemented photocatalysts flows through the glass tubes of the solar receiver and is cleaned using solar radiation in the RayWOx solar photocatalytic water purification device. KACO has adopted this photocatalytic device and tested this device with nano-TiO$_2$ catalyst.[136]

Besides water purification, nanobased surface-modified catalysts are being imposed widely around the globe for hydrocarbon production. With the alarming rate of environmental pollution, the search for alternatives of fossil fuel such as biodiesel or monoalkyl esters of fatty acids is being considered as the main research focus in renewable energy sectors. Some research work based on nanocatalysts for biofuel production is shown in Table 5.2.

Developed countries like the USA, Germany, Italy, Japan, and others are significantly investing in the development of nanocatalyst-based technology. Global companies are shown in Table 5.3, involved in the development of nanomaterials-based catalysts to eliminate pollutants during wastewater treatment and hydrocarbon production.

Table 5.2 Recent advancement in biofuel production using surface-modified nanocatalysts.

No.	Catalysts	The efficiency of biodiesel yields (%)	References
1.	TiO$_2$/rGO	98	[137]
2.	Mn-doped ZnO	97	[138]
3.	CaO–MgO	98.5	[139]
4.	Sulfonated graphene	98	[140]
5.	CaO/SiO$_2$	95.2	[141]
6.	ZrO$_2$@SiO$_2$	98.6	[142]
7.	CaO/Zn	93.2	[143]

Table 5.3 Global companies involved in the development of nanomaterials-based catalysts to eliminate pollutants and hydrocarbon production.

Company/Institution	Origin	Homepage
Toshiba materials Co., Ltd.	Tokyo, Japan	www.toshiba-water.com
Daicel Corporation	Osaka, Japan	www.daicel.com
Crystal	New York, USA	www.crystalwatercare.com
Ishihara Sangyo Kaisha, Ltd.	Osaka, Japan	www.iskweb.co.jp
KRONOS Worldwide, Inc.	Dallas, USA	www.kronostio2.com
Huber technology	Berching, Germany	www.huber.de
Chriwa water treatment technology	Hambühren, Germany	www.chriwa.de
HUBER environmental technology (China) Co., Ltd.	China	www.huber.cn.com
Ovivo	Canada	www.ovivowater.ca
Ontario	Canada	www.investontario.ca

6. Conclusion

Nanocatalysts with modified and improved structures have shown high efficacy for the direct remediation of environmental pollutants and conversion to other valuable hydrocarbons. Along with the metal and metal oxide—derived nanocatalysts, nonmetal-doped and carbon nanostructure-doped catalysts can selectively diminish pollutants from the environment. However, challenges and chances coexist to explore the solutions for converting wastes into valuable products. Designing and preparing highly efficient catalytic materials are still a major requirement in the removal of pollutants and hydrocarbon generation. These pave the way for future research, where modification of nanocatalysts with desired functionalities is necessary to further increase the rate of pollutants removal capacity as well as accelerate the catalytic rate of hydrocarbon production. We recommend such functionalities could be stimuli-responsive polymer, two-dimensional transition metal dichalcogenide, biomaterials, and so on.

Acknowledgments

Support from the University Grants Commission (UGC) of Bangladesh and Bangladesh University of Engineering and Technology (BUET) is highly appreciated.

References

[1] Chaudhry FN, Malik MF. Factors affecting water pollution: a review. *J Ecosyst Ecography* March 2017;**07**(01):1—3. https://doi.org/10.4172/2157-7625.1000225.

[2] Martins SCS, Martin CM, Fiúza LMCG, Santaella ST. Immobilization of microbial cells: a promising tool for treatment of toxic pollutants in industrial wastewater. *Afr J Biotechnol* July 2013;**12**(28):4412—8. https://doi.org/10.5897/ajb12.2677.

[3] Yaqoob AA, Parveen T, Umar K, Mohamad Ibrahim MN. Role of nanomaterials in the treatment of wastewater: a review. *Water* Feb. 2020;**12**(2):495. https://doi.org/10.3390/w12020495.

[4] Li W, et al. A short review of recent advances in CO_2 hydrogenation to hydrocarbons over heterogeneous catalysts. *RSC Adv* Feb. 14, 2018;**8**(14):7651—69. https://doi.org/10.1039/c7ra13546g. Royal Society of Chemistry.

[5] Shi Z, et al. Direct conversion of CO_2 to long-chain hydrocarbon fuels over K-promoted CoCu/TiO_2 catalysts. *Catal Today* Aug. 2018;**311**:65—73. https://doi.org/10.1016/j.cattod.2017.09.053.

[6] Somwanshi SB, Somvanshi SB, Kharat PB. Nanocatalyst: a brief review on synthesis to applications. In: *Journal PhysConf Ser*, **1644**; Oct. 2020. p. 12046. https://doi.org/10.1088/1742-6596/1644/1/012046.

[7] Rajabi HR, Shahrezaei F, Farsi M. Zinc sulfide quantum dots as powerful and efficient nanophotocatalysts for the removal of industrial pollutant. *J Mater Sci Mater Electron* Sep. 2016;**27**(9):9297—305. https://doi.org/10.1007/s10854-016-4969-4.

[8] Wacławek S, Padil VVT, Černík M. Major advances and challenges in heterogeneous catalysis for environmental applications: a review. *Ecol. Chem. Eng. S* Mar. 2018;**25**(1):9—34. https://doi.org/10.1515/eces-2018-0001.

[9] Farooq MH, Aslam I, Shuaib A, Anam HS, Rizwan M, Kanwal Q. Band gap engineering for improved photocatalytic performance of CuS/TiO_2 composites under solar light irradiation. *Bull Chem Soc Ethiop* 2019;**33**(3):561—71. https://doi.org/10.4314/bcse.v33i3.16.

[10] Shang J, Chai M, Zhu Y. Photocatalytic degradation of polystyrene plastic under fluorescent light. *Environ Sci Technol* Oct. 2003;**37**(19):4494–9. https://doi.org/10.1021/es0209464.

[11] Zhang YH, et al. Highly enhanced photocatalytic H2 evolution of Cu_2O microcube by coupling with TiO_2 nanoparticles. *Nanotechnology* Feb. 2019;**30**(14):145401. https://doi.org/10.1088/1361-6528/aafccb.

[12] Kalpana K, Selvaraj V. Thiourea assisted hydrothermal synthesis of $ZnS/CdS/Ag_2S$ nanocatalysts for photocatalytic degradation of Congo red under direct sunlight illumination. *RSC Adv* Jan. 2016;**6**(5):4227–36. https://doi.org/10.1039/c5ra16242d.

[13] Awfa D, Ateia M, Fujii M, Johnson MS, Yoshimura C. Photodegradation of pharmaceuticals and personal care products in water treatment using carbonaceous-TiO_2 composites: a critical review of recent literature. *Water Res* Oct. 01, 2018;**142**:26–45. https://doi.org/10.1016/j.watres.2018.05.036. Elsevier Ltd.

[14] Daer S, Kharraz J, Giwa A, Hasan SW. Recent applications of nanomaterials in water desalination: a critical review and future opportunities. *Desalination* Jul. 01, 2015;**367**:37–48. https://doi.org/10.1016/j.desal.2015.03.030. Elsevier B.V.

[15] Wu Y, et al. Environmental remediation of heavy metal ions by novel-nanomaterials: a review. *Environ Pollut* Mar. 01, 2019;**246**:608–20. https://doi.org/10.1016/j.envpol.2018.12.076. Elsevier Ltd.

[16] Tang WW, et al. Impact of humic/fulvic acid on the removal of heavy metals from aqueous solutions using nanomaterials: a review. *Sci Total Environ* Jan. 15, 2014;**468–469**:1014–27. https://doi.org/10.1016/j.scitotenv.2013.09.044. Elsevier.

[17] Narayanan R, El-Sayed MA. Catalysis with transition metal nanoparticles in colloidal solution: nanoparticle shape dependence and stability. *J Phys Chem B* Jul. 2005;**109**(26):12663–76. https://doi.org/10.1021/jp051066p.

[18] Liu Y, Li J, Qiu X, Burda C. Novel TiO_2 nanocatalysts for wastewater purification: tapping energy from the sun. In: *Water Sci Technol*, **54**; Oct. 2006. p. 47–54. https://doi.org/10.2166/wst.2006.733. 8.

[19] Langeslay RR, Kaphan DM, Marshall CL, Stair PC, Sattelberger AP, Delferro M. Catalytic applications of vanadium: a mechanistic perspective. *Chem Rev* Feb. 27, 2019;**119**(4):2128–91. https://doi.org/10.1021/acs.chemrev.8b00245. American Chemical Society.

[20] Taourati R, Khaddor M, El Kasmi A. Stable ZnO nanocatalysts with high photocatalytic activity for textile dye treatment. *Nano Struct Nano-Objects* Apr. 2019;**18**:100303. https://doi.org/10.1016/j.nanoso.2019.100303.

[21] Suthakaran S, Dhanapandian S, Krishnakumar N, Ponpandian N. Surfactants assisted SnO_2 nanoparticles synthesized by a hydrothermal approach and potential applications in water purification and energy conversion. *J Mater Sci Mater Electron* Jul. 2019;**30**(14):13174–90. https://doi.org/10.1007/s10854-019-01681-7.

[22] Widiyandari H, Purwanto A, Balgis R, Ogi T, Okuyama K. CuO/WO_3 and Pt/WO_3 nanocatalysts for efficient pollutant degradation using visible light irradiation. *Chem Eng J* Jan. 2012;**180**:323–9. https://doi.org/10.1016/j.cej.2011.10.095.

[23] Rohokale MS, Dhabliya D, Sathish T, Vijayan V, Senthilkumar N. A novel two-step co-precipitation approach of $CuS/NiMn_2O_4$ heterostructured nanocatalyst for enhanced visible light driven photocatalytic activity via efficient photo-induced charge separation properties. *Phys B Condens Matter* Feb. 2021:412902. https://doi.org/10.1016/j.physb.2021.412902.

[24] Devendran P, Alagesan T, Nallamuthu N, Asath Bahadur S, Pandian K. Single-precursor synthesis of sub-10 nm CdS nanoparticles embedded on graphene sheets nanocatalyst for active photodegradation under visible light. *Appl Surf Sci* Dec. 2020;**534**:147614. https://doi.org/10.1016/j.apsusc.2020.147614.

[25] Wang ZQ, et al. High-performance and long-lived Cu/SiO_2 nanocatalyst for CO_2 hydrogenation. *ACS Catal* Jun. 2015;**5**(7):4255–9. https://doi.org/10.1021/acscatal.5b00682.

[26] Das R, et al. Recent advances in nanomaterials for water protection and monitoring. *Chem Soc Rev* Nov. 21, 2017;**46**(22):6946–7020. https://doi.org/10.1039/c6cs00921b. Royal Society of Chemistry.

[27] Shanshool HM, Yahaya M, Yunus WMM, Abdullah IY. Investigation of energy band gap in polymer/ZnO nanocomposites. *J Mater Sci Mater Electron* Sep. 2016;**27**(9):9804—11. https://doi.org/10.1007/s10854-016-5046-8.

[28] Alipour A, Mansour Lakouraj M, Tashakkorian H. Study of the effect of band gap and photoluminescence on biological properties of polyaniline/CdS QD nanocomposites based on natural polymer. *Sci Rep* Dec. 2021;**11**(1):1913. https://doi.org/10.1038/s41598-020-80038-1.

[29] Tandon PK, Bahadur Singh S, Kumar Tandon P. *Catalysis: a brief review on nano-catalyst.* 2014. Accessed: Mar. 03, 2021. [Online]. Available: https://www.researchgate.net/publication/284727255.

[30] Prasad Rao J, Gruenberg P, Geckeler KE. Magnetic zero-valent metal polymer nanoparticles: current trends, scope, and perspectives. *Prog Polym Sci* Jan. 01, 2015;**40**(1):138—47. https://doi.org/10.1016/j.progpolymsci.2014.07.002. Elsevier Ltd.

[31] Tratnyek PG, Salter AJ, Nurmi JT, Sarathy V. Environmental applications of zerovalent metals: iron vs. zinc. In: *ACS Symposium Series*, **1045**; Aug. 2010. p. 165—78. https://doi.org/10.1021/bk-2010-1045.ch009.

[32] Dong H, et al. An overview on limitations of TiO2-based particles for photocatalytic degradation of organic pollutants and the corresponding countermeasures. *Water Res* Aug. 01, 2015;**79**:128—46. https://doi.org/10.1016/j.watres.2015.04.038. Elsevier Ltd.

[33] Umar K, Dar AA, Haque MM, Mir NA, Muneer M. Photocatalysed decolourization of two textile dye derivatives, Martius Yellow and Acid Blue 129, in UV-irradiated aqueous suspensions of Titania. *Desalination Water Treat* 2012;**46**(1—3):205—14. https://doi.org/10.1080/19443994.2012.677527.

[34] Ali A, et al. Synthesis, characterization, applications, and challenges of iron oxide nanoparticles. *Nanotechnol Sci Appl* Aug. 2016;**9**:49—67. https://doi.org/10.2147/NSA.S99986.

[35] Lee C. Oxidation of organic contaminants in water by iron-induced oxygen activation: a short review. *Environ. Eng. Res.* Sep. 2015;**20**(3):205—11. https://doi.org/10.4491/eer.2015.051.

[36] Pawar M, Sendoğdular ST, Gouma P. A brief overview of TiO$_2$ photocatalyst for organic dye remediation: case study of reaction mechanisms involved in Ce-TiO$_2$ photocatalysts system. *J Nanomater* 2018;**2018**. https://doi.org/10.1155/2018/5953609. Hindawi Limited.

[37] Umar K, Haque MM, Mir NA, Muneer M, Farooqi IH. Titanium dioxide-mediated photocatalysed mineralization of two selected organic pollutants in aqueous suspensions. *J Adv Oxid Technol* Jul. 2013;**16**(2):252—60. https://doi.org/10.1515/jaots-2013-0205.

[38] Yu J, Nguyen CTK, Lee H. Preparation of blue TiO$_2$ for visible-light-driven photocatalysis. In: *Titanium dioxide - material for a sustainable environment.* InTech; 2018.

[39] Borgarello E, Kiwi J, Grätzel M, Pelizzetti E, Visca M. Visible light induced water cleavage in colloidal solutions of chromium-doped titanium dioxide particles. *J Am Chem Soc* 1982;**104**(11):2996—3002. https://doi.org/10.1021/ja00375a010.

[40] Mishra M, Chun DM. α-Fe$_2$O$_3$ as a photocatalytic material: a review. *Appl Catal Gen* Jun. 05, 2015;**498**:126—41. https://doi.org/10.1016/j.apcata.2015.03.023. Elsevier B.V.

[41] Zhou X, et al. Visible light induced photocatalytic degradation of rhodamine B on one-Dimensional iron oxide particles. *J Phys Chem C* Oct. 2010;**114**(40):17051—61. https://doi.org/10.1021/jp103816e.

[42] Ristić M, Musić S, Godec M. Properties of γ-FeOOH, α-FeOOH and α-Fe$_2$O$_3$ particles precipitated by hydrolysis of Fe^{3+} ions in perchlorate containing aqueous solutions. *J Alloys Compd* Jun. 2006;**417**(1—2):292—9. https://doi.org/10.1016/j.jallcom.2005.09.043.

[43] Lu H, Wang J, Stoller M, Wang T, Bao Y, Hao H. An overview of nanomaterials for water and wastewater treatment. *Adv Mater Sci Eng* 2016;**2016**. https://doi.org/10.1155/2016/4964828.

[44] Daneshvar N, Salari D, Khataee AR. Photocatalytic degradation of azo dye acid red 14 in water on ZnO as an alternative catalyst to TiO2. *J Photochem Photobiol A Chem* Mar. 2004;**162**(2—3):317—22. https://doi.org/10.1016/S1010-6030(03)00378-2.

[45] Shahpal A, Aziz Choudhary M, Ahmad Z. An investigation on the synthesis and catalytic activities of pure and Cu-doped zinc oxide nanoparticles. *Cogent Chem.* Jan. 2017;**3**(1):1301241. https://doi.org/10.1080/23312009.2017.1301241.

[46] Yi G, Li X, Yuan Y, Zhang Y. Redox active Zn/ZnO duo generating superoxide (O_2^-) and H_2O_2 under all conditions for environmental sanitation. *Environ Sci Nano* Jan. 2019;**6**(1):68–74. https://doi.org/10.1039/c8en01095a.

[47] Das R, Abd Hamid SB, Ali ME, Ismail AF, Annuar MSM, Ramakrishna S. Multifunctional carbon nanotubes in water treatment: the present, past and future. *Desalination* Dec. 01, 2014;**354**:160–79. https://doi.org/10.1016/j.desal.2014.09.032. Elsevier.

[48] Ren X, Chen C, Nagatsu M, Wang X. Carbon nanotubes as adsorbents in environmental pollution management: a review. *Chem Eng J* Jun. 01, 2011;**170**(2–3):395–410. https://doi.org/10.1016/j.cej.2010.08.045.

[49] Mauter MS, Elimelech M. Environmental applications of carbon-based nanomaterials. *Environ Sci Technol* Aug. 15, 2008;**42**(16):5843–59. https://doi.org/10.1021/es8006904. American Chemical Society.

[50] Das R, Abd Hamid SB, Ali ME, Annuar MSM, Samsudin EMB, Bagheri S. Covalent functionalization schemes for tailoring solubility of multi-walled carbon nanotubes in water and acetone solvents. *Sci Adv Mater* 2015;**7**(12):2726–37. https://doi.org/10.1166/sam.2015.2694.

[51] Ali ME, Das R, Maamor A, Hamid SBA. Multifunctional carbon nanotubes (CNTs): a new dimension in environmental remediation. In: *Adv Mater Res*, **832**; 2014. p. 328–32. https://doi.org/10.4028/www.scientific.net/AMR.832.328.

[52] Das R. *Nanohybrid catalyst based on carbon nanotube*. Cham: Springer International Publishing; 2017.

[53] Li Y, Chen J, Liu J, Ma M, Chen W, Li L. Activated carbon supported TiO_2-photocatalysis doped with Fe ions fo continuous treatment of dye wastewater in a dynamic reactor. *J Environ Sci* Aug. 2010;**22**(8):1290–6. https://doi.org/10.1016/S1001-0742(09)60252-7.

[54] Singh SB. Iron and iron oxide-based eco-nanomaterials for catalysis and water remediation. In: *Handbook of ecomaterials*. Springer International Publishing; 2018. p. 1–21.

[55] Das R. Carbon nanotube in water treatment. In: *Carbon nanostructures*, **0**. Springer International Publishing; 2017. p. 23–54. 9783319581507.

[56] Bahuguna A, Kumar A, Krishnan V. Carbon-Support-based heterogeneous nanocatalysts: synthesis and applications in organic reactions. *Asian J Org Chem* Aug. 2019;**8**(8):1263–305. https://doi.org/10.1002/ajoc.201900259.

[57] Smith AT, LaChance AM, Zeng S, Liu B, Sun L. Synthesis, properties, and applications of graphene oxide/reduced graphene oxide and their nanocomposites. *Nano Mater. Sci.* Mar. 2019;**1**(1):31–47. https://doi.org/10.1016/j.nanoms.2019.02.004.

[58] Axet MR, Durand J, Gouygou M, Serp P. Surface coordination chemistry on graphene and two-dimensional carbon materials for well-defined single atom supported catalysts. In: *Advances in organometallic chemistry*, **71**. Academic Press Inc.; 2019. p. 53–174.

[59] Khan M, et al. Graphene based metal and metal oxide nanocomposites: synthesis, properties and their applications. *J Mater Chem A* Jun. 2015;**3**(37):18753–808. https://doi.org/10.1039/c5ta02240a.

[60] Saifuddin N, Raziah AZ, Junizah AR. Carbon nanotubes: a review on structure and their interaction with proteins. *J Chem* 2013. https://doi.org/10.1155/2013/676815.

[61] Hamid SBA, Das R, Ali ME. Photoconductive carbon nanotube (CNT): a potential candidate for future renewable energy. *Adv Mater Res* 2014;**925**:48–51. https://doi.org/10.4028/www.scientific.net/AMR.925.48.

[62] Rakov EG, Baronin IV, Anoshkin IV. Carbon nanotubes for catalytic applications. *Catal Ind* Mar. 2010;**2**(1):26–8. https://doi.org/10.1134/S2070050410010046.

[63] Shaban M, Ashraf AM, Abukhadra MR. $TiO2$ nanoribbons/carbon nanotubes composite with enhanced photocatalytic activity; fabrication, characterization, and application. *Sci Rep* Dec. 2018;**8**(1):781. https://doi.org/10.1038/s41598-018-19172-w.

[64] Das R. *Introduction*. 2019. p. 1–5.

[65] Zinatloo-Ajabshir S, Salavati-Niasari M, Hamadanian M. Praseodymium oxide nanostructures: novel solvent-less preparation, characterization and investigation of their optical and photocatalytic properties. *RSC Adv* Apr. 2015;**5**(43):33792–800. https://doi.org/10.1039/c5ra00817d.

[66] Soofivand F, Mohandes F, Salavati-Niasari M. Silver chromate and silver dichromate nanostructures: sonochemical synthesis, characterization, and photocatalytic properties. *Mater Res Bull* Jun. 2013;**48**(6):2084–94. https://doi.org/10.1016/j.materresbull.2013.02.025.

[67] Geise GM, Lee HS, Miller DJ, Freeman BD, McGrath JE, Paul DR. Water purification by membranes: the role of polymer science. *J Polym Sci Part B Polym Phys* Aug. 2010;**48**(15):1685–718. https://doi.org/10.1002/polb.22037.

[68] Pandey N, Shukla SK, Singh NB. Water purification by polymer nanocomposites: an overview. *Nanocomposites* Apr. 03, 2017;**3**(2):47–66. https://doi.org/10.1080/20550324.2017.1329983. Taylor and Francis Inc.

[69] Jawaid M, Ahmad A, Ismail N, Rafatullah M, editors. *Environmental remediation through carbon based nano composites*. Singapore: Springer Singapore; 2021.

[70] Dhakshinamoorthy A, Asiri AM, Garcia H. Catalysis by metal-organic frameworks in water. *Chem Commun* Sep. 2014;**50**(85):12800–14. https://doi.org/10.1039/c4cc04387a.

[71] De Vietro N, et al. Photocatalytic inactivation of *Escherichia coli* bacteria in water using low pressure plasma deposited TiO_2 cellulose fabric. *Photochem Photobiol Sci* Sep. 2019;**18**(9):2248–58. https://doi.org/10.1039/c9pp00050j.

[72] Murillo-Sierra JC, Hernández-Ramírez A, Hinojosa-Reyes L, Guzmán-Mar JL. A review on the development of visible light-responsive WO_3-based photocatalysts for environmental applications. *Chem Eng J Adv* Mar. 2021;**5**:100070. https://doi.org/10.1016/j.ceja.2020.100070.

[73] Fedotova MP, Voronova GA, Emelyanova EY, Vodyankina OV. A method of preparation of active TiO_2-SiO_2 photocatalysts for water purification. *Stud Surf Sci Catal* Jan. 2010;**175**:723–6. https://doi.org/10.1016/S0167-2991(10)75145-4.

[74] Rao CNR. Perovskites. In: *Encyclopedia of physical science and technology*. Elsevier; 2003. p. 707–14.

[75] Peña MA, Fierro JLG. Chemical structures and performance of perovskite oxides. *Chem Rev* Jul. 2001;**101**(7):1981–2017. https://doi.org/10.1021/cr980129f. American Chemical Society.

[76] Phraewphiphat T, et al. Syntheses, structures, and ionic conductivities of perovskite-structured lithium-strontium-aluminum/gallium-tantalum-oxides. *J Solid State Chem* May 2015;**225**:431–7. https://doi.org/10.1016/j.jssc.2015.01.007.

[77] Alvarez G, Conde-Gallardo A, Montiel H, Zamorano R. About room temperature ferromagnetic behavior in $BaTiO_3$ perovskite. *J Magn Magn Mater* Mar. 2016;**401**(401):196–9. https://doi.org/10.1016/j.jmmm.2015.10.031.

[78] Ichimura K, Inoue Y, Yasumori I. Hydrogenation and hydrogenolysis of hydrocarbons on perovskite oxides. *Catal Rev* Dec. 1992;**34**(4):301–20. https://doi.org/10.1080/01614949208016314.

[79] Wang GJ, Qin YN, Ma Z, Qi XZ, Ding T. Study on the catalytic reduction mechanism of SO_2 by CO over doped copper perovskite catalyst in presence of oxygen. *React Kinet Catal Lett* Aug. 2006;**89**(2):229–36. https://doi.org/10.1007/s11144-006-0132-1.

[80] Yang X, Luo L, Zhong H. Catalytic properties of $LnSrCoO_4$ (ln = La, Sm) in the oxidation of CO and C_3H_8. *React Kinet Catal Lett* 2004;**81**(2):219–27. https://doi.org/10.1023/B:REAC.0000019426.76399.e9.

[81] Tavakkoli H, Yazdanbakhsh M. Fabrication of two perovskite-type oxide nanoparticles as the new adsorbents in efficient removal of a pesticide from aqueous solutions: kinetic, thermodynamic, and adsorption studies. *Micropor Mesopor Mater* Aug. 2013;**176**:86–94. https://doi.org/10.1016/j.micromeso.2013.03.043.

[82] Das N, Kandimalla S. Application of perovskites towards remediation of environmental pollutants: an overview: a review on remediation of environmental pollutants using perovskites. *Int J Environ Sci Technol* Jul. 01, 2017;**14**(7):1559–72. https://doi.org/10.1007/s13762-016-1233-7. Center for Environmental and Energy Research and Studies.

[83] Chen YZ, Zhang R, Jiao L, Jiang HL. Metal–organic framework-derived porous materials for catalysis. *Coord Chem Rev* May 01, 2018;**362**:1–23. https://doi.org/10.1016/j.ccr.2018.02.008. Elsevier B.V.

[84] Xu GR, An ZH, Xu K, Liu Q, Das R, Zhao HL. Metal organic framework (MOF)-based micro/nanoscaled materials for heavy metal ions removal: the cutting-edge study on designs, synthesis, and applications. *Coord Chem Rev* Jan. 15, 2021;**427**:213554. https://doi.org/10.1016/j.ccr.2020.213554. Elsevier B.V.

[85] Huang H, Shen K, Chen F, Li Y. *Metal–Organic frameworks as a good platform for the fabrication of single-atom catalysts.* 2020. https://doi.org/10.1021/acscatal.0c01459.

[86] Sikarwar VS, Zhao M, Fennell PS, Shah N, Anthony EJ. Progress in biofuel production from gasification. *Prog Energy Combust Sci* Jul. 01, 2017;**61**:189–248. https://doi.org/10.1016/j.pecs.2017.04.001. Elsevier Ltd.

[87] Verma M, Godbout S, Brar SK, Solomatnikova O, Lemay SP, Larouche JP. Biofuels production from biomass by thermochemical conversion technologies. *Int J Chem Eng* 2012. https://doi.org/10.1155/2012/542426.

[88] Uddin MN, et al. An overview of recent developments in biomass pyrolysis technologies. *Energies* Nov. 01, 2018;**11**(11):3115. https://doi.org/10.3390/en11113115. MDPI AG.

[89] Refaat AA. Biodiesel production using solid metal oxide catalysts. *Int J Environ Sci Technol* Dec. 2011;**8**(1):203–21. https://doi.org/10.1007/BF03326210.

[90] Yigezu ZD, Muthukumar K. Catalytic cracking of vegetable oil with metal oxides for biofuel production. *Energy Convers Manag* Aug. 2014;**84**:326–33. https://doi.org/10.1016/j.enconman.2014.03.084.

[91] Balasundram V, et al. Thermogravimetric studies on the catalytic Pyrolysis of rice husk. *Chem Eng Trans* 2017;**56**:427–32. https://doi.org/10.3303/CET1756072.

[92] Zhang L, Bao Z, Xia S, Lu Q, Walters K. Catalytic pyrolysis of biomass and polymer wastes. *Catalysts* Dec. 2018;**8**(12):659. https://doi.org/10.3390/catal8120659.

[93] Peterson AA, Vogel F, Lachance RP, Fröling M, Antal MJ, Tester JW. Thermochemical biofuel production in hydrothermal media: a review of sub- and supercritical water technologies. *Energy Environ Sci* Jul. 23, 2008;**1**(1):32–65. https://doi.org/10.1039/b810100k. Royal Society of Chemistry.

[94] Zhang Y, Chen WT. 5 - hydrothermal liquefaction of protein-containing feedstocks. In: *Direct thermochemical liquefaction for energy applications.* Elsevier; 2018. p. 127–68.

[95] Widayat W, Darmawan T, Rosyid RA, Hadiyanto H. Biodiesel production by using CaO catalyst and ultrasonic assisted. *J Phys* Aug. 2017;**877**(1):12037. https://doi.org/10.1088/1742-6596/877/1/012037.

[96] Dalai AK, Kulkarni MG, Meher LC. *Biodiesel productions from vegetable oils using heterogeneous catalysts and their applications as lubricity additives.* 2006. https://doi.org/10.1109/EICCCC.2006.277228.

[97] Zhao X, Wei L, Cheng S, Julson J. Review of heterogeneous catalysts for catalytically upgrading vegetable oils into hydrocarbon biofuels. *Catalysts* Mar. 2017;**7**(12):83. https://doi.org/10.3390/catal7030083.

[98] Shen Y, Yoshikawa K. Recent progresses in catalytic tar elimination during biomass gasification or pyrolysis - a review. *Renew Sustain Energy Rev* May 01, 2013;**21**:371–92. https://doi.org/10.1016/j.rser.2012.12.062. Pergamon.

[99] Sana SS, et al. *Nanocatalysts to improve the production of microbial fuel applications.* Singapore: Springer; 2020. p. 229–47.

[100] Richardson Y, Blin J, Volle G, Motuzas J, Julbe A. In situ generation of Ni metal nanoparticles as catalyst for H_2-rich syngas production from biomass gasification. *Appl Catal A Gen* Jul. 2010;**382**(2):220–30. https://doi.org/10.1016/j.apcata.2010.04.047.

[101] Abu El-Rub Z, Bramer EA, Brem G. Review of catalysts for tar elimination in biomass gasification processes. *Ind Eng Chem Res* Oct. 27, 2004;**43**(22):6911–9. https://doi.org/10.1021/ie0498403. American Chemical Society.

[102] Kaur R, Gera P, Jha MK, Bhaskar T. Thermochemical route for biohydrogen production. In: *Biohydrogen.* Elsevier; 2019. p. 187–218.

[103] Han J, Kim H. The reduction and control technology of tar during biomass gasification/pyrolysis: an overview. *Renew Sustain Energy Rev* 2008;**12**(2):397–416. Accessed: Mar. 03, 2021. [Online]. Available, https://ideas.repec.org/a/eee/rensus/v12y2008i2p397-416.html.

[104] Nordgreen T, Liliedahl T, Sjöström K. Metallic iron as a tar breakdown catalyst related to atmospheric, fluidised bed gasification of biomass. *Fuel* Mar. 2006;**85**(5–6):689–94. https://doi.org/10.1016/j.fuel.2005.08.026.

[105] Balat M, Balat M, Kirtay E, Balat H. Main routes for the thermo-conversion of biomass into fuels and chemicals. Part 2: gasification systems. *Energy Convers Manag* Dec. 2009;**50**(12):3158–68. https://doi.org/10.1016/j.enconman.2009.08.013.

[106] Balat M, Balat M, Kirtay E, Balat H. Main routes for the thermo-conversion of biomass into fuels and chemicals. Part 1: pyrolysis systems. *Energy Convers Manag* Dec. 2009;**50**(12):3147–57. https://doi.org/10.1016/j.enconman.2009.08.014.

[107] Adegoke KA, Adegoke RO, Ibrahim AO, Adegoke SA, Bello OS. Electrocatalytic conversion of CO_2 to hydrocarbon and alcohol products: realities and prospects of Cu-based materials. *Sustain Mater Technol* Sep. 01, 2020;**25**:e00200. https://doi.org/10.1016/j.susmat.2020.e00200. Elsevier B.V.

[108] Wei J, et al. Directly converting CO_2 into a gasoline fuel. *Nat Commun* May 2017;**8**(1):1–9. https://doi.org/10.1038/ncomms15174.

[109] Yang H, et al. A review of the catalytic hydrogenation of carbon dioxide into value-added hydrocarbons. *Catal Sci Technol* Oct. 16, 2017;**7**(20):4580–98. https://doi.org/10.1039/c7cy01403a. Royal Society of Chemistry.

[110] Evans G, Smith C. Biomass to liquids technology. In: *Comprehensive renewable energy*, **5**. Elsevier Ltd; 2012. p. 155–204.

[111] Ye RP, et al. CO_2 hydrogenation to high-value products via heterogeneous catalysis. *Nat Commun* Dec. 01, 2019;**10**(1):1–15. https://doi.org/10.1038/s41467-019-13638-9. Nature Research.

[112] Yao B, et al. Transforming carbon dioxide into jet fuel using an organic combustion-synthesized Fe-Mn-K catalyst. *Nat Commun* Dec. 2020;**11**(1):1–12. https://doi.org/10.1038/s41467-020-20214-z.

[113] Hu J, Yu F, Lu Y. Application of Fischer–tropsch synthesis in biomass to liquid conversion. *Catalysts* Jun. 2012;**2**(2):303–26. https://doi.org/10.3390/catal2020303.

[114] Rauch R, Kiennemann A, Sauciuc A. Fischer-tropsch synthesis to biofuels (BtL process). In: *The role of catalysis for the sustainable production of bio-fuels and bio-chemicals*. Elsevier Inc.; 2013. p. 397–443.

[115] Takenaka S, Shimizu T, Otsuka K. Complete removal of carbon monoxide in hydrogen-rich gas stream through methanation over supported metal catalysts. *Int J Hydrogen Energy* Aug. 2004;**29**(10):1065–73. https://doi.org/10.1016/j.ijhydene.2003.10.009.

[116] Rodemerck U, Holeňa M, Wagner E, Smejkal Q, Barkschat A, Baerns M. Catalyst development for CO_2 hydrogenation to fuels. *ChemCatChem* Jul. 2013;**5**(7):1948–55. https://doi.org/10.1002/cctc.201200879.

[117] *Latest water purification technologies - top five - Water Technology*. https://www.water-technology.net/features/latest-water-purification-technologies-top-five.

[118] Namshah KS, Mohamed RM. WO3–TiO2 nanocomposites for paracetamol degradation under visible light. *Appl Nanosci* Nov. 2018;**8**(8):2021–30. https://doi.org/10.1007/s13204-018-0888-4.

[119] Chhabra T, Kumar A, Bahuguna A, Krishnan V. Reduced graphene oxide supported MnO_2 nanorods as recyclable and efficient adsorptive photocatalysts for pollutants removal. *Vacuum* Feb. 2019;**160**:333–46. https://doi.org/10.1016/j.vacuum.2018.11.053.

[120] Montalvo-Herrera T, Sánchez-Martínez D, Torres-Martínez LM. Sonochemical synthesis of $CaBi_6O_{10}$ nanoplates: photocatalytic degradation of organic pollutants (ciprofloxacin and methylene blue) and oxidizing species study (h^+, $OH\cdot$, H_2O_2 and O_2^-). *J Chem Technol Biotechnol* Jul. 2017;**92**(7):1496–502. https://doi.org/10.1002/jctb.5252.

[121] Muraro PCL, et al. Iron oxide nanocatalyst with titanium and silver nanoparticles: synthesis, characterization and photocatalytic activity on the degradation of Rhodamine B dye. *Sci Rep* Dec. 2020;**10**(1):1–9. https://doi.org/10.1038/s41598-020-59987-0.

[122] Nazim M, Khan AAP, Asiri AM, Kim JH. Exploring rapid photocatalytic degradation of organic pollutants with porous CuO nanosheets: synthesis, dye removal, and kinetic studies at room temperature. *ACS Omega* Feb. 2021;**6**(4):2601–12. https://doi.org/10.1021/acsomega.0c04747.

[123] Yang X, Duan L, Ran X. Photocatalytic degradation of organic dyes by a donor–acceptor type conjugated polymer: poly(thiophene-1,3,4-oxadiazole) and its photocatalytic mechanism. *Polym Int* Sep. 2018;**67**(9):1282–90. https://doi.org/10.1002/pi.5652.

[124] Su F, Li P, Huang J, Gu M, Liu Z, Xu Y. Photocatalytic degradation of organic dye and tetracycline by ternary Ag$_2$O/AgBr−CeO$_2$ photocatalyst under visible-light irradiation. *Sci Rep* Dec. 2021;**11**(1): 85. https://doi.org/10.1038/s41598-020-76997-0.

[125] Huo J, Yu D, Li H, Luo B, Arulsamy N. Mechanistic investigation of photocatalytic degradation of organic dyes by a novel zinc coordination polymer. *RSC Adv* Nov. 2019;**9**(67):39323−31. https://doi.org/10.1039/c9ra07821e.

[126] Zhao J, Biswas MRUD, Oh WC. A novel BiVO$_4$-GO-TiO$_2$-PANI composite for upgraded photocatalytic performance under visible light and its non-toxicity. *Environ Sci Pollut Res* Apr. 2019;**26**(12): 11888−904. https://doi.org/10.1007/s11356-019-04441-6.

[127] Kratofil Krehula L, et al. Conducting polymer polypyrrole and titanium dioxide nanocomposites for photocatalysis of RR45 dye under visible light. *Polym Bull* Apr. 2019;**76**(4):1697−715. https://doi.org/10.1007/s00289-018-2463-2.

[128] Jana B, Bhattacharyya S, Patra A. Conjugated polymer P3HT-Au hybrid nanostructures for enhancing photocatalytic activity. *Phys Chem Chem Phys* Jun. 2015;**17**(23):15392−9. https://doi.org/10.1039/c5cp01769f.

[129] Kozhevnikova NS, et al. Nanocrystalline TiO$_2$ doped by small amount of pre-synthesized colloidal CdS nanoparticles for photocatalytic degradation of 1,2,4-trichlorobenzene. *Sustain Chem Pharm* Mar. 2019;**11**:1−11. https://doi.org/10.1016/j.scp.2018.11.004.

[130] Yao P, Liu H, Wang D, Chen J, Li G, An T. Enhanced visible-light photocatalytic activity to volatile organic compounds degradation and deactivation resistance mechanism of titania confined inside a metal-organic framework. *J Colloid Interface Sci* Jul. 2018;**522**:174−82. https://doi.org/10.1016/j.jcis.2018.03.075.

[131] Zhang Y, Zhou J, Feng Q, Chen X, Hu Z. Visible light photocatalytic degradation of MB using UiO-66/g-C$_3$N$_4$ heterojunction nanocatalyst. *Chemosphere* Dec. 2018;**212**:523−32. https://doi.org/10.1016/j.chemosphere.2018.08.117.

[132] He L, Dong Y, Zheng Y, Jia Q, Shan S, Zhang Y. A novel magnetic MIL-101(Fe)/TiO$_2$ composite for photo degradation of tetracycline under solar light. *J Hazard Mater* Jan. 2019;**361**:85−94. https://doi.org/10.1016/j.jhazmat.2018.08.079.

[133] Zhang M, et al. Two pure MOF-photocatalysts readily prepared for the degradation of methylene blue dye under visible light. *Dalton Trans* 2018;**47**(12):4251−8. https://doi.org/10.1039/c8dt00156a.

[134] Wang C, Xue Y, Wang P, Ao Y. Effects of water environmental factors on the photocatalytic degradation of sulfamethoxazole by AgI/UiO-66 composite under visible light irradiation. *J Alloys Compd* Jun. 2018;**748**:314−22. https://doi.org/10.1016/j.jallcom.2018.03.129.

[135] Gehrke I, Geiser A, Somborn-Schulz A. Innovations in nanotechnology for water treatment. *Nanotechnol Sci Appl* Jan. 06, 2015;**8**:8−9. https://doi.org/10.2147/NSA.S43773. Dove Medical Press Ltd.

[136] *Inverters for solar PV systems battery storage | KACO new energy*. https://kaco-newenergy.com/home.

[137] Baskar G, Aberna Ebenezer Selvakumari I, Aiswarya R. Biodiesel production from castor oil using heterogeneous Ni doped ZnO nanocatalyst. *Bioresour Technol* Feb. 2018;**250**:793−8. https://doi.org/10.1016/j.biortech.2017.12.010.

[138] Baskar G, Gurugulladevi A, Nishanthini T, Aiswarya R, Tamilarasan K. Optimization and kinetics of biodiesel production from Mahua oil using manganese doped zinc oxide nanocatalyst. *Renew Energy* Apr. 2017;**103**:641−6. https://doi.org/10.1016/j.renene.2016.10.077.

[139] Tahvildari K, Anaraki YN, Fazaeli R, Mirpanji S, Delrish E. The study of CaO and MgO heterogenic nano-catalyst coupling on transesterification reaction efficacy in the production of biodiesel from recycled cooking oil. *J Environ Heal Sci Eng* Oct. 2015;**13**(1):1−9. https://doi.org/10.1186/s40201-015-0226-7.

[140] Borah MJ, Devi A, Saikia RA, Deka D. Biodiesel production from waste cooking oil catalyzed by in-situ decorated TiO$_2$ on reduced graphene oxide nanocomposite. *Energy* Sep. 2018;**158**:881−9. https://doi.org/10.1016/j.energy.2018.06.079.

[141] Samart C, Chaiya C, Reubroycharoen P. Biodiesel production by methanolysis of soybean oil using calcium supported on mesoporous silica catalyst. *Energy Convers Manag* Jul. 2010;**51**(7):1428–31. https://doi.org/10.1016/j.enconman.2010.01.017.

[142] Ibrahim MM, Mahmoud HR, El-Molla SA. Influence of support on physicochemical properties of ZrO_2 based solid acid heterogeneous catalysts for biodiesel production. *Catal Commun* Mar. 2019;**122**:10–5. https://doi.org/10.1016/j.catcom.2019.01.008.

[143] Rahman WU, et al. Biodiesel synthesis from eucalyptus oil by utilizing waste egg shell derived calcium based metal oxide catalyst. *Process Saf Environ Protect* Feb. 2019;**122**:313–9. https://doi.org/10.1016/j.psep.2018.12.015.

CHAPTER 6

Surface-modified nanomaterial-based catalytic materials for the production of liquid fuels

Indrajeet R. Warkad, Hanumant B. Kale and Manoj B. Gawande

Department of Industrial and Engineering Chemistry, Institute of Chemical Technology, Mumbai Marathwada Campus, Jalna, Maharashtra, India

1. Introduction

The rising population upsurges the world's energy demands because of reformed lifestyle and other requirements, i.e., fuels for transportation and industrialization.[1] The major concern of the 21st century is the need to quest highly impressive means to alter crude oil to fulfill the global energy demands. Researchers are looking for a more sustainable way for the production of fuels to fulfill the world energy demands. To overcome the two major concerns, i.e., diminishing the fossil fuels and raising concerns about global climate change, the scientific community moves toward the expansion of pathways for fuel production depends on renewable resources, such as liquid fuels synthesis from bio-derived precursors for the transportation and energy source so the cost of these fuels should be low as compared with petroleum fuels.[2,3] In consequence of this, the design and development of energy sources, utilization, and conversion of biomass to liquid fuels, biodiesel has fascinated widespread attention due to its rich source of energy.[4,5] Synthesis of nonoxygenated fuels are requisite for the gasoline and transportation; most of the time it is highly necessary to design a catalyst system to remove the oxygen from biomass and forms a long-chain C_9 to C_{15} fuel.[6]

A variety of transition metal-based nanomaterials (NMs) are notified for the manufacturing the liquid fuels. In particular some of the surface engineered nanocatalyst having better control of their surface morphology, chemical, and electronic properties at the nanoscale.[7,8] These surface-modified nanomaterials (SMNs) exhibited higher catalytic activity and reactivity toward biomass valorization, transformation of carbon dioxide (CO_2) to liquid fuels because of the high surface area, robustness, tunable morphology, unique chemical and physical characteristics relative to their bulk counterparts, and the ability to modify their surfaces, offering unique opportunities for the catalytic converter.[9,10] The SMNs can be entail of (a) passivating very reactive nanoparticles (NPs), (b) stabilizing a very aggregative NPs in a medium (which may be a solvent, i.e., ionic liquids, or a polymer or any other support like graphene, carbon nanotubes, metal oxides, etc.) where the NPs are to be dispersed, (c) functionalizing the NPs for the applications of molecular recognition, and (d) promoting the assembly of NPs. The primary evidence that the world utilized various nonfossil resources, particularly wind, solar, and geothermal sources, and biomass that can cooperatively meeting energy requisites globally. To achieve the goal of energy fulfillment needs to be develop an efficient and functional wide range of "biomass refineries" for the transformation of biomass.[11]

Gasification, pyrolysis, and chemical hydrolysis are three major routes so far demonstrated for the production of biofuels from the natural organic carbon source.[12,13] To date various SMNs are reported for the production of liquid fuels like 5-hydroxymethylfurfural (HMF), furfural, alcohols, and polyols, oils, and acids from biomass as a starting material.[14–21] Carbohydrates, lignocellulose-based biomass utilized for the production of fuels, among them lignocellulose is nonedible biomass stated as "no competition with food" widely utilized as a raw material for the synthesizing various liquid chemicals.[5,22] Apart from this, biodiesel (fatty acid methyl esters) is an alternative way for the conventional diesel derived from petroleum, obtained from animal fats and plant-based oils.[23] Renewable biomass feedstock used for the production of biodiesel and it blended with petroleum diesel acts as a fuel or in diesel engines consisting of mono-alkyl esters acquired from the plant oil transesterification process in ethanol or methanol as a reaction medium.[24,25]

Apart from the aforementioned liquid fuels, CO_2 is one of the feedstock utilized for C_1 and C_2 fuels manufacturing along with hydrocarbons.[8,26] Transformation of CO_2 to chemical fuels is the sustainable ways to defeat the energy crisis arising due to the decreasing of fossil fuels because of a growing population and advanced lifestyle.[27] It is well-known that a huge amount of (33 gigatons) CO_2 is emitted globally; which is very threatening to environment-related concerns i.e., air pollution, acid rain, etc. CO_2 and carbon monoxide (CO) are the major sources of greenhouse gases that can be upgraded to liquid fuels and feedstocks through electrocatalysis.[28] Various SMNs are utilized for the capture and utilization of CO_2 to methanol, formic acid, ethanol, and propanol production.[29–32] Among them methanol is a viable and clean energy source for gasoline, diesel, and other important chemical feedstocks.[33] In the view of reduction of carbon footprints, CO_2 conversion to methanol is a key reaction in the recently proposed "methanol economy"[34] and "liquid sunshine" vision.[35] Apart from this other alternative ways also reported for the methanol production from biomass.[36] In this book chapter, we summarized some reported protocols for synthesis and applications of SMNs for the manufacturing of liquid fuels like alcohols and polyols i.e., 1,2-propanediol, 1,4-butanediol, ethylene glycol, glycerol, 1,3-propanediol, also furanics compounds (furfural and 5-hydroxymethylfurfural), biodiesel (fatty acid methyl esters, free fatty acid, etc.) from natural occurring organic biomass as a precursor.[37–50] Herein, we also summarized the synthesis of energy fuels like methanol, ethanol, propanol, and formic acid from hydrogenation and electrocatalytic CO_2 reduction in presence of SMNs (Fig. 6.1).[38–41,51–55]

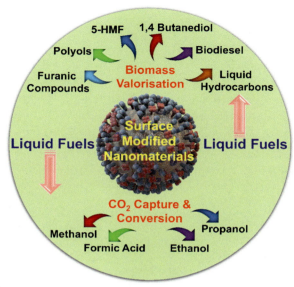

Figure 6.1 Schematic representation of different types of liquid fuels obtained from biomass-derived precursors and CO$_2$ sequestration.

2. Surface modified nanomaterials (SMNs) for biomass conversion to liquid fuels

Lignocellulose and lignin are the renewable nonwood biomass feedstock efficiently utilized for the manufacturing of liquid fuels and chemical precursors via catalytic transformations.[56,57] It will reduce the dependence on nonrenewable fossil fuels resources and overcome the upcoming global energy crisis. Since the evolution of NMs for catalytic application, to date, several SMNs reported for the production of liquid fuels through biomass valorization.[25,37,44,58–67] The single-pot conversion of cellulose into polyols attracted much attention in academics and industries because of ease in process for liquid fuels productions.[56] The production of glycols from cellulosic biomass is one of the viable ways for liquid fuel formation via biomass valorization because of the high efficiency of product output along with diols having large market demand.[68] Pang et al. have demonstrated ball milling pretreatment is one of the sustainable approaches for the synthesis of ethylene glycol (EG) via catalytic upgradation of lignocellulosic biomass.[69] Lignin is another source of biomass made up of amorphous three-dimensional polymer made up of methoxylated phenylpropane components that can be catalytically upgraded into aromatics and subsequently valuable chemicals.[70–73] Several SMNs demonstrated for the production of liquids from upgrading biomass are described below.

2.1 Production of 1, 2-propylene glycol and ethylene glycol

Lignocellulosic biomass consists of cellulose as a major component, various heterogeneous, bimetallic catalytic nanocomposites are developed for transformation of cellulose to value-added chemicals and fuels.[74–76] Among them glycols like 1,2-propylene glycol (1, 2-PG) and EG obtained via direct hydrogenolysis of cellulose-based biomass are the promising and efficient ways for liquid fuel production.[50,77–80] Several noble metals (Pt, Pd) supported on solid Lewis acids support, i.e., AlW, Al_2O_3 also demonstrated for the production of polyols but they exhibited comparative low yield and selectivity of products.[81–84] To achieve high selectivity, Liu et al. have demonstrated the surface-modified silica-supported Ru–W bifunctional catalyst for conversion of glucose to small diols i.e., butanediol (BDO), EG, and 1, 2-PG.[85] Silica surface-modified with polyethylene glycol followed by Ru–W adsorbed on silica surface by wet impregnation process. The prepared silica-supported Ru-WO_x catalysts exhibited high selectivity and catalytic efficiency toward the production of diols (91.7% diol formation) from glucose (98.2% of glucose conversion).

Recently Zhang et al. have synthesized bimetallic Ni–Sn catalysts by hydrothermal process.[50] Ni–Sn catalysts employed for the EG and 1,2-PG production selectively from cellulose (Fig. 6.2A). Ni–Sn alloy NPs evenly dispersed on activated carbon (AC) with the size of 10–150 nm revealed by transmission electron microscopy (TEM) and scanning tunneling electron microscopy (STEM) images (Fig. 6.2B–D). Ni–Sn bimetallic catalyst achieved 86.6% selectivity for total polyol production in that EG obtained was 57.6%. Also, 32.2% of 1,2-PG yield reported in a mixture of Ni/AC and SnO nanocomposite. The Ni–Sn catalyst switches the selectivity from EG to 1,2-PG due to a change in valency of Sn bound on the surface of Ni/AC in the transformation of cellulose. The Ni–Sn/AC catalyst exhibited two metallic sites, i.e., alloyed Sn species which are involved in the conversion of glucose to glycolaldehyde through retro-aldol condensation, and the other was unalloyed metallic Ni for the transformation of glycolaldehyde to EG via hydrogenation. The Ni–Sn nanocatalyst forms an alloy so that exhibited long-term stability and selectivity for glycol formation. To date, various bimetallic Ni-based composites reported for energy and environmental applications.[86] Apart from the above-reported protocols various Ni-based nanocomposites, for example, bimetallic Ni–La(III) catalyst,[87] Ni–W/M catalysts,[88] also employed for the transformation of lignocellulosic biomass to glycols.

Core-shell NPs having several applications in the area of biology, catalysis, material chemistry, and sensors due to its unique electronic and structural properties.[89] Catalyst recyclability and reusability significantly reduce the consumption of energy, use of noble metals as well as cost of the catalyst.[61] Till date several magnetically recoverable catalyst reported for the production of glucose, sorbitol, 1,2-PG, and EG from cellulosic biomass but every catalyst has its own drawbacks like low selectivity, recyclability.[63,90–92] Very

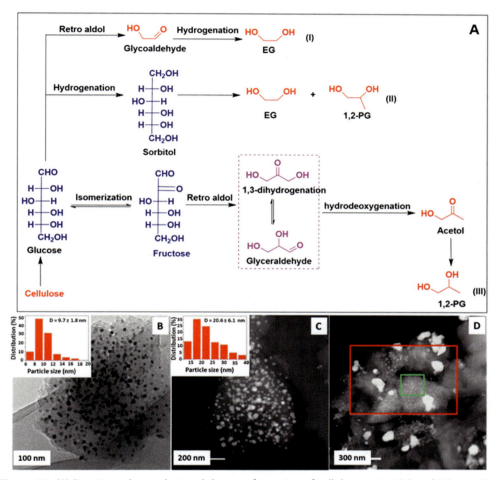

Figure 6.2 (A) Reaction scheme depicted the transformation of cellulose to 1, 2-PG and EG over Ni–Sn/AC nanocatalysts; TEM image of (B) Ni/AC; STEM images of (C, D) Ni–Sn(90)/AC. *(Reproduced with permission from Ref. 50 Copyright 2016 American Chemical Society.)*

recently Lv et al. have synthesized the magnetically recoverable core-shell $Fe_3O_4@SiO_2/Ru-WO_x$ nanocatalyst via the chemical reduction method.[46] The bifunctional $Fe_3O_4@SiO_2/Ru-WO_x$ catalyst was synthesized by using Ru NPs and WO_x (common solid acid) species on a surface of $Fe_3O_4@SiO_2$ nanospheres as support (Fig. 6.3A). The core-shell $Fe_3O_4@SiO_2/Ru-WO_x$ nanostructures employed for the production of EG, glycerol, and 1, 2-PG from cellulose. These core-shell NPs reported 96.8% of cellulose conversion to liquid fuels along with 32.4% of 1, 2-PG selectivity. The higher catalytic selectivity for the 1,2-PG formation was achieved in 2 h at 245°C and 5 MPa. The $Fe_3O_4@SiO_2$ nanospheres exhibited a Lewis acid site responsible for the glucose to

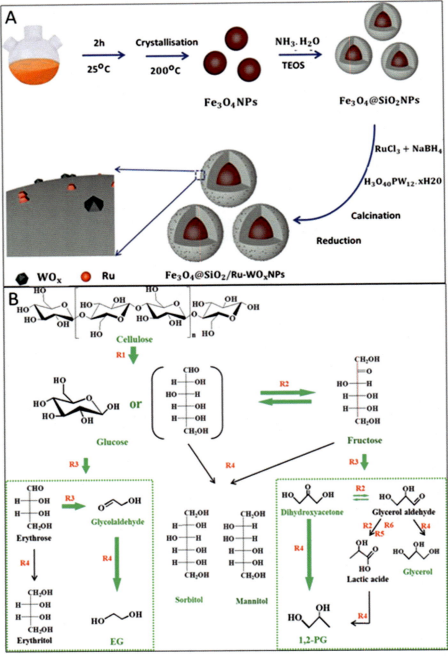

Figure 6.3 (A) Schematic fabrication of Fe_3O_4@SiO_2/Ru-WO_x; (B) Steps involved in the synthesis of EG, glycerol, and 1, 2-PG from lignocellulosic biomass. *(Reproduced with permission from Ref. 46 Copyright 2020 American Chemical Society.)*

fructose isomerization before the condensation of glucose via retro-aldol pathway. Whereas the Ru acts as a metal active site for the hydrogenation of C_3 intermediate and converted to 1,2-PG (Fig. 6.3B). The Fe_3O_4@SiO_2 and Ru-WO_x based core-shell NPs exhibited long-term catalytic stability after the magnetic recovery also makes this catalyst most prominent for the conversion of cellulose to liquid fuels.

Several tungsten-based catalysts were reported for hydrogenolysis of biomass-derived from lignocellulose to EG with high conversion rate and selectivity but due to low catalytic stability there use was limited.[56] Li et al. have reported tungsten-based hydrogenolysis catalysts (Ni–W/M) for the catalytic synthesis of EG from cellulose with good catalytic stability.[88] The Ni–W/M nanocatalyst was prepared by using MIL-125 (Ti) (Materials of Institut Lavoisier), as a template via the impregnation method. Ni–W/M catalyst enhanced the 100% rate of cellulose conversion and also reported 68.7% selectivity to EG production. SEM and TEM/HR-TEM images revealed that the small cylinder shape morphology of catalyst along with uniform dispersion of Ni NPs and TiO_2 NPs around the porous carbon. Among the library of catalysts, i.e., Ni–W/MO, Ni–W/T, and Ni–W/M, the Ni–W/M nanocatalyst manifested good catalytic efficiency and stability (seven runs) for the transformation of cellulose to EG (Fig. 6.4A–C). Transformation of cellulose to EG proceeded via three different catalytic steps, i.e., cellulose hydrolysis and retro-aldol condensation, followed by hydrogenation (Fig. 6.4D). During the catalytic transformation of cellulose, active tungsten component is retained due to the binding of tungsten with the catalyst support. The high stability of tungsten-based catalyst arises due to Ti–O–W bonds present in the moiety.

2.2 Production of 1,4-butanediol

The US Department of Energy's report, in 2004 was stated that succinic acid (SA) is the most significant C_4 feedstock utilized for the manufacturing of several valuable chemicals obtained through the fermentation process of renewable biomass sources in presence of CO_2 and water.[93–95] The biomass-derived SA used as a precursor for preparation of several smaller chemicals like tetrahydrofuran (THF), g-butyrolactone (GBL), and 1,4-butanediol (BDO), via several noble and non-noble metal-based catalytic hydrogenation.[59,96–103] The C_4/C_5 compounds and alcohols like BDO not only employed in the polymer-based industry for the preparation of polyurethanes, polyethers, and polyesters but also utilized as a fuel additive in fuel industry to increase the gasoline octane number.[101,104]

Among the list of transition metals, iron (Fe) is a very inexpensive metal firstly reported for the synthesis of BDO, GBL, and THF by Liu et al.[103] They reported FeO_x-promoted Pd/C catalysts for the SA hydrogenation in an aqueous medium to GBL, BDO, and THF. The Pd–FeO_x/C catalyst was prepared via the impregnation technique with a fixed content of Pd(3%) followed by varying Fe (1%, 5%, and 10%) composition.[103] Among the prepared Pd–10FeO_x/C, Pd–5FeO_x/C, and Pd–1FeO_x/C, the Pd and Fe NPs are uniformly distributed on active carbon support with some

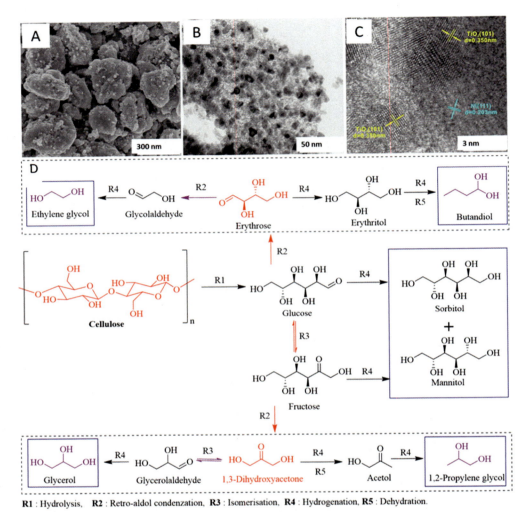

Figure 6.4 Schematic of Ni–W/M catalyst (A) SEM image; (B, C) TEM and HRTEM images; (D) a plausible mechanism for the production of EG, glycerol, and 1,2-PG from biomass-based cellulose. *(Reproduced with permission from Ref. 88 Copyright 2020 American Chemical Society.)*

agglomerates in Pd−5FeO$_x$/C nanocomposite revealed by STEM images and its elemental mapping (Fig. 6.5A−D). Also, Pd−5FeO$_x$/C showed high catalytic conversion of aqueous SA. Fe improved the activity as well as tuned the distribution of product and it was also acting as an efficient promoter for the SA hydrogenation. Aqueous SA converted into BDO with a 70% yield catalyzed by Pd−5FeO$_x$/C under the optimum reaction environment, i.e., 200°C and 5 MPa of H$_2$ pressure (Fig. 6.5E). Because of uniform dispersion of Pd and the synergistic effect between Fe and Pd species high

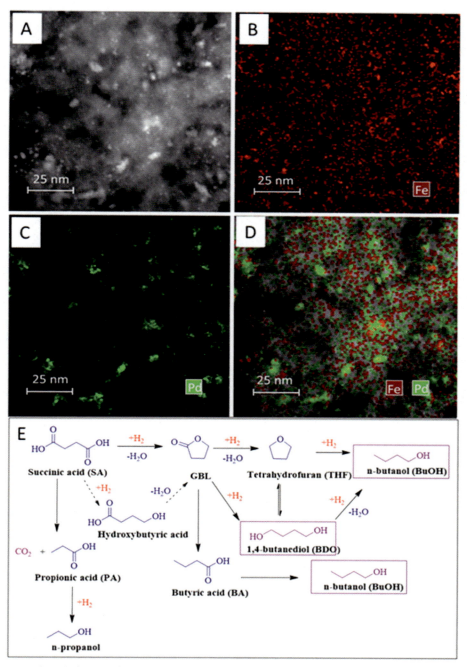

Figure 6.5 (A–D) Elemental mapping STEM and STEM images of Pd-5FeO$_x$/C catalyst; (E) Plausible reaction pathways for hydrogenation of SA catalyzed by Pd-5FeO$_x$/C. *(Reprinted with permission from Ref. 103 Copyright 2015 Royal Society of Chemistry.)*

catalytic efficiency and selectivity reported for the conversion of SA into BDO using Pd—5FeO$_x$/C catalyst was ascribed to the higher acidity of the catalyst. But unfortunately, it is observed that the catalytic activity was decreased after several runs because of the lower stability of the Pd-FeO$_x$/C catalyst hydrothermally. Later on, Huang et al. have demonstrated a more stable bimetallic titania-supported CuCo (CuCo/TiO$_2$) nanocomposite for hydrogenation of GBL to BDO.[105] The high catalytic efficiency demonstrated by Cu$_{0.1}$Co$_{0.9}$/TiO$_2$ nanocatalyst consists of a 1:9 atomic ratio of Cu—Co manifested 95% of product, i.e., BDO from GBL. The Cu—Co bimetallic core-shell catalyst manifested high selectivity (80%) and operation stability (150 h) in continuous flow conditions.

In line with this work, Nishimura et al. have developed Cu-based bimetallic catalyst, i.e., Cu—Pd NPs supported hydroxyapatite (HAP) catalysts (Cu—Pd/HAP) by coimpregnation method for BDO synthesis.[106] Herein the monometallic Cu/HAP or Pd/HAP catalyst individually not achieved the high selectivity of BDO formation, but after the addition of 8 wt% Cu in 2 wt% Pd resulted in bimetallic catalyst facilitated the hydrogenation of SA to BDO afforded 82% yield. Notably the monometallic Pd/HAP catalyst achieved the higher selectivity for production of butyric acid (BA) as a major product and GBL as a minor product as well as Cu/HAP resulted in 16% conversion of SA to GBL. The BA formation reduced by adding Cu in the Pd/HAP catalyst i.e., 8Cu—2Pd/HAP catalyst reported 100% conversion of SA into liquid products via hydrogenation. Apart from this, the very interesting protocol for preparation of 1,3-butanediol through hydrogenolysis of biomass-derived precursor 1,4-anhydroerythritol reported by Liu et al.[107] The author prepared tungsten-modified platinum supported on silica (Pt-WO$_x$/SiO$_2$) catalyst for selectively 1,4-anhydroerythritol hydrogenolysis to 1,3-butanediol via ring-opening followed by elimination of 2° hydroxy groups in 1,2,3-butanediol. The 54% yield of 1,3-butanediol was obtained in one-pot hydrogenolysis of 1,4-anhydroerythritol.

Very recently Le et al.[108] have demonstrated that the catalytic activity and product selectivity directly depends on the electronic behavior as well as metal-support interaction of the supporting medium. Cu—Pd bimetallic nanoalloy adsorbed on SiO$_2$, γ-Al$_2$O$_3$, and TiO$_2$ synthesized through impregnation process attained more than 60% selectivity for hydrogenation of SA. The average size of CuPd NPs uniformly dispersed on γ-Al$_2$O$_3$, SiO$_2$, and TiO$_2$ supports investigated by TEM (Fig. 6.6A—C). The CuPd/SiO$_2$, CuPd/TiO$_2$, and CuPd/γ-Al$_2$O$_3$ catalyst exhibited 86%, 97%, as well as 90% selectivity for production of BDO, THF, and GBL from SA hydrogenation, respectively (Fig. 6.6D). Herein they reported that the THF could be produced from either GBL hydrogenation or BDO dehydration because of the characteristics of catalyst and reaction conditions. Here author proved that the structure of the supported CuPd NPs controlled by metal-support interactions also catalytic efficiency and selectivity enhanced due to the small size of CuPd NPs, Lewis acidity of support, along with the favorable reaction conditions.

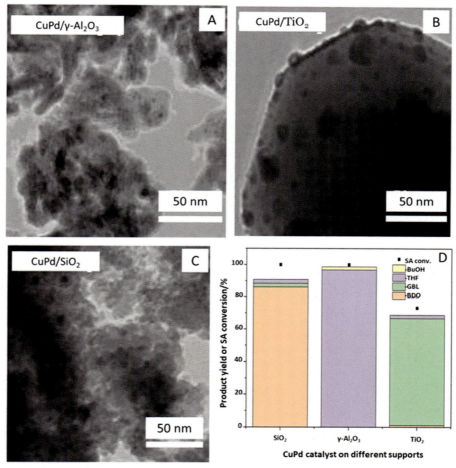

Figure 6.6 (A–C) TEM images of CuPd NPs adsorbed on γ-Al$_2$O$_3$, SiO$_2$, and TiO$_2$ supports; (D) % Conversion of SA to different liquid fuels over a CuPd NPs adsorbed on SiO$_2$, γ-, TiO$_2$ and Al$_2$O$_3$ catalyst. *(Reprinted with permission from Ref. 108 Copyright 2021 Elsevier.)*

2.3 Production of furfural alcohol and related liquid fuels

Catalytic conversion of lignocellulosic biomass, i.e., hemicellulose into biochemicals and biofuels like furfural (FAL) attracted immense attention nowadays as a fuel source to overcome the global energy crisis.[60,64,109] Depolymerization of biomass into furfurals for example, 5-hydroxymethylfurfural, furfuryl alcohol, was the key step in catalytic up-gradation of FALs toward the targeted liquid fuels.[110] However several homo- and heterogeneous catalytic methodologies reported for the transformation of FALs to C$_4$ and C$_5$ chemicals like furfuryl alcohol (FOL), alkyl levulinate (AL), 2,5-dimethylfuran, 2-methyl furan, γ-valerolactone (GVL), 5-ethoxy methyl furfural, levulinic acid (LA) as well as long-chain hydrocarbons.[20,66,111–114] Among the list of biochemicals, FOL was the key component in chemicals and polymer industry utilized for resins, vitamin C, lysine, synthetic fibers, and lubricants production.[115] By considering the significance of FOL, Zhao et al. have reported modified Cu-based catalyst, i.e., Cu-supported sulfonate group modified active carbon (Cu/AC-SO$_3$H) for the conversion of FAL through hydrogenation to FOL.[43] The Cu/AC-SO$_3$H nanostructures were synthesized via a liquid-phase chemical reduction process. The modified copper nanocatalyst enhanced the catalytic efficiency for the transformation of FAL to FOL in the liquid phase via hydrogenation where 2-propanol was utilized as a hydrogen source. The Cu/AC-SO$_3$H catalyst was selectively hydrogenated FAL to FOL with 100% conversion in 2 h at 378 K along with 0.4 MPa of H$_2$ pressure. The plausible reaction mechanism for the synthesis of FOL demonstrated in Fig. 6.7. The support of —SO$_3$H group grafted AC would

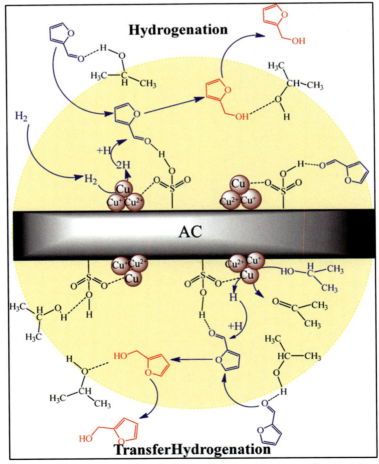

Figure 6.7 Mechanistic pathway for hydrogenation of FAL to FOL. *(Reproduced with permission from Ref. 43 Copyright 2017 American Chemical Society.)*

be manifested as an effective candidate for the synthesis of Cu-supported catalysts attributed to greater catalytic efficiency for the FAL hydrogenation along with long-term operational stability.

Also in other protocols, Neeli et al. have reported Rh/ED-KIT-6 catalyst containing Rh NPs enclosed in KIT-6 viz., mesoporous silica featured with N1-[3-(trimethoxysilyl)propyl]ethane-1,2-diamine prepared via Rh^{3+} adsorption followed by liquid–phase chemical reduction.[116] The Rh/ED-KIT-6 catalyst employed for the conversion of FAL to FOL via hydrogenation in presence of hydrogen source such as formic acid. The Rh/ED-KIT-6 catalyst displayed a 204 h^{-1} turnover frequency (TOF) higher than the Pd, Ru, or Ni-based KIT-6 catalyst. The Rh/ED-KIT-6 demonstrated the catalytic activity up to three cycles after that Rh species gets oxidized.

The platform chemicals like FOL were further converted into highly desirable building blocks like GVL, AL, and LA by stepwise hydrogenation process.[12,65,117,118] In this regard one very interesting protocol i.e., zeolite-based bifunctional magnetic ZSM-5 catalyst reported by Varma et al.[112] The bifunctional core-shell type magnetic ZSM-5 zeolite structures employed for the selective production of GVL from FOL through a tandem alcoholysis/hydrogenation/cyclization pathways. Herein various valuable liquid fuels can be prepared by replacing the reaction solvents, such as 2-butyl levulinate obtained in presence of two butanol as well as a mixture of solvents like water/2-butanol selectively produced LA over the zeolite-based magnetic ZSM-5 nanocomposite. The magnetic core-shell ZSM-5 zeolite was prepared via embedding magnetite particles into the ZSM-5 zeolite grains (Si/Al = 14) by utilizing a cationic polymer, subsequently by calcination as well as ion-exchange method to convert the zeolite from its sodic structures (NaZSM-5) to the ammoniacal structure (NH_4ZSM-5). The HRTEM images revealed that the pure magnetite shows microspheres with a 200 nm size, as well as the pictures of the magnetic zeolite revealed that the presence of large HZSM-5 grain (Fig. 6.8A–D). The overall transformation pathway for the production of GVL from FOL comprises a single-pot process (Fig. 6.8E). The magnetic ZSM-5 zeolite catalyst displayed high catalytic stability and selectivity for the hydrogenation of FOL.

Along with the above protocols in recent times, Ma et al. have synthesized easily recoverable magnetic Fe_3O_4-12 NPs by coprecipitation method for the transformation of FAL to FOL.[47] The heterogeneous Fe_3O_4 NPs exhibited higher catalytic efficiency in the catalytic transfer hydrogenation of lignocellulosic FAL in presence of 2-propanol as a hydrogen precursor under optimum reaction conditions (Fig. 6.9A and B). Fe_3O_4-12 catalyst displayed high catalytic performance and 90.1% yield of FOL from FAL with a conversion rate of FAL was 97.5% at 160°C for 5 h. The prepared magnetic Fe_3O_4-12 NPs exhibited more acid-base active sites along with high catalytic efficiency has a large potential toward the catalytic conversion of xylose into FOL. Apart from the above reports, very recently ZrO_2, Al_2O_3, and ZrO_2/Al_2O_3 catalytic systems were reported for hydrogenation of FAL.[119] The ZrO_2/Al_2O_3 catalyst exhibited high

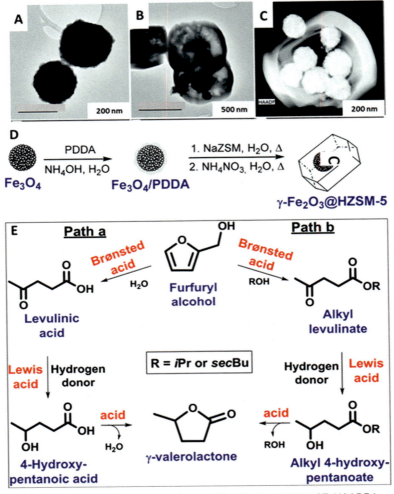

Figure 6.8 HRTEM image of (A) magnetite spheres; (B) γ-Fe$_2$O$_3$-HZSM5; (C) HAADF image (γ-Fe$_2$O$_3$ microspheres encapsulated in HZSM-5 grains); (D) Synthesis of the HZSM-5 catalyst; (E) Synthesis of GVL from FOL. *(Reproduced with permission from Ref. 112 Copyright 2016 Royal Society of Chemistry.)*

surface area, small crystal size, and Lewis acidity responsible for the catalytic transformation of FAL to FOL. The ZrO$_2$ catalyst selectively produced i-propyl levulinate and i-propylfurfuryl ether via etherification of FAL. Along with this, the direct transformation of FAL to LA reported by Nandiwale et al.[120] The bifunctional H$_3$PW$_{12}$O$_{40}$ (heteropolyacid) supported over a SiO$_2$ exhibited Lewis and Brønsted acid active sites can be employed for the catalytic hydrogenation of FAL to LA in presence of isopropyl alcohol as a reaction solvent as well as a hydrogen source. The H$_3$PW$_{12}$O$_{40}$/SiO$_2$ catalyst obtained a nearly 51% yield of LA via direct conversion of FAL.

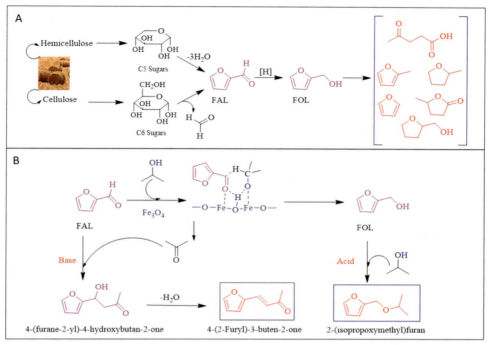

Figure 6.9 (A) Schematic synthesis of FOL and conversion of FOL to other liquid fuels from biomass; (B) the reaction pathway for catalytic transfer hydrogenation of FAL to FOL over Fe$_3$O$_4$-12 NPs. (Reprinted with permission from Ref. 47 Copyright 2020 Elsevier.)

2.4 Production of 5-hydroxymethylfurfural and related liquid derivatives

Among the list of valuable platform fuels, 5-hydroxymethylfurfural (HMF) has been served as an intermediate chemical for the synthesis of several essential industrial chemicals like adipic acid, furan dicarboxylic acid, and fuels.[121] HMF is a main intermediate connecting link between the biomass and biofuels obtained from cellulose valorization to carbohydrates i.e., glucose or fructose followed by synthesis of HMF from glucose.[122] By considering the industrial importance of HMF, Zhang et al. have reported magnetically recoverable porous Fe$_3$O$_4$@SiO$_2$@TiO$_2$-HPW NPs by adsorbing H$_3$PW$_{12}$O$_{40}$ (HPW) on the surfaces of TiO$_2$ and SiO$_2$-coated Fe$_3$O$_4$ NPs via sonication method.[123] The synthesized magnetically-recoverable Fe$_3$O$_4$@SiO$_2$@TiO$_2$-HPW catalyst exhibited high catalytic efficiency via single-step conversion of inulin-based biomass to HMF. The production of HMF achieved from inulin and fructose in presence of a solvent such as dimethyl sulfoxide that influenced the reaction time as well as temperature. Hence, the HMF yields derived from inulin and fructose reached 54% and 83%, respectively. However, apart from the above-reported protocol, recent studies targeted the conversion of HMF to furan-ring-retaining as well as furan-ring-opening

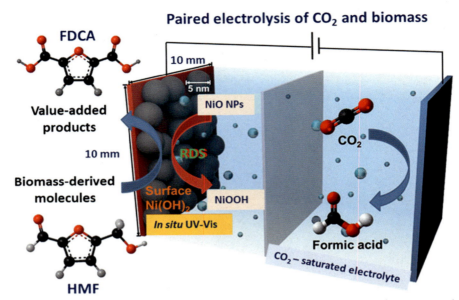

Figure 6.10 Schematic of HMF oxidation. *(Reproduced with permission from Ref. 126 Copyright 2020 American Chemical Society.)*

products.[124–128] The nickel oxide (NiO) NPs also reported for the HMF oxidation to 2,5-furan dicarboxylic acid (FDCA) by Choi et al. (Fig. 6.10).[126]

NiO nanocatalysts provided surface phase transformation and also uniform surface structure manifested the high volume ratio, surface area, which can make NiO NPs an efficient catalyst in neutral reaction conditions. The HMF oxidation mechanism of NiO NPs has been investigated in near-neutral and CO_2 saturated electrolytes by mechanistically and electrokinetic analysis combined with *in-situ* UV-visible spectroscopy. Also, Ge et al. have discovered the nickel-molybdenum sulfide supported alumina (Ni–MoS_2/mAl_2O_3) catalyst for the production of 2,5-dimethylfuran (DMF) from HMF through hydrogenolysis.[127] The catalyst was synthesized via the evaporation-induced self-assembly technique. The polar protic solvent, i.e., 2-propanol is utilized as a hydrogen source forms ether intermediate due to this the catalytic performance improved significantly and the highest DMF yield of 95% was obtained. Coordinated unsaturated site in Ni–MoS_2/mAl_2O_3 catalyst replaced by Mo raised the turn over frequency as well as lower the activation energy barrier.

In another protocol, Ordomsky et al. have demonstrated Br–Pd/Al_2O_3 nanocatalyst for the transformation of HMF to DMF selectively via hydrodeoxygenation (Fig. 6.11A).[129] The designed catalyst was synthesized by the wet impregnation process. The Pd combined with bromine created Brønsted acid sites, which are adsorbed on the

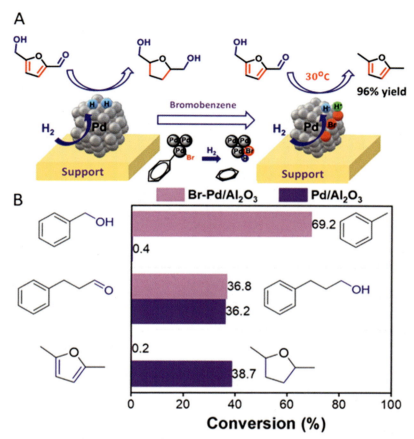

Figure 6.11 (A) HMF deoxygenation from DMF over a Pd–Br dual metal-acid catalyst; (B) 3-phenylpropanol, DMF, and benzyl alcohol hydrogenation catalyzed by Pd/Al$_2$O$_3$ as well as Br–Pd/Al$_2$O$_3$ nanocatalysts. *(Reproduced with permission from Ref. 129 Copyright 2020 American Chemical Society.)*

catalytic metal surface directly. The catalytic efficiency of Pd–Br bifunctional catalyst increased due to Brønsted sites created by bromine on the surface of Pd manifested an extremely higher yield (96%) for DMF production. Pd surface with Pd–Br sites enhanced the heterolytic dissociation of hydrogen for deoxygenation of the hydroxyl group. The Pd/Al$_2$O$_3$ catalyst exhibited high catalytic hydrogenation activity and high hydrodeoxygenation activity exhibited by Br– Pd/Al$_2$O$_3$ demonstrated in Fig. 6.11B.

Apart from several nanomaterial-based catalysts, single-atom catalyst (SACs) is one of the emerging fields that provide maximum metal surface, well-defined active sites so that it can increase the 100% metal-atom utilization for obtaining high catalytic efficiency, and selectivity in various catalytic transfer reactions.[130–132] To enhance the stability of single metal active sites several bimetallic systems are demonstrated with well-defined

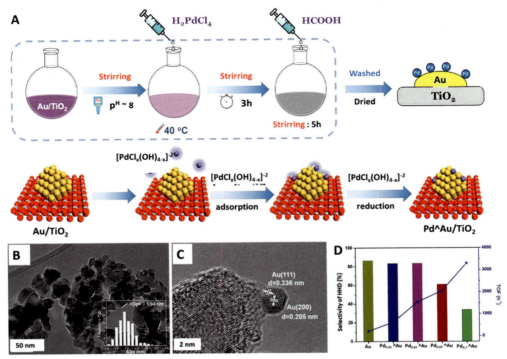

Figure 6.12 (A) Schematic pathway for the Pd—Au/TiO$_2$ synthesis; (B, C) AC-HRTEM and STEM-HAADF images of Pd—Au/TiO$_2$; (D) Effect of Pd composition in Pd—Au/TiO$_2$ catalyst on the TOF and products selectivity. *(Reproduced with permission from Ref. 124 Copyright 2019 American Chemical Society.)*

on-surface nanoarchitectures.[127,129] In this regard, Zhu et al. have demonstrated Au—Pd bimetallic single-site catalysts for the furan-ring-opening transformation of HMF to di and triketone products.[124] The Pd atom decorated on the surface of Au NPs supported by titania(Pd—Au/TiO$_2$) by deposition-precipitation technique (Fig. 6.12A). The Pd—Au/TiO$_2$ catalyst exhibited 90% selectivity along with 100% conversion of HMF due to strong Lewis acidity. The AC-HRTEM and STEM-HAADF images of Pd—Au/TiO$_2$ showed the homogeneous dispersion of Pd and Au on the surface of TiO$_2$ (Fig. 6.12B—D). The high catalytic stability proved by a large turnover number which controlled the ring-opening of HMF to keto products. Similarly, in recent protocols, bifunctional nickel phosphide NPs were also reported for the production of diketones from furanic aldehydes.[133] Herein the hydrogen-activating ability and surface acidity enhanced the conversion of biofuranic aldehydes. Apart from this Co-alloyed Pt (Pt$_1$/Co) single-atom catalysts also reported for the conversion of HMF to DMF through hydrodeoxygenation.[128]

2.5 Production of biodiesel

The production of biodiesel also known as fatty acid methyl ester (FAME) from biomass has been one of the most extensive research areas.[134,135] By notifying the importance of biofuels as a nonrenewable energy source, Chiang et al. have reported core-shell Fe_3O_4@silica magnetic NPs activated with a triazabicyclodecene (TBD) as a strong base for biodiesel production from microalgae harvesting.[37] The TBD-Fe_3O_4@silica NPs manifested large catalytic efficiency than the acidic conventional catalyst i.e., Sulfuric acid, Amberlyst-15 in single-step conversion of algae oil to biodiesel via transesterification. In this protocol, TBD-functionalized Fe_3O_4@silica NPs showed the effective conversion of biodiesel from microalgae harvesting with a yield of 97.1%. Core-shell TBD-functionalized magnetic Fe_3O_4@silica NPs are very efficient and recyclable microalgae harvester can be utilized for producing fuels. From this report, we can conclude that the functionalized core-shell NPs can be effectively utilized for the synthesis of biodiesel and various useful chemicals from algal biomass. Coequally, Jiang et al. also prepared CaO coated magnetic NPs ($MgFe_2O_4$@CaO) by a precipitation method in the presence of sodium dodecylbenzene sulfonate for the biodiesel production.[136] The TEM and HRTEM images discovered that the core-shell morphology of $MgFe_2O_4$@CaO (Fig. 6.13A and B). The heterogeneous $MgFe_2O_4$@CaO catalyst was effectively utilized for the synthesis of biodiesel from methanol and soybean oil as precursors with a 98.3% yield of FAME as well as the molar ratio of methanol and soybean oil affected the yield of biodiesel (Fig. 6.13C). The FAME yield was improved up to 90% by adding 1 wt% water in reaction mass in presence of $MgFe_2O_4$@CaO catalyst (Fig. 6.13D). The pure CaO catalyst decreased the FAME yield up to 15.7% so it is proved that the $MgFe_2O_4$@CaO nanocomposite synergistically enhanced the rate of reaction and product output. As compared with the pure CaO catalyst and $MgFe_2O_4$@CaO nanostructures, $MgFe_2O_4$@CaO exhibited strong water resistance, higher catalytic activity, and better acid-resistant. These pieces of evidence indicated that the $MgFe_2-O_4$@CaO catalyst was highly efficient and recyclable alkaline catalyst that would be open an environmentally friendly route for the manufacturing of biodiesel.

2.6 Production of liquid hydrocarbons

The transformation of biomass-derived resources into fuels for transportation and valuable chemical products is drawing more attention because of limited fossil fuels and rising energy demand.[137] Esterification is the key reaction in the utilization of biomass into liquid fuels. In this regards, Song et al. have synthesized dendritic fibrous N-doped carbon support (DFNC-[C_3N][OTf]) through the *in-situ* water-in-oil microemulsion method employed for the esterification reaction.[138] The particle size and morphology perfection of the support can be tuned by changing the composition of urea in the preparation process (Fig. 6.14A). The N-doped carbon support exhibited chemical bonding

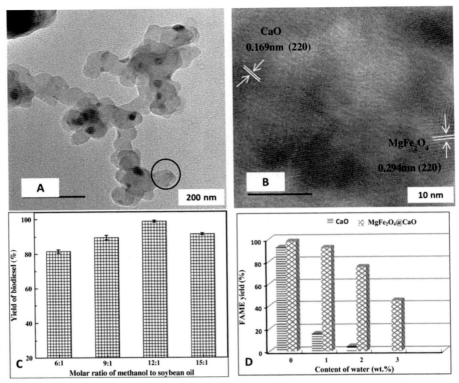

Figure 6.13 TEM and HRTEM images of (A, B) MgFe$_2$O$_4$@CaO; (C) Effect of the molar ratio of methanol to soybean oil on biodiesel production; (D) Impact of water content on a yield of FAME. *(Reproduced with permission from Ref. 136 Copyright 2016 Elsevier.)*

with Brønsted acidic ionic liquids (BAILs) is one of the significant strategies to preparing highly active solid acids for the catalytic transformation of biomass-based organic matter to liquid fuels and platform chemicals (Fig. 6.14B). The high catalytic efficiency of DFNC–[C$_3$N][OTf] catalysts attributed to hierarchical porosity, superstrong Brønsted acidity, as well as chemical adsorption of the -[C$_3$][OTf-] sites on the wall of wrinkled fibers of the supporting DFNC. The unique 3D dendritic fibrous nonspherical structure displayed excellent catalytic reusability and activity for biomass transformation via esterification. Similarly, silica nanospheres adsorbed on BAIL recently reported for the valorization of lignocellulosic biomass for LA and ethyl levulinate.[139]

Long-chain hydrocarbons obtained from biomass after some modifications are also utilized as transportation fuels.[6] Moussa et al. have reported a graphene-supported, iron-based NPs (Fe$_{15}$K$_5$-G) catalyst for the synthesis of liquid hydrocarbons (C$_{16}$H$_{34}$ and C$_{20}$H$_{44}$) from syngas by Fischer-Tropsch synthesis (FTS).[140] The Fe$_{15}$K$_5$-G catalyst was synthesized through microwave irradiation reduction technique. The less valuable side products such as methane and CO$_2$ are reduced by graphene uniquely as compared to other carbon-based supports like carbon nanotubes. The catalytic stability, selectivity,

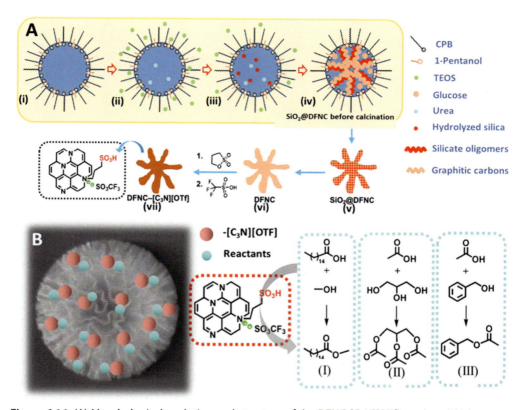

Figure 6.14 (A) Morphological evolution and structure of the DFNC-[C$_3$N][OTf] catalyst; (B) The esterification of (I) palmitic acid (II) glycerol, and (III) acetic acid in presence of DFNC-[C$_3$N][OTf] catalyst. *(Reproduced with permission from Ref. 138 Copyright 2019 American Chemical Society.)*

and activity of graphene-supported Fe-based NPs exhibited a promising industrial application for the transformation of syn gases such as CO. Another FTS was studied by Li et al. for the synthesis of liquid hydrocarbon.[141] The nano-ZSM-5 zeolite (U—Co/H-ZSM-5) catalyst was prepared via the wet impregnation process. The prepared cobalt crystal exhibited a small size increased the large contact surface area, which is beneficial for synthesizing the C$_{5-12}$ products. The U—Co/H-ZSM-5 catalyst displayed superior catalytic performance toward the CO conversion and reported higher C$_{5-12}$ selectivity. The strong interaction between ZSM-5 zeolite and cobalt crystal enhanced the CO transformation. The larger contact surface of U—Co/H-ZSM-5 zeolite reduced the number of C$_{20+}$ products. In another protocol, conversion of CO$_2$ to C$_{2+}$ hydrocarbons selectively reported by using copper–polyaniline (Cu-PANI) interfaces.[142] In Fig. 6.15 it is observed that the large difference in the catalytic selectivity of Cu-PANI and pristine Cu due to the surface modifications. Herein the current density of CO$_2$RR was enhanced at the Cu/PANI surface, subsequently, the C$_{2+}$ hydrocarbons yielded as a major product. The catalytic selectivity and activity of Cu surface toward CO$_2$RR can be

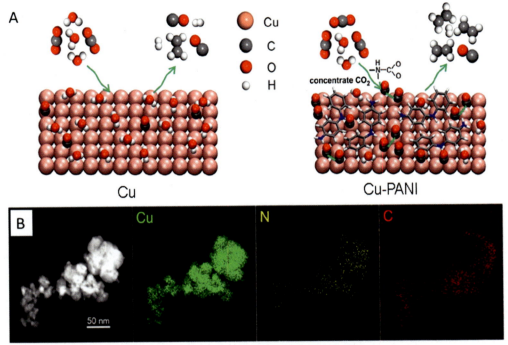

Figure 6.15 (A) Schematic representation of Cu and Cu-PANI electrodes for the conversion of CO_2; Elemental mapping and TEM image of (B) Cu-PANI catalyst Cu (green), N (cyan), and C (red). (Reproduced with permission from Ref. 142 Copyright 2020 American Chemical Society.)

greatly modified by a coating of a thin film of PANI, which is not electronically conductive in weak alkaline conditions such as a solution of $KHCO_3$ but can enriched the CO_2 adsorption via the binding between the —NH— group on the CO_2 molecule and PANI.

3. Surface-modified nanomaterials for the transformation of carbon dioxide to liquid fuels

CO_2 is one of the major greenhouse gas responsible for global warming, earth temperature rises 2°C every year that will be impacted on the environment so it is necessary to take any action against it.[143] Capture, conversion, and utilization of CO_2 by photocatalysis, artificial photosynthesis, electrochemical, chemical, and biological approaches help to lower the increasing level of CO_2 in the atmosphere and also provides alternative fuel and value-added product to fulfill the demand of energy as a renewable source of energy.[144,145] It would be desirable to utilize the captured CO_2 as a precursor for the synthesis of valuable chemical fuels. CO_2 utilization signifies the CO_2 is either used directly or transformed into high-value liquid fuels, is likely to be one way to reduce CO_2 emissions.[146,147] To date, several protocols reported for the manufacturing of liquid fuels like methanol, ethanol, and propanol from CO_2[38–41,51–55] are briefly summarized in below sections.

3.1 Reduction of CO₂ to methanol

The increasing demands for fuels and continuous reduction in fossil resources reserves CO_2 a greenhouse gas responsible for climate change can be utilized as a safe and inexpensive single carbon source for renewable energy feedstock for fuel production.[20] Nowadays methanol is produced on a large amount (scale = >40 million tons, globally in 2009), mostly from the syngas (CO/H₂).[148] Chemical recycling of captured CO_2 to methanol is denoted as methanol economy due to several applications of methanol in chemical industries.[149] In this regard, Witoon et al. developed CuO—ZnO nanocomposites for the hydrogenation of CO_2 to methanol.[38] The catalyst was manufactured via a chitosan-based coprecipitation process. Chitosan acted as a coordination compound for uniform combination and also a soft template for the preparation of nanospheres for CuO—ZnO nanocomposites. The small crystallite size of hollow CuO and ZnO nanospheres was observed as compared with a catalyst that has an unmodified surface. The catalyst synthesized by using precipitating agent i.e., chitosan manifested a large space time yield for production of methanol compared with the unmodified surface catalyst. This activity enhancement due to the synergetic effect of CuO NPs embedded in the CuO/ZnO nanostructures. Also in another report by Li et al. have described the catalytic transformation of CO_2 to methanol using acid-etched oyster shells supported CuO/ZnO/OS nanocatalyst.[52] Herein, Oyster shells (OS) has been used as a template or precursor for preparing functionalised NMs. The CuO/ZnO/OS catalyst was developed by an efficient route using the inexpensive and eco-friendly supporting medium i.e., acid-etched oyster shells (Fig. 6.16A). Similarly, the activated catalyst, i.e., CuO/ZnO/a-OS synthesized by calcination of Cu(Zn)-BTC/a-OS in air, and employed for the CO_2 hydrogenation reaction for increasing activity and efficiency of nanocatalyst. Surprisingly, the rate of reduction of CO_2 over CuO/ZnO/a-OS raised from 1.9% to 6.6% in comparison with CuO/ZnO/OS, and the yield of methanol also enhanced from 1.3% to 4.2%. In this work, the acid-etched OS (a-OS) enhanced the catalytic performance of CuO/ZnO/a-OS nanocatalysts due to the higher interfacial spacing between the two active components i.e., Cu and Zn. Cu and Zn-based active sites on OS manifested the higher catalytic efficiency for the synthesis of methanol from CO_2 hydrogenation. SEM and TEM images confirmed that the homogeneous dispersion of Cu(Zn)-BTC NPs on a-OS having the nanoscale size (Fig. 6.16B—G). The reported results suggested that the unused OS on acid-etching process can be utilized as a support for the synthesis of nanocatalytic materials.

In another report, Imyen et al. have prepared Fe-ZSM-5@ZIF-8 nanocomposite using calcination followed by a multilayer deposition method for the production of methanol from methane adsorption at low temperature in one-pot (Fig. 6.17A).[53] SEM images (Fig. 6.17B), revealed that the Fe-ZSM-5@ZIF-8 composite having a particle size of 750 ± 300 nm and the TEM image of Fe-ZSM-5@ZIF-8 (Fig. 6.17C) confirmed

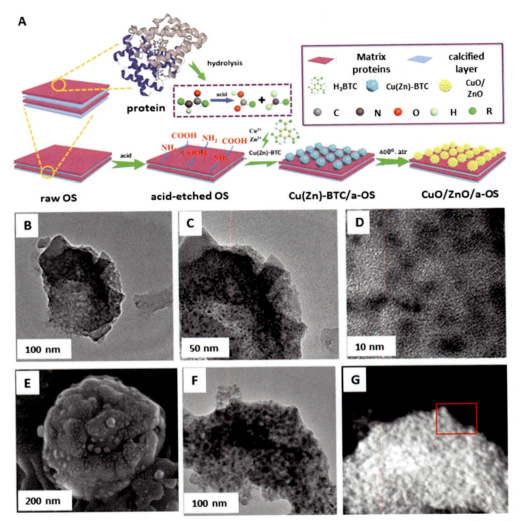

Figure 6.16 (A) Synthesis of CuO/ZnO/a-OS nanocatalysts; TEM images of (B–D) Cu(Zn)-BTC/a-OS; SEM images of the (E) CuO/ZnO/a-OS composites; TEM images of (F, G) CuO/ZnO/a-OS. *(Reproduced with permission from Ref. 52 Copyright 2020 American Chemical Society.)*

that the nanocomposite is composed of stacking NPs at mostly outer surfaces of the Fe-ZSM-5 in the range of 640 ± 190 nm. The designed Fe-ZSM-5@ZIF-8 catalyst exhibited very good performance in terms of higher methanol yield and higher adsorption of methane i.e., 4 mL/min of 3% CH_4/He (Fig. 6.17D and E) as compared to the isolated materials, for example, Fe-ZSM-5, and ZIF-8. Due to synergism between Fe-ZSM-5 and ZIF-8, the catalytic efficiency, as well as gas adsorption affinity, increases

Surface-modified nanomaterial-based catalytic materials for the production of liquid fuels 155

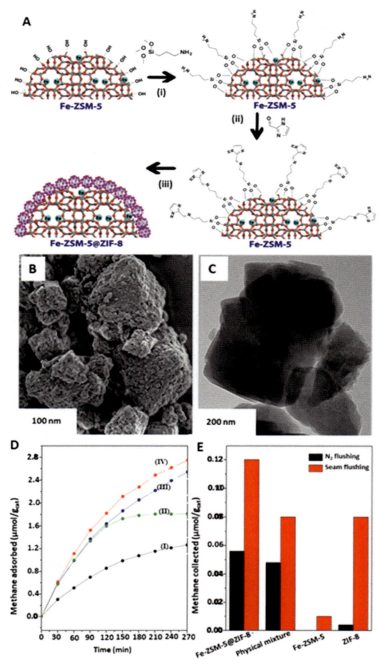

Figure 6.17 (A) Steps involved in the synthesis of Fe-ZSM-5@ZIF-8 nanocomposite; SEM image of (B) Fe-ZSM-5@ZIF-8 composite; and TEM images of (C) Fe-ZSM-5; (D) adsorption of methane on a various catalyst surface; and (E) comparison of production of methanol from methane over different types of catalysts. *(Reproduced with permission from Ref. 53 Copyright 2020 American Chemical Society.)*

tremendously. The adsorbed methane in the Fe-ZSM-5@ZIF-8 nanocomposite has been converted into methanol at 150°C for 1 atm within 30 min. Moreover, this strategy can facilitate the desorption of methanol because of the hydrophobic nature of ZIF-8 at the outer surface of the catalyst.

3.2 Reduction of CO_2 to ethanol

Ethanol is also a valuable fuel like methanol and it has large demand in chemical industries as well as in transportation.[150] Ethanol is utilized as a fuel additive and for preparing various industrial intermediate considered to be one of the most preferable chemical.[151] The higher selectivity of ethanol synthesis is still a great challenge because the mixture of alcohols or hydrocarbons is the main product reported via the CO_2 hydrogenation, while the direct synthesis of ethanol is rarely reported.[152] Herein, some advanced NMs are described for the synthesis of ethanol from CO_2 reduction reaction. The metal-organic framework (MOF) derived NMs provide abundant catalytic active sites for the reduction of CO_2 to liquid fuels due to their tunable geometric framework, electrocatalytic stability, tunable pore size, and large surface area.[153] In this context, Zhang et al. have synthesized Cu/Cu_2O NPs adsorbed on zeolitic imidazolate framework-L (ZIF-L)-coated N-doped graphene nanosheets (Cu-GNC-VL) for the production of ethanol from CO_2 reduction (Fig. 6.18A).[41] The Zn and Cu NPs homogeneously dispersed on graphene revealed in STEM and SEM images of Cu-GNC-VL catalyst (Fig. 6.18B–D). The GO/ZIF-L nanocomposites having homogeneous and vertical growth exhibited high catalytic surface area, excellent electrical conductivity, and operational stability. However, Cu GNC-VL nanostructures displayed 70.52% faradaic efficiency (FE) for the production of ethanol (Fig. 6.18E and F). The collaborative effect of MOF-derived N-doped porous carbon as well as surface-modified graphene enhanced the physico-chemical properties of Cu GNC-VL nanocomposites.

Similarly, Yang et al. have synthesized Au NPs in Cu_2O nanocavities (Au@Cu_2O yolk-shell) by using the hydrazine hydration reduction method for the electroreduction of CO_2 through the confinement of the CO intermediate at the low potential for production of ethanol.[54] The Au-core embedded with copper nanocavity was reduced the CO_2 to CO, obtained a high concentrated CO, and Cu-shell to transforms CO into ethanol. Au embedded in porous Cu_2O shell structure with uneven cavities responsible for boosting the catalytic activity (Fig. 6.19A–E). The high concentration of CO in the nanocavity is utilized for enhancing the Cu_2O nanocavity hollow size. The yolk-shell Au@Cu_2O nanostructures demonstrated 52.3% FE at −0.30 V (vs. RHE) for ethanol production through electrocatalytic CO_2 reduction (Fig. 6.19F–G). Nanocavities in Cu-shell retained the large concentration of CO subsequently enhanced the ethanol production (Fig. 6.19H).

3.3 Reduction of CO_2 to propanol

The production of C_{2+} alcohols from the CO_2 reduction, which was recently described as the next main challenge in the CO_2 reduction reaction.[154] Among the several C_{2+}

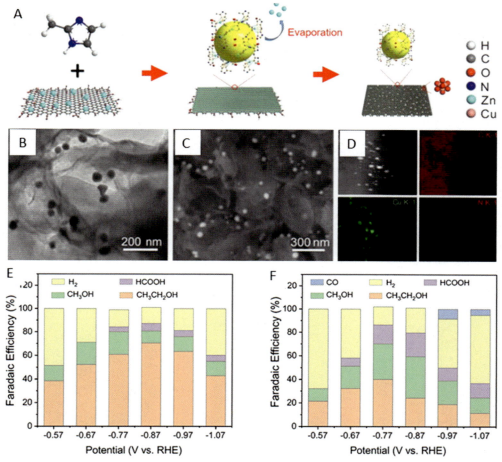

Figure 6.18 Schematic of Cu GNC-VL (A) Synthesis of nanocatalyst; (B, C) HAADF-STEM and SEM images; (D) Homogeneous dispersion of Cu and N by energy dispersive spectroscopy; The FE of (E) Cu GNC-VL and (F) Cu ZIF-L@GO catalyst. *(Reproduced with permission from Ref. 41 Copyright 2020 American Chemical Society.)*

alcohols, propanol was utilized as a fuel to alcohol-based fuel cells and it exhibited high specific energy density than ethanol and methanol. Propanol can be easily transformed into propylene, which is one of the valuable chemical feedstock.[39] Recently, Kondratenko and his group have demonstrated the propanol production from catalytic hydrogenation CO_2 in the H_2 atmosphere over a K doped Au NPs supported on SiO_2 and TiO_2 -based catalysts.[39] The K—Au/SiO$_2$ and K—Au/TiO$_2$ catalyst was synthesized by deposition-precipitation technique followed by impregnation process. The K—Au/SiO$_2$ catalyst property was tuned by the K promoter in the surface of the nanocatalyst. However Au-related TOF values decreased at higher loading of K. The promoter,

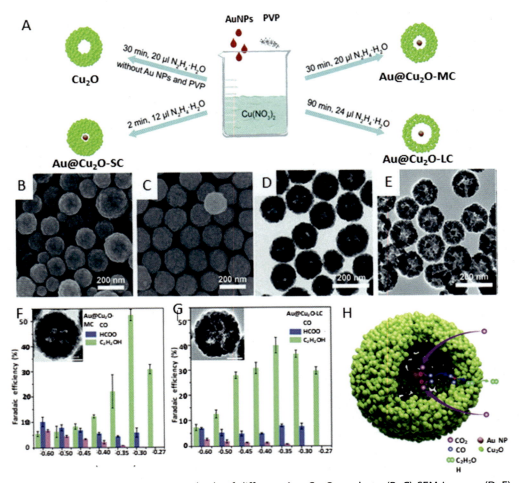

Figure 6.19 (A) The preparation methods of different Au@Cu$_2$O catalysts; (B, C) SEM images; (D, E) TEM pictures of Cu$_2$O and Au@Cu$_2$O-small cavity (SC), middle cavity (MC), large cavity (LC); (F, G) The FEs of CO, formic acid, and ethanol; (H) nanocavity of Au@Cu$_2$O catalyst. *(Reproduced with permission from Ref. 54 Copyright 2020 Royal Society of Chemistry.)*

i.e., K affected the kinetics of CO$_2$ transformation to CO and subsequently propanol formation. CO$_2$ has weakly adhered on K$_2$SiO$_3$ than K$_2$O and therefore the CO$_2$ is quickly reduced to CO using Au NPs supported on silicate. Apart from the above protocols, several noble/non-noble metal-based nanocatalysts were also reported for the production of propanol from CO$_2$ catalytic as well as an electrolytic pathway.[155–158]

3.4 Reduction of CO$_2$ to formic acid

Among the list of renewable hydrogen sources, formic acid is utilized as a hydrogen source obtained from electrocatalytic CO$_2$ reduction.[159] Since the decades, formic acid is utilized as a fine chemical as well in a direct formic acid fuel cell as a fuel notified

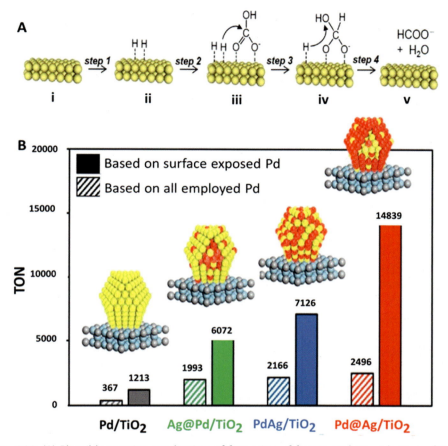

Figure 6.20 (A) Plausible reaction mechanism of formation of formic acid via reduction of CO_2; (B) Comparison of the turnover number and catalytic activities of Pd and Ag supported TiO_2 catalysts with a change in metallic compositions along with Pd/TiO_2 for hydrogenation of CO_2. *(Reproduced with permission from Ref. 40 Copyright 2018 American Chemical Society.)*

as an efficient energy source.[160] Within this perspective, Yamashita and his group have synthesized PdAg bimetallic NPs adsorbed on TiO_2 through wet impregnation followed by a reduction method for the selective synthesis of formic acid from the hydrogenation of CO_2 under optimum reaction environment (2.0 MPa, at 100°C) (Fig. 6.20A–B).[40] This study demonstrated the advanced designing of catalytically active sites for hydrogenation of CO_2 and the enhancement in the efficiency of the target reaction by atomically precise PdAg alloy NPs through surface engineering. Synergistically noble Pd atoms and neighboring Ag atoms, at even a lower Pd/Ag ratio, boosted the catalytic reduction of CO_2. Several other supports, such as layered double hydroxide, CeO_2, Al_2O_3, and MgO were also utilized for the adsorption of Pd/Ag bimetallic NPs among them PdAg alloy NPs were immobilized on TiO_2 reported a high turnover number. Notably, Ag and Pd were located in the same place, demonstrated that the formation of PdAg alloy, however in the core region Pd atoms were preferentially located, while in the shell region Ag atoms were located in the Pd@Ag/TiO$_2$ catalyst. Also, Nematollahi et al. have described a Pt nanowire (Pt NW) for the synthesis of formic acid from the electrocatalytic reduction of CO$_2$.[55] Pt NW was synthesized by the solvothermal method. The Pt NW assembly employed as a cocatalyst for electrocatalytic reduction of CO$_2$ to methanol as well as formic acid herein the CO$_2$ was activated by pyridinium (PyrH$^+$). Morphological modifications of Pt NW accumulated higher hydride resulted in superior electrochemical performance in CO$_2$ electroreduction. Apart from the above protocols, recently very fascinating NMs were also demonstrated for the formic acid production as well as some bimetallic nanocatalyst also described for the conversion of formic acid into hydrogen as a energy source in direct formic acid fuel cell.[161–165]

4. Future perspectives and conclusion

Nowadays liquid fuels produced from biomass valorization are a very fascinating field due to their high abundance in nature as a renewable organic-based substrate. For decades scientific community devoted to the design and development of sustainable functional materials for a breakdown of biomass into its simple monomeric units as renewable future power. Several SMNs have been demonstrated for the production of liquid fuels such as HMF, furfural, polyols, alcohols, oils, and acids from biomass. But still, there is a lack of product selectivity as well as low catalytic surface area so this area requires more improvement toward the systematic design of nano- or single-atom based catalysts that offer maximum atom efficiency, high selectivity, the stability of catalyst at higher temperature as well as pressure, along with the potential of the nanocomposites to predefine and regulates the mechanistic pathways to get maximum yield of products. While designing the catalyst for biomass transformations, the surface of NMs modifies with Lewis acid as well as Brønsted acid so that the catalytic efficiency and product selectivity 10 times increased as compared with bare NMs. Previously reported data have been extensively showed a suite of catalytic materials, which are capable of transforming biomass-derived components into different chemical compounds ranging from hydrocarbons to bio-aromatics. However, there will be a need to develop an environmentally benign methodology for large scale synthesis of fuels from biomass. In this context, mechanochemical strategies such as ball milling and microwave irradiation-assisted techniques, flow chemistry and photoflow methodologies need to be explored for biomass valorization because of inexpensive and low solvent consumption.[166]

The overall transformation of biobased organic matter takes place at higher pressure and temperature of reaction over a nanocatalyst. Besides, the number of reactions suffers from low conversions or selectivities, which then needs to separate several intermediates from reaction mass to obtain the final product.

Effective use of nanomaterials is crucial to meet the current and future challenges due to changes in feedstock supply (biomass-derived feeds), product demand (diesel versus gasoline), and environmental concerns. Recent developments in this field and the progression of surface-modified nanomaterial catalysts and additives throughout recent history demonstrate how a mature field can continuously evolve through the successful utilization of nanomaterials to overcome the challenges it faces. In consequence of this, there will be scope for a single-atom catalyst that provides high atom efficiency and large surface area to increases the rate of reaction and we get desired product selectivity.

In this book chapter, we have summarized various previously reported SMNs for liquid fuel productions from biomass precursors. The composition of biomass is very diverse such as carbohydrates and cellulosic materials as well as lignin as biopolymers. The high catalytic transformation of the organic-based substrate into fuels attributed to surface acidity and size of NMs. In SMNs, the size and functionality of NMs directly affected the selectivity and yield of the product. High emission of carbon dioxide to the environment is threatened because CO_2 is directly responsible for global warming. It is highly necessary to develop strategies for the sequestration of CO_2. In this book chapter, we described some SMNs reported for electro- as well as the thermocatalytic transformation of CO_2 to methanol, ethanol, formic acid, and propanol. These chemicals have extensive applications in industry as well as fuel cell devices.

Acknowledgments

I. R. W and H. B. K would like to gratefully acknowledge the Institute of Chemical Technology, Mumbai-Marathwada Campus, Jalna, India for providing doctoral fellowship.

References

1. Serrano-Ruiz JC, Dumesic JA. Catalytic routes for the conversion of biomass into liquid hydrocarbon transportation fuels. *Energy Environ Sci* 2011;**4**(1):83–99.
2. Braden DJ, Henao CA, Heltzel J, Maravelias CC, Dumesic JA. Production of liquid hydrocarbon fuels by catalytic conversion of biomass-derived levulinic acid. *Green Chem* 2011;**13**(7).
3. West RM, Kunkes EL, Simonetti DA, Dumesic JA. Catalytic conversion of biomass-derived carbohydrates to fuels and chemicals by formation and upgrading of mono-functional hydrocarbon intermediates. *Catal Today* 2009;**147**(2):115–25.
4. Shuttleworth PS, De bruyn M, Parker HL, Hunt AJ, Budarin VL, Matharu AS, Clark JH. Applications of nanoparticles in biomass conversion to chemicals and fuels. *Green Chem* 2014;**16**(2):573–84.
5. Gerardy R, Debecker DP, Estager J, Luis P, Monbaliu JM. Continuous flow upgrading of selected C_2-C_6 platform chemicals derived from biomass. *Chem Rev* 2020;**120**(15):7219–347.
6. Kunkes EL, Simonetti DA, West RM, Serrano-Ruiz JC, Gärtner CA, Dumesic JA. Catalytic conversion of biomass to monofunctional hydrocarbons and targeted liquid-fuel classes. *Science* 2008;**322**(5900):417.
7. Pomerantseva E, Bonaccorso F, Feng X, Cui Y, Gogotsi Y. Energy storage: the future enabled by nanomaterials. *Science* 2019;**366**(6468):eaan8285.
8. Selva M, Luque R. Benign-by-design advanced nanomaterials for environmental and energy-related applications. *Curr Opin Green Sustain Chem* 2019;**15**:98–102.
9. Mark LO, Zhu C, Medlin JW, Heinz H. Understanding the surface reactivity of ligand-protected metal nanoparticles for biomass upgrading. *ACS Catal* 2020;**10**(10):5462–74.
10. Muñoz-Batista MJ, Rodriguez-Padron D, Puente-Santiago AR, Luque R. Mechanochemistry: toward sustainable design of advanced nanomaterials for electrochemical energy storage and catalytic applications. *ACS Sustain Chem Eng* 2018;**6**(8):9530–44.

11. Dapsens PY, Mondelli C, Pérez-Ramírez J. Biobased chemicals from conception toward industrial reality: lessons learned and to be learned. *ACS Catal* 2012;**2**(7):1487–99.
12. Alonso DM, Bond JQ, Dumesic JA. Catalytic conversion of biomass to biofuels. *Green Chem* 2010;**12**(9):1493–513.
13. Corma A, Iborra S, Velty A. Chemical routes for the transformation of biomass into chemicals. *Chem Rev* 2007;**107**(6):2411–502.
14. Nasrollahzadeh M, Soheili Bidgoli NS, Shafiei N, Soleimani F, Nezafat Z, Luque R. Low-cost and sustainable (nano)catalysts derived from bone waste: catalytic applications and biofuels production. *Biofuel Bioprod Biorefin* 2020;**14**(6):1197–227.
15. Akhundi A, Badiei A, Ziarani GM, Habibi-Yangjeh A, Muñoz-Batista MJ, Luque R. Graphitic carbon nitride-based photocatalysts: toward efficient organic transformation for value-added chemicals production. *Mol Catal* 2020;**488**:110902.
16. Portilla-Zuñiga OM, Martínez JJ, Casella M, Lick DI, Sathicq ÁG, Luque R, Romanelli GP. Etherification of 5-hydroxymethylfurfural using a heteropolyacid supported on a silica matrix. *Mol Catal* 2020;**494**:111125.
17. Su T, Zhao D, Wang Y, Lü H, Varma RS, Len C. Innovative protocols in the catalytic oxidation of 5-hydroxymethylfurfural. *ChemSusChem* 2020;**14**(1):266–80.
18. Jorge EYC, Lima CGS, Lima TM, Marchini L, Gawande MB, Tomanec O, Varma RS, Paixão MW. Sulfonated dendritic mesoporous silica nanospheres: a metal-free Lewis acid catalyst for the upgrading of carbohydrates. *Green Chem* 2020;**22**(5):1754–62.
19. Verma S, Baig RBN, Nadagouda MN, Len C, Varma RS. Sustainable pathway to furanics from biomass via heterogeneous organo-catalysis. *Green Chem* 2017;**19**(1):164–8.
20. Kumar S, Gawande MB, Kopp J, Kment S, Varma RS, Zbořil R. P- and F-co-doped carbon nitride nanocatalysts for photocatalytic CO_2 reduction and thermocatalytic furanics synthesis from sugars. *ChemSusChem* 2020;**13**(19):5231–8.
21. Zhao D, Su T, Wang Y, Varma RS, Len C. Recent advances in catalytic oxidation of 5-hydroxymethylfurfural. *Mol Catal* 2020;**495**:111133.
22. Costa FF, Oliveira DTD, Brito YP, Rocha Filho GND, Alvarado CG, Balu AM, Luque R, Nascimento LASD. Lignocellulosics to biofuels: an overview of recent and relevant advances. *Curr Opin Green Sustain Chem* 2020;**24**:21–5.
23. Rajabi F, Luque R. Highly ordered mesoporous functionalized pyridinium protic ionic liquids framework as efficient system in esterification reactions for biofuels production. *Mol Catal* 2020;**498**:111238.
24. Pramanik K. Properties and use of jatropha curcas oil and diesel fuel blends in compression ignition engine. *Renew Energy* 2003;**28**(2):239–48.
25. Deng X, Fang Z, Liu Y-H, Yu C-L. Production of biodiesel from Jatropha oil catalyzed by nanosized solid basic catalyst. *Energy* 2011;**36**(2):777–84.
26. Zheng Y, Zhang W, Li Y, Chen J, Yu B, Wang J, Zhang L, Zhang J. Energy related CO_2 conversion and utilization: advanced materials/nanomaterials, reaction mechanisms and technologies. *Nano Energy* 2017;**40**:512–39.
27. Monga Y, Kumar P, Sharma RK, Filip J, Varma RS, Zboril R, Gawande MB. Sustainable synthesis of nanoscale zerovalent iron particles for environmental remediation. *ChemSusChem* 2020;**13**(13):3288–305.
28. Ma W, He X, Wang W, Xie S, Zhang Q, Wang Y. Electrocatalytic reduction of CO_2 and CO to multi-carbon compounds over Cu-based catalysts. *Chem Soc Rev* 2021;**50**(23):12897–914.
29. Galadima A, Muraza O. Catalytic thermal conversion of CO_2 into fuels: perspective and challenges. *Renew Sust Energ Rev* 2019;**115**:109333.
30. Ra EC, Kim KY, Kim EH, Lee H, An K, Lee JS. Recycling carbon dioxide through catalytic hydrogenation: recent key developments and perspectives. *ACS Catal* 2020;**10**(19):11318–45.
31. Gao P, Li S, Bu X, Dang S, Liu Z, Wang H, Zhong L, Qiu M, Yang C, Cai J, Wei W, Sun Y. Direct conversion of CO_2 into liquid fuels with high selectivity over a bifunctional catalyst. *Nat Chem* 2017;**9**(10):1019–24.
32. Benson EE, Kubiak CP, Sathrum AJ, Smieja JM. Electrocatalytic and homogeneous approaches to conversion of CO_2 to liquid fuels. *Chem Soc Rev* 2009;**38**(1):89–99.

33. Sen R, Goeppert A, Kar S, Prakash GKS. Hydroxide based integrated CO_2 capture from air and conversion to methanol. *J Am Chem Soc* 2020;**142**(10):4544−9.
34. Olah G. Beyond oil and gas: the methanol economy. *Angew Chem Int Ed* 2005;**44**:2636−9.
35. Shih CF, Zhang T, Li J, Bai C. Powering the future with liquid sunshine. *Joule* 2018;**2**(10):1925−49.
36. Wu CT, Yu KM, Liao F, Young N, Nellist P, Dent A, Kroner A, Tsang SC. A non-syn-gas catalytic route to methanol production. *Nat Commun* 2012;**3**:1050.
37. Chiang YD, Dutta S, Chen CT, Huang YT, Lin KS, Wu JC, Suzuki N, Yamauchi Y, Wu KC. Functionalized Fe_3O_4@silica core-shell nanoparticles as microalgae harvester and catalyst for biodiesel production. *ChemSusChem* 2015;**8**(5):789−94.
38. Witoon T, Permsirivanich T, Donphai W, Jaree A, Chareonpanich M. CO_2 hydrogenation to methanol over Cu/ZnO nanocatalysts prepared via a chitosan-assisted co-precipitation method. *Fuel Process Technol* 2013;**116**:72−8.
39. Mavlyankariev SA, Ahlers SJ, Kondratenko VA, Linke D, Kondratenko EV. Effect of support and promoter on activity and selectivity of gold nanoparticles in propanol synthesis from CO_2, C_2H_4, and H_2. *ACS Catal* 2016;**6**(5):3317−25.
40. Mori K, Sano T, Kobayashi H, Yamashita H. Surface engineering of a supported PdAg catalyst for hydrogenation of CO_2 to formic acid: elucidating the active Pd atoms in alloy nanoparticles. *J Am Chem Soc* 2018;**140**(28):8902−9.
41. Zhang Y, Li K, Chen M, Wang J, Liu J, Zhang Y. Cu/Cu_2O nanoparticles supported on vertically ZIF-L-coated nitrogen-doped graphene nanosheets for electroreduction of CO_2 to ethanol. *ACS Appl Nano Mater* 2019;**3**(1):257−63.
42. Mittal N, Nisola GM, Malihan LB, Seo JG, Kim H, Lee S-P, Chung W-J. One-pot synthesis of 2,5-diformylfuran from fructose using a magnetic bi-functional catalyst. *RSC Adv* 2016;**6**(31):25678−88.
43. Gong W, Chen C, Zhang Y, Zhou H, Wang H, Zhang H, Zhang Y, Wang G, Zhao H. Efficient synthesis of furfuryl alcohol from H_2-hydrogenation/transfer hydrogenation of furfural using sulfonate group modified Cu catalyst. *ACS Sustain Chem Eng* 2017;**5**(3):2172−80.
44. He J, Schill L, Yang S, Riisager A. Catalytic transfer hydrogenation of bio-based furfural with NiO nanoparticles. *ACS Sustain Chem Eng* 2018;**6**(12):17220−9.
45. Koley P, Chandra Shit S, Joseph B, Pollastri S, Sabri YM, Mayes ELH, Nakka L, Tardio J, Mondal J. Leveraging $Cu/CuFe_2O_4$-catalyzed biomass-derived furfural hydrodeoxygenation: a nanoscale metal-organic-framework template is the prime key. *ACS Appl Mater Interfaces* 2020;**12**(19):21682−700.
46. Lv M, Xin Q, Yin D, Jia Z, Yu C, Wang T, Yu S, Liu S, Li L, Liu Y. Magnetically recoverable bifunctional catalysts for the conversion of cellulose to 1,2-propylene glycol. *ACS Sustain Chem Eng* 2020;**8**(9):3617−25.
47. Ma M, Hou P, Zhang P, Cao J, Liu H, Yue H, Tian G, Feng S. Magnetic Fe_3O_4 nanoparticles as easily separable catalysts for efficient catalytic transfer hydrogenation of biomass-derived furfural to furfuryl alcohol. *Appl Catal A Gen* 2020;**602**:117709.
48. Shao Y, Wang J, Du H, Sun K, Zhang Z, Zhang L, Li Q, Zhang S, Liu Q, Hu X. Importance of magnesium in Cu-based catalysts for selective conversion of biomass-derived furan compounds to diols. *ACS Sustain Chem Eng* 2020;**8**(13):5217−28.
49. Di X, Shao Z, Li C, Li W, Liang C. Hydrogenation of succinic acid over supported rhenium catalysts prepared by the microwave-assisted thermolytic method. *Catal Sci Technol* 2015;**5**(4):2441−8.
50. Sun R, Zheng M, Pang J, Liu X, Wang J, Pan X, Wang A, Wang X, Zhang T. Selectivity-switchable conversion of cellulose to glycols over Ni−Sn catalysts. *ACS Catal* 2016;**6**(1):191−201.
51. Pieta IS, Kadam RG, Pieta P, Mrdenovic D, Nowakowski R, Bakandritsos A, Tomanec O, Petr M, Otyepka M, Kostecki R, Khan MAM, Zboril R, Gawande MB. The hallmarks of copper single atom catalysts in direct alcohol fuel cells and electrochemical CO_2 fixation. *Adv Mater Interfaces* 2021: 2001822. https://doi.org/10.1002/admi.202001822.
52. Liu X, Zhan G, Wu J, Li W, Du Z, Huang J, Sun D, Li Q. Preparation of integrated CuO/ZnO/OS nanocatalysts by using acid-etched oyster shells as a support for CO_2 hydrogenation. *ACS Sustain Chem Eng* 2020;**8**(18):7162−73.

53. Imyen T, Znoutine E, Suttipat D, Iadrat P, Kidkhunthod P, Bureekaew S, Wattanakit C. Methane utilization to methanol by a hybrid Zeolite@Metal-organic framework. *ACS Appl Mater Interfaces* 2020;**12**(21):23812−21.
54. Zhang B-B, Wang Y-H, Xu S-M, Chen K, Yang Y-G, Kong Q-H. Tuning nanocavities of Au@Cu$_2$O yolk−shell nanoparticles for highly selective electroreduction of CO$_2$ to ethanol at low potential. *RSC Adv* 2020;**10**(33):19192−8.
55. Rabiee A, Nematollahi D. Pyridinium-facilitated CO$_2$ electroreduction on Pt nanowire: enhanced electrochemical performance in CO$_2$ conversion. *Environ Prog Sustain Energy* 2019;**38**(1):112−7.
56. Wang A, Zhang T. One-pot conversion of cellulose to ethylene glycol with multifunctional tungsten-based catalysts. *Acc Chem Res* 2013;**46**(7):1377−86.
57. Barta K, Ford PC. Catalytic conversion of nonfood woody biomass solids to organic liquids. *Acc Chem Res* 2014;**47**(5):1503−12.
58. Boz N, Degirmenbasi N, Kalyon DM. Conversion of biomass to fuel: transesterification of vegetable oil to biodiesel using KF loaded nano-γ-Al$_2$O$_3$ as catalyst. *Appl Catal B* 2009;**89**(3−4):590−6.
59. Luque R, Clark JH, Yoshida K, Gai PL. Efficient aqueous hydrogenation of biomass platform molecules using supported metal nanoparticles on Starbons®. *ChemComm* 2009;(35):5305−7.
60. Dutta S, De S, Saha B, Alam MI. Advances in conversion of hemicellulosic biomass to furfural and upgrading to biofuels. *Catal Sci Technol* 2012;**2**(10):2025−36.
61. Gawande MB, Branco PS, Varma RS. Nano-magnetite (Fe$_3$O$_4$) as a support for recyclable catalysts in the development of sustainable methodologies. *Chem Soc Rev* 2013;**42**(8):3371−93.
62. van Putten RJ, van der Waal JC, de Jong E, Rasrendra CB, Heeres HJ, de Vries JG. Hydroxymethylfurfural, a versatile platform chemical made from renewable resources. *Chem Rev* 2013;**113**(3):1499−597.
63. Zhang C, Wang H, Liu F, Wang L, He H. Magnetic core−shell Fe$_3$O$_4$@C-SO$_3$H nanoparticle catalyst for hydrolysis of cellulose. *Cellulose* 2013;**20**(1):127−34.
64. Delidovich I, Leonhard K, Palkovits R. Cellulose and hemicellulose valorization: an integrated challenge of catalysis and reaction engineering. *Energy Environ Sci* 2014;**7**(9):2803−30.
65. Démolis A, Essayem N, Rataboul F. Synthesis and applications of alkyl levulinates. *ACS Sustain Chem Eng* 2014;**2**(6):1338−52.
66. Li X, Jia P, Wang T. Furfural: a promising platform compound for sustainable production of C$_4$ and C$_5$ chemicals. *ACS Catal* 2016;**6**(11):7621−40.
67. Calvo-Serrano R, Guo M, Pozo C, Galán-Martín Á, Guillén-Gosálbez G. Biomass conversion into fuels, chemicals, or electricity? A network-based life cycle optimization approach applied to the European union. *ACS Sustain Chem Eng* 2019;**7**(12):10570−82.
68. Zheng M, Pang J, Sun R, Wang A, Zhang T. Selectivity control for cellulose to diols: dancing on eggs. *ACS Catal* 2017;**7**(3):1939−54.
69. Pang J, Zheng M, Li X, Sebastian J, Jiang Y, Zhao Y, Wang A, Zhang T. Unlock the compact structure of lignocellulosic biomass by mild ball milling for ethylene glycol production. *ACS Sustain Chem Eng* 2019;**7**(1):679−87.
70. Zakzeski J, Bruijnincx PCA, Jongerius AL, Weckhuysen BM. The catalytic valorization of lignin for the production of renewable chemicals. *Chem Rev* 2010;**110**(6):3552−99.
71. Li C, Zhao X, Wang A, Huber GW, Zhang T. Catalytic transformation of lignin for the production of chemicals and fuels. *Chem Rev S* 2015;**115**(21):11559−624.
72. Merino O, Fundora-Galano G, Luque R, Martínez-Palou R. Understanding microwave-assisted lignin solubilization in protic ionic liquids with multiaromatic imidazolium cations. *ACS Sustain Chem Eng* 2018;**6**(3):4122−9.
73. Li H, Riisager A, Saravanamurugan S, Pandey A, Sangwan RS, Yang S, Luque R. Carbon-increasing catalytic strategies for upgrading biomass into energy-intensive fuels and chemicals. *ACS Catal* 2018;**8**(1):148−87.
74. Yabushita M, Kobayashi H, Fukuoka A. Catalytic transformation of cellulose into platform chemicals. *Appl Catal B* 2014;**145**:1−9.

75. Zhu S, Xu J, Cheng Z, Kuang Y, Wu Q, Wang B, Gao W, Zeng J, Li J, Chen K. Catalytic transformation of cellulose into short rod-like cellulose nanofibers and platform chemicals over lignin-based solid acid. *Appl Catal B* 2020;**268**:118732.
76. Song J, Fan H, Ma J, Han B. Conversion of glucose and cellulose into value-added products in water and ionic liquids. *Green Chem* 2013;**15**(10):2619−35.
77. De Clercq R, Dusselier M, Sels BF. Heterogeneous catalysis for bio-based polyester monomers from cellulosic biomass: advances, challenges and prospects. *Green Chem* 2017;**19**(21):5012−40.
78. Yazdani P, Wang B, Du Y, Kawi S, Borgna A. Lanthanum oxycarbonate modified Cu/Al_2O_3 catalysts for selective hydrogenolysis of glucose to propylene glycol: base site requirements. *Catal Sci Technol* 2017;**7**(20):4680−90.
79. Li C, Xu G, Li K, Wang C, Zhang Y, Fu Y. A weakly basic Co/CeO_x catalytic system for one-pot conversion of cellulose to diols: Kungfu on eggs. *ChemComm* 2019;**55**(53):7663−6.
80. Pang J, Zheng M, Li X, Jiang Y, Zhao Y, Wang A, Wang J, Wang X, Zhang T. Selective conversion of concentrated glucose to 1,2-propylene glycol and ethylene glycol by using RuSn/AC catalysts. *Appl Catal B* 2018;**239**:300−8.
81. Chambon F, Rataboul F, Pinel C, Cabiac A, Guillon E, Essayem N. Cellulose conversion with tungstated-alumina-based catalysts: influence of the presence of platinum and mechanistic studies. *ChemSusChem* 2013;**6**(3):500−7.
82. Besson M, Gallezot P, Pinel C. Conversion of biomass into chemicals over metal catalysts. *Chem Rev* 2014;**114**(3):1827−70.
83. Lazaridis PA, Karakoulia SA, Teodorescu C, Apostol N, Macovei D, Panteli A, Delimitis A, Coman SM, Parvulescu VI, Triantafyllidis KS. High hexitols selectivity in cellulose hydrolytic hydrogenation over platinum (Pt) vs. ruthenium (Ru) catalysts supported on micro/mesoporous carbon. *Appl Catal B* 2017;**214**:1−14.
84. Liu C, Zhang C, Sun S, Liu K, Hao S, Xu J, Zhu Y, Li Y. Effect of WOx on bifunctional $Pd-WO_x/Al_2O_3$ catalysts for the selective hydrogenolysis of glucose to 1,2-propanediol. *ACS Catal* 2015;**5**(8):4612−23.
85. Liu Y, Liu Y, Wu Q, Zhang Y. Catalytic conversion of glucose into lower diols over highly dispersed SiO_2-supported Ru-W. *Catal Commun* 2019;**129**:105731.
86. De S, Zhang J, Luque R, Yan N. Ni-based bimetallic heterogeneous catalysts for energy and environmental applications. *Energy Environ Sci* 2016;**9**(11):3314−47.
87. Sun R, Wang T, Zheng M, Deng W, Pang J, Wang A, Wang X, Zhang T. Versatile nickel−lanthanum(III) catalyst for direct conversion of cellulose to glycols. *ACS Catal* 2015;**5**(2):874−83.
88. Li N, Liu X, Zhou J, Ma Q, Liu M, Chen W. Enhanced Ni/W/Ti catalyst stability from Ti−O−W linkage for effective conversion of cellulose into ethylene glycol. *ACS Sustain Chem Eng* 2020;**8**(26):9650−9.
89. Gawande MB, Goswami A, Asefa T, Guo H, Biradar AV, Peng D-L, Zboril R, Varma RS. Core−shell nanoparticles: synthesis and applications in catalysis and electrocatalysis. *Chem Soc Rev* 2015;**44**(21):7540−90.
90. Podolean I, Negoi A, Candu N, Tudorache M, Parvulescu VI, Coman SM. Cellulose capitalization to bio-chemicals in the presence of magnetic nanoparticle catalysts. *Top Catal* 2014;**57**(17):1463−9.
91. Zhang J, Wu S-b, Liu Y. Direct conversion of cellulose into sorbitol over a magnetic catalyst in an extremely low concentration acid system. *Energy Fuels* 2014;**28**(7):4242−6.
92. Manaenkov OV, Mann JJ, Kislitza OV, Losovyj Y, Stein BD, Morgan DG, Pink M, Lependina OL, Shifrina ZB, Matveeva VG, Sulman EM, Bronstein LM. Ru-containing magnetically recoverable catalysts: a sustainable pathway from cellulose to ethylene and propylene glycols. *ACS Appl Mater Interfaces* 2016;**8**(33):21285−93.
93. Werpy TA, Holladay JE, White JF. *Top value added chemicals from biomass: I. Results of screening for potential candidates from sugars and synthesis gas*. Richland, WA (USA): Pacific Northwest National Laboratory (PNNL); 2004.
94. Delhomme C, Weuster-Botz D, Kühn FE. Succinic acid from renewable resources as a C_4 building-block chemical—a review of the catalytic possibilities in aqueous media. *Green Chem* 2009;**11**(1):13−26.

95. Koutinas AA, Vlysidis A, Pleissner D, Kopsahelis N, Lopez Garcia I, Kookos IK, Papanikolaou S, Kwan TH, Lin CSK. Valorization of industrial waste and by-product streams via fermentation for the production of chemicals and biopolymers. *Chem Soc Rev* 2014;**43**(8):2587−627.
96. Deshpande RM, Buwa VV, Rode CV, Chaudhari RV, Mills PL. Tailoring of activity and selectivity using bimetallic catalyst in hydrogenation of succinic acid. *Catal Commun* 2002;**3**(7):269−74.
97. Tapin B, Epron F, Especel C, Ly BK, Pinel C, Besson M. Study of monometallic Pd/TiO$_2$ catalysts for the hydrogenation of succinic acid in aqueous phase. *ACS Catal* 2013;**3**(10):2327−35.
98. Tapin B, Epron F, Especel C, Ly BK, Pinel C, Besson M. Influence of the Re introduction method onto Pd/TiO$_2$ catalysts for the selective hydrogenation of succinic acid in aqueous-phase. *Catal Today* 2014;**235**:127−33.
99. Kang KH, Hong UG, Bang Y, Choi JH, Kim JK, Lee JK, Han SJ, Song IK. Hydrogenation of succinic acid to 1,4-butanediol over Re−Ru bimetallic catalysts supported on mesoporous carbon. *Appl Catal A Gen* 2015;**490**:153−62.
100. Ly BK, Tapin B, Aouine M, Delichere P, Epron F, Pinel C, Especel C, Besson M. Insights into the oxidation state and location of rhenium in Re-Pd/TiO$_2$ catalysts for aqueous-phase selective hydrogenation of succinic acid to 1,4-butanediol as a function of palladium and rhenium deposition methods. *ChemCatChem* 2015;**7**(14):2161−78.
101. Sun D, Sato S, Ueda W, Primo A, Garcia H, Corma A. Production of C$_4$ and C$_5$ alcohols from biomass-derived materials. *Green Chem* 2016;**18**(9):2579−97.
102. Ly BK, Tapin B, Epron F, Pinel C, Especel C, Besson M. In situ preparation of bimetallic ReO$_x$-Pd/TiO$_2$ catalysts for selective aqueous-phase hydrogenation of succinic acid to 1,4-butanediol. *Catal Today* 2020;**355**:75−83.
103. Liu X, Wang X, Xu G, Liu Q, Mu X, Liu H. Tuning the catalytic selectivity in biomass-derived succinic acid hydrogenation on FeO$_x$-modified Pd catalysts. *J Mater Chem A* 2015;**3**(46):23560−9.
104. Li F, Lu T, Chen B, Huang Z, Yuan G. Pt nanoparticles over TiO$_2$−ZrO$_2$ mixed oxide as multifunctional catalysts for an integrated conversion of furfural to 1,4-butanediol. *Appl Catal A Gen* 2014;**478**:252−8.
105. Huang Z, Barnett KJ, Chada JP, Brentzel ZJ, Xu Z, Dumesic JA, Huber GW. Hydrogenation of γ-butyrolactone to 1,4-butanediol over CuCo/TiO$_2$ bimetallic catalysts. *ACS Catal* 2017;**7**(12):8429−40.
106. Le SD, Nishimura S. Highly selective synthesis of 1,4-butanediol via hydrogenation of succinic acid with supported Cu−Pd alloy nanoparticles. *ACS Sustain Chem Eng* 2019;**7**(22):18483−92.
107. Liu L, Asano T, Nakagawa Y, Tamura M, Tomishige K. One-pot synthesis of 1,3-butanediol by 1,4-anhydroerythritol hydrogenolysis over a tungsten-modified platinum on silica catalyst. *Green Chem* 2020;**22**(8):2375−80.
108. Le SD, Nishimura S. Effect of support on the formation of CuPd alloy nanoparticles for the hydrogenation of succinic acid. *Appl Catal B* 2021;**282**:119619.
109. Climent MJ, Corma A, Iborra S. Conversion of biomass platform molecules into fuel additives and liquid hydrocarbon fuels. *Green Chem* 2014;**16**(2):516−47.
110. Bohre A, Dutta S, Saha B, Abu-Omar MM. Upgrading furfurals to drop-in biofuels: an overview. *ACS Sustain Chem Eng* 2015;**3**(7):1263−77.
111. Mariscal R, Maireles-Torres P, Ojeda M, Sádaba I, López Granados M. Furfural: a renewable and versatile platform molecule for the synthesis of chemicals and fuels. *Energy Environ Sci* 2016;**9**(4):1144−89.
112. Lima TM, Lima CGS, Rathi AK, Gawande MB, Tucek J, Urquieta-González EA, Zbořil R, Paixão MW, Varma RS. Magnetic ZSM-5 zeolite: a selective catalyst for the valorization of furfuryl alcohol to γ-valerolactone, alkyl levulinates or levulinic acid. *Green Chem* 2016;**18**(20):5586−93.
113. Tadele K, Verma S, Gonzalez MA, Varma RS. A sustainable approach to empower the bio-based future: upgrading of biomass via process intensification. *Green Chem* 2017;**19**(7):1624−7.
114. Jorge EYC, Lima TdM, Lima CGS, Marchini L, Castelblanco WN, Rivera DG, Urquieta-González EA, Varma RS, Paixão MW. Metal-exchanged magnetic β-zeolites: valorization of lignocellulosic biomass-derived compounds to platform chemicals. *Green Chem* 2017;**19**(16):3856−68.

115. Villaverde MM, Bertero NM, Garetto TF, Marchi AJ. Selective liquid-phase hydrogenation of furfural to furfuryl alcohol over Cu-based catalysts. *Catalysis Today* 2013;**213**:87−92.
116. Neeli CKP, Chung Y-M, Ahn W-S. Catalytic transfer hydrogenation of furfural to furfuryl alcohol by using ultrasmall Rh nanoparticles embedded on diamine-functionalized KIT-6. *ChemCatChem* 2017;**9**(24):4570−9.
117. Lange J-P. Lignocellulose conversion: an introduction to chemistry, process and economics. *Biofuel Bioprod Biorefin* 2007;**1**(1):39−48.
118. Lange J-P, vanderHeide E, van Buijtenen J, Price R. Furfural—a promising platform for lignocellulosic biofuels. *ChemSusChem* 2012;**5**(1):150−66.
119. García-Sancho C, Jiménez-Gómez CP, Viar-Antuñano N, Cecilia JA, Moreno-Tost R, Mérida-Robles JM, Requies J, Maireles-Torres P. Evaluation of the ZrO_2/Al_2O_3 system as catalysts in the catalytic transfer hydrogenation of furfural to obtain furfuryl alcohol. *Appl Catal A Gen* 2021;**609**:117905.
120. Nandiwale KY, Vishwakarma M, Rathod S, Simakova I, Bokade VV. One-pot cascade conversion of renewable furfural to levulinic acid over a bifunctional $H_3PW_{12}O_{40}/SiO_2$ catalyst in the absence of external H_2. *Energy Fuels* 2021;**35**(1):539−45.
121. Rosatella AA, Simeonov SP, Frade RFM, Afonso CAM. 5-Hydroxymethylfurfural (HMF) as a building block platform: biological properties, synthesis and synthetic applications. *Green Chem* 2011;**13**(4):754−93.
122. Teong SP, Yi G, Zhang Y. Hydroxymethylfurfural production from bioresources: past, present and future. *Green Chem* 2014;**16**(4):2015−26.
123. Zhang L, Shah A, Michel Jr FC. Synthesis of 5-hydroxymethylfurfural from fructose and inulin catalyzed by magnetically-recoverable $Fe_3O_4@SiO_2@TiO_2$−HPW nanoparticles. *J Chem Technol Biotechnol* 2019;**94**(10):3393−402.
124. Zhu M-M, Du X-L, Zhao Y, Mei B-B, Zhang Q, Sun F-F, Jiang Z, Liu Y-M, He H-Y, Cao Y. Ring-opening transformation of 5-hydroxymethylfurfural using a golden single-atomic-site palladium catalyst. *ACS Catal* 2019;**9**(7):6212−22.
125. Yang Y, Xie Y, Deng D, Li D, Zheng M, Duan Y. Highly selective conversion of HMF to 1-hydroxy- 2,5-hexanedione on Pd/MIL-101(Cr). *ChemistrySelect* 2019;**4**(37):11165−71.
126. Choi S, Balamurugan M, Lee K-G, Cho KH, Park S, Seo H, Nam KT. Mechanistic investigation of biomass oxidation using nickel oxide nanoparticles in a CO_2-saturated electrolyte for paired electrolysis. *J Phys Chem Lett* 2020;**11**(8):2941−8.
127. Han W, Tang M, Li J, Li X, Wang J, Zhou L, Yang Y, Wang Y, Ge H. Selective hydrogenolysis of 5-hydroxymethylfurfural to 2,5-dimethylfuran catalyzed by ordered mesoporous alumina supported nickel-molybdenum sulfide catalysts. *Appl Catal B* 2020;**268**:118748.
128. Gan T, Liu Y, He Q, Zhang H, He X, Ji H. Facile synthesis of Kilogram-scale Co-alloyed Pt single-atom catalysts via ball milling for hydrodeoxygenation of 5-hydroxymethylfurfural. *ACS Sustain Chem Eng* 2020;**8**(23):8692−9.
129. Wu D, Zhang S, Hernández WY, Baaziz W, Ersen O, Marinova M, Khodakov AY, Ordomsky VV. Dual metal—acid Pd-Br catalyst for selective hydrodeoxygenation of 5-hydroxymethylfurfural (HMF) to 2,5-dimethylfuran at ambient temperature. *ACS Catal* 2020;**11**(1):19−30.
130. Bakandritsos A, Kadam RG, Kumar P, Zoppellaro G, Medved M, Tuček J, Montini T, Tomanec O, Andrýsková P, Drahoš B, Varma RS, Otyepka M, Gawande MB, Fornasiero P, Zbořil R. Mixed-valence single-atom catalyst derived from functionalized graphene. *Adv Mater* 2019;**31**(17):1900323.
131. Hannagan RT, Giannakakis G, Flytzani-Stephanopoulos M, Sykes ECH. Single-atom alloy catalysis. *Chem Rev* 2020;**120**(21):12044−88.
132. Wang Y, Chu F, Zeng J, Wang Q, Naren T, Li Y, Cheng Y, Lei Y, Wu F. Single atom catalysts for fuel cells and rechargeable batteries: principles, advances, and opportunities. *ACS Nano* 2021;**15**(1):210−39.
133. Fujita S, Nakajima K, Yamasaki J, Mizugaki T, Jitsukawa K, Mitsudome T. Unique catalysis of nickel phosphide nanoparticles to promote the selective transformation of biofuranic aldehydes into diketones in water. *ACS Catal* 2020;**10**(7):4261−7.

134. Zhang Q, Yang T, Lei D, Wang J, Zhang Y. Efficient production of biodiesel from esterification of lauric acid catalyzed by ammonium and silver Co-doped phosphotungstic acid embedded in a zirconium metal-organic framework nanocomposite. *ACS Omega* 2020;**5**(22):12760−7.
135. Zhong L, Feng Y, Wang G, Wang Z, Bilal M, Lv H, Jia S, Cui J. Production and use of immobilized lipases in/on nanomaterials: a review from the waste to biodiesel production. *Int J Biol Macromol* 2020;**152**:207−22.
136. Liu Y, Zhang P, Fan M, Jiang P. Biodiesel production from soybean oil catalyzed by magnetic nanoparticle $MgFe_2O_4@CaO$. *Fuel* 2016;**164**:314−21.
137. Shylesh S, Gokhale AA, Ho CR, Bell AT. Novel strategies for the production of fuels, lubricants, and chemicals from biomass. *Acc Chem Res* 2017;**50**(10):2589−97.
138. Sun Y, Zhang Q, Zhang C, Liu J, Guo Y, Song D. In situ approach to dendritic fibrous nitrogen-doped carbon nanospheres functionalized by brønsted acidic ionic liquid and their excellent esterification catalytic performance. *ACS Sustain Chem Eng* 2019;**7**(17):15114−26.
139. Song D, Liu J, Zhang C, Guo Y. Design of Brønsted acidic ionic liquid functionalized mesoporous organosilica nanospheres for efficient synthesis of ethyl levulinate and levulinic acid from 5-hydroxymethylfurfural. *Catal Sci Technol* 2021. https://doi.org/10.1039/D0CY01941K.
140. Moussa SO, Panchakarla LS, Ho MQ, El-Shall MS. Graphene-supported, iron-based nanoparticles for catalytic production of liquid hydrocarbons from synthesis gas: the role of the graphene support in comparison with carbon nanotubes. *ACS Catal* 2014;**4**(2):535−45.
141. Liu C, Chen Y, Zhao Y, Lyu S, Wei L, Li X, Zhang Y, Li J. Nano-ZSM-5-supported cobalt for the production of liquid fuel in Fischer-Tropsch synthesis: effect of preparation method and reaction temperature. *Fuel* 2020;**263**:116619.
142. Wei X, Yin Z, Lyu K, Li Z, Gong J, Wang G, Xiao L, Lu J, Zhuang L. Highly selective reduction of CO_2 to C_{2+} hydrocarbons at copper/polyaniline interfaces. *ACS Catal* 2020;**10**(7):4103−11.
143. IPCC fifth assessment report: climate change 2014, synthesis report. Summary for policymaker. Cambridge, U.K: Cambridge University Press; 2014.
144. Mac Dowell N, Fennell PS, Shah N, Maitland GC. The role of CO_2 capture and utilization in mitigating climate change. *Nat Clim Change* 2017;**7**(4):243−9.
145. von der Assen N, Jung J, Bardow A. Life-cycle assessment of carbon dioxide capture and utilization: avoiding the pitfalls. *Energy Environ Sci* 2013;**6**(9):2721−34.
146. Burkart MD, Hazari N, Tway CL, Zeitler EL. Opportunities and challenges for catalysis in carbon dioxide utilization. *ACS Catal* 2019;**9**(9):7937−56.
147. Kar S, Goeppert A, Prakash GKS. Integrated CO_2 capture and conversion to formate and methanol: connecting two threads. *Acc Chem Res* 2019;**52**(10):2892−903.
148. Brown NJ, García-Trenco A, Weiner J, White ER, Allinson M, Chen Y, Wells PP, Gibson EK, Hellgardt K, Shaffer MSP, Williams CK. From organometallic zinc and copper complexes to highly active colloidal catalysts for the conversion of CO_2 to methanol. *ACS Catal* 2015;**5**(5):2895−902.
149. Kar S, Sen R, Goeppert A, Prakash GKS. Integrative CO_2 capture and hydrogenation to methanol with reusable catalyst and amine: toward a carbon neutral methanol economy. *J Am Chem Soc* 2018;**140**(5):1580−3.
150. Yue H, Ma X, Gong J. An alternative synthetic approach for efficient catalytic conversion of syngas to ethanol. *Acc Chem Res* 2014;**47**(5):1483−92.
151. Bai S, Shao Q, Wang P, Dai Q, Wang X, Huang X. Highly active and selective hydrogenation of CO_2 to ethanol by ordered Pd−Cu nanoparticles. *J Am Chem Soc* 2017;**139**(20):6827−30.
152. Goeppert A, Czaun M, Jones JP, Surya Prakash GK, Olah GA. Recycling of carbon dioxide to methanol and derived products - closing the loop. *Chem Soc Rev* 2014;**43**(23):7995−8048.
153. Al-Rowaili FN, Jamal A, Ba Shammakh MS, Rana A. A review on recent advances for electrochemical reduction of carbon dioxide to methanol using metal−organic framework (MOF) and non-MOF catalysts: challenges and future prospects. *ACS Sustain Chem Eng* 2018;**6**(12):15895−914.
154. Jouny M, Luc W, Jiao F. General techno-economic analysis of CO_2 electrolysis systems. *Ind Eng Chem Res* 2018;**57**(6):2165−77.
155. Geioushy RA, Khaled MM, Alhooshani K, Hakeem AS, Rinaldi A. Graphene/ZnO/Cu_2O electrocatalyst for selective conversion of CO_2 into n-propanol. *Electrochim Acta* 2017;**245**:456−62.

156. Francis SA, Velazquez JM, Ferrer IM, Torelli DA, Guevarra D, McDowell MT, Sun K, Zhou X, Saadi FH, John J, Richter MH, Hyler FP, Papadantonakis KM, Brunschwig BS, Lewis NS. Reduction of aqueous CO_2 to 1-propanol at MoS_2 electrodes. *Chem Mater* 2018;**30**(15):4902–8.
157. Rahaman M, Kiran K, Montiel IZ, Grozovski V, Dutta A, Broekmann P. Selective n-propanol formation from CO_2 over degradation-resistant activated PdCu alloy foam electrocatalysts. *Green Chem* 2020;**22**(19):6497–509.
158. Rahaman M, Dutta A, Zanetti A, Broekmann P. Electrochemical reduction of CO_2 into multicarbon alcohols on activated Cu mesh catalysts: an identical location (IL) study. *ACS Catal* 2017;**7**(11):7946–56.
159. Kortlever R, Peters I, Koper S, Koper MTM. Electrochemical CO_2 reduction to formic acid at low overpotential and with high faradaic efficiency on carbon-supported bimetallic Pd–Pt nanoparticles. *ACS Catal* 2015;**5**(7):3916–23.
160. Yu X, Pickup PG. Recent advances in direct formic acid fuel cells (DFAFC). *J Power Sources* 2008;**182**(1):124–32.
161. Sun Q, Fu X, Si R, Wang C-H, Yan N. Mesoporous silica-encaged ultrafine bimetallic nanocatalysts for CO_2 hydrogenation to formates. *ChemCatChem* 2019;**11**(20):5093–7.
162. Yang G, Kuwahara Y, Masuda S, Mori K, Louis C, Yamashita H. PdAg nanoparticles and aminopolymer confined within mesoporous hollow carbon spheres as an efficient catalyst for hydrogenation of CO_2 to formate. *J Mater Chem A* 2020;**8**(8):4437–46.
163. Sun Q, Chen BWJ, Wang N, He Q, Chang A, Yang C-M, Asakura H, Tanaka T, Hülsey MJ, Wang C-H, Yu J, Yan N. Zeolite-encaged Pd–Mn nanocatalysts for CO_2 hydrogenation and formic acid dehydrogenation. *Angew Chem Int Ed* 2020;**59**(45):20183–91.
164. Liu J, Lan L, Liu X, Yang X, Wu X. Facile synthesis of agglomerated Ag–Pd bimetallic dendrites with performance for hydrogen generation from formic acid. *Int J Hydrog Energy* 2021;**46**(9):6395–403.
165. Kuwahara Y, Fujie Y, Mihogi T, Yamashita H. Hollow mesoporous organosilica spheres encapsulating PdAg nanoparticles and poly(ethyleneimine) as reusable catalysts for CO_2 hydrogenation to formate. *ACS Catal* 2020;**10**(11):6356–66.
166. Szcześniak B, Phuriragpitikhon J, Choma J, Jaroniec M. Recent advances in the development and applications of biomass-derived carbons with uniform porosity. *J Mater Chem A* 2020;**8**(36):18464–91.

CHAPTER 7

SMN-based catalytic membranes for environmental catalysis

Nilesh R. Manwar and Manoj B. Gawande

Department of Industrial and Engineering Chemistry, Institute of Chemical Technology, Mumbai Marathwada Campus, Jalna, Maharashtra, India

1. Introduction

Advances in functional nanomaterials have received enormous attention from researchers in various fields of catalysis. As a result, a reasonable number of reviews, books, patents, and research papers are published, emphasizing multidisciplinary applications.[1–9] In the view of rapidly evolved chemical industrialization for the production of various chemicals is allied with product separation problems, which reduces the purity of required chemicals.[2,10] To improve these product qualities, essence considers clean and safe green chemistry approaches for energy-effective technology.[11]

At the one end, material scientists, chemists, and chemical engineers are working toward catalytic membrane reactors (MR), where membrane-based separation and chemical reaction are performed in one single unit. MR is associated with multifunctional features of catalytic reactors, and it has been recognized as a promising approach for industries (chemical and petroleum), environmental protection, and energy conversions through catalysis.[12–20] IUPAC defines a catalytic MR as a single physical device that simultaneously performs a chemical reaction and membrane-based separation.[21] Thus, the catalytic MR is well conceptualized; however, membranes are crucial and play a role of a separator and its involvement in the reaction system.

On the other end, engineered/functionalized nanomaterials are an emerging class of potential materials with a broad range of practical applications.[10,22–25] These include ammonia (NH_3) and hydrogen (H_2) production,[26] environmental protection,[27] photocatalysis,[28] waste removal,[29] fiber, and industrial mechanical processes.[3,30] Engineered nanoparticle was defined as "any intentionally produced particle that has a characteristic dimension from 1 to 100 nm and has properties that are not shared by non-nano scale particles with the same chemical composition." Here the second part of the definition unveils the sizable impact of nanoparticles (NPs) on the environment and health.[27] Despite this, nanomaterials have colossal demand to empower the research advances with innovative solutions for environmental remediations.[6,27,31] Nanomaterials with a large surface area (surface-to-volume ratio) provide high catalytic sites in catalysis; however, microscopic NPs have a tendency to agglomerate, which reduces the catalyst stability reusability and overall catalytic performance. Therefore, functionalizing these NPs without diminishing the surface properties is necessary, and the added functional groups tailor the surface of nanomaterials, which improves the specific sites on the surface.[3,4,32]

Surface-modified nanomaterials (SMNs) which are typically produced from inorganic silica/silicates, noble metals, semiconductors, metal oxides, polymers, and carbonaceous materials (cellulose, chitosan) are the most appealing choices for environmental industries.[4] These functional nanomaterials are extensively employed for environmental applications (remediation of air, wastewater, and soil), and catalytic processes include adsorption, photodegradation, alcohol oxidation, nanofiltration, disinfection, and pollutant-monitoring sensors.[3,4,8] Because of these versatile applications of SMNs, the selection of fabrication methods and functionalization is of significant interest.[3] In recent years, catalytic MR engineering[16,18,33] has shown a prime role in reconceptualizing/redesigning industrial processes, which includes the use of membrane units during various processes of petrochemical industry, sector of biotechnology, environmental remediation (wastewater treatment), and energy conversions (NH_3 and H_2 production), where catalytic MR accomplish the demand of process intensification.

In the context of these innovative developments of SMNs and catalytic MR, the functionalized SMNs based catalytic membrane is an appropriate solution to overcome the associated challenges in NPs (Section 7.2). The activity of this hybrid system is majorly dependent on available different types of catalytic membranes; however, their choice is based on specific characteristic parameters like product selectivity, membrane lifetime, mechanic and physicochemical properties of MR, and the operating cost. The primary catalytic membrane was broadly categorized into organic (polymeric) and inorganic (ceramic/metallic) membranes based on these known properties and chemical processes of interest(Section 7.3).[34] Optimizing this organic/inorganic membrane improves the reactant conversions (product yield) and reduces the separation cost for further commercial development. Indeed, an appropriate design for catalytic organic (polymeric) membranes is more advantageous and attractive among MRs. However, inorganic catalytic MR has the main concern about operative high-temperature (>500°C), which is more challenging for large-scale development of MR. Therefore, only the polymeric MRs are overviewed here (Section 7.4). Surface functionalization of nanomaterials with advanced polymeric membranes possesses unique properties as SMNs based catalytic membranes for developing innovative environmental industry (Section 7.5). Multifunctional features of these materials are a prime focus of this chapter (Section 7.6). Herein, we have overviewed the recent advances of SMN-based catalytic membrane and will deliver certain new concepts for their rationale designs and the selection of polymeric membranes for environmental catalysis.

This book chapter discusses the context of SMNs and engineering catalytic MRs for the environmental industry. The biological membranes have minimal industrial applications, and inorganic (silica) membrane also restricts its use for large-scale dynamic processes. This chapter only covers nonbiological catalysts, and special attention has been paid to environmental catalytic applications of functionalized polymeric membranes. Short limitations of nanomaterials are also given; however, to overcome this, we have specified the advantageous composition of polymeric membranes, their selection criteria, and more specific application of SMN-based polymeric catalytic MRs are discussed here.

2. Challenges in SMNs and catalytic MRs

The remarkable properties of SMNs include a high surface area for adsorption, high activity, and reasonable product selectivity for environmental catalysis, antimicrobial/antifungal properties for disinfection, magnetic properties for particle separations, and other optoelectronic properties useful for photonics and sensing applications (water/air quality monitoring).[6] However, SMNs face some challenges that hinder their overall performance of environmental catalysis. Although the microscopic SMNs provide significant surface area for their high reactivity, NPs aggregate in water and disprove environmental benefits. TiO_2/ZnO NPs aggregates severely once added to the water. Additional difficulties in the retention and reuse of nanomaterials influence economic costs related to their loss and negatively impact ecosystems. Therefore, these issues are crucial, debated reasonably in the present time and regarding same, as it has toxic, adverse impacts on environment and health, which increases the human concern (Fig. 7.1).[6,27,35−37]

The retention and dissolution ability of SMNs is another challenging aspect. Understanding of the effective and reliable synthetic approach is very much essential to retaining suspended NPs. In one aspect, nanomagnetic cores are created by mounting nanomaterials, which have magnetic properties, are adopted widely because of their accessible separation features. However, specific metal-based magnetic nanomaterials are not stable because of their physicochemical properties.[38] It can cause corrosion under harsh environmental conditions, indirectly releases secondary components of pollution. In another aspect, they were using catalytic MRs with functional properties allowing them to be anchored in the nanomaterials, including NPs, nanotubes, nanofibers, metal-organic frameworks, and nanosheets.[2,39] The overall process of these membrane-associated materials also became challenging; however, Fig. 7.2 shows the most prominent issues, consequences, and possible solutions for nanomaterial-based polymeric membranes for the distillation process.[40]

Furthermore, technical bottlenecks of essential catalytic membranes (organic/inorganic)[4] are also listed in Table 7.1. Despite these challenges and limitations of SMNs and catalytic MRs, we introduce a practical incorporation approach for SMN-based

Figure 7.1 Opportunities and challenges of SMNs.

polymeric membranes that could overcome the drawbacks mentioned earlier. The suggested modifications will extend the research directions and encourage the incorporation of methodologies given under Section 7.5. Tailored SMN-based polymeric nanocomposite membranes are helpful for the desired application. It also aids in the regulation of the physicochemical features listed earlier as well as the negative/toxic effects of some nanomaterials on the environment.

3. Types of catalytic membrane reactors (MRs)

Typical multifunctional features of catalytic MRs are of various types, and several different criteria have been suggested for its classification (Fig. 7.3). There are four different types of catalytic MR categorization criteria: (1) transport function of the membrane (includes extractor, distributor, and contactor type of MRs); (2) the nature of catalytic materials (artificial/biological); (3) the nature of the catalytic membrane is the most important criterion used to distinguish organic/polymeric MRs and inorganic/ceramic MRs; and (4) the role of the catalytic membranes is inert or reactive (Fig. 7.3A).[33] Analytic membranes are broadly classified into inorganic (ceramic/metallic) membranes and organic (polymeric) membranes (Fig. 7.3B).

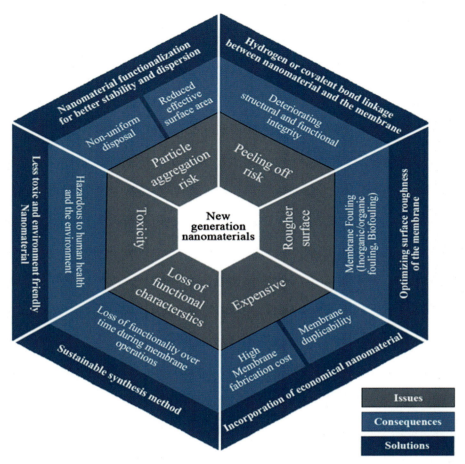

Figure 7.2 Summary of severe issues in the manufacture of nanomaterial-based polymeric membrane for the distillation process. *(Reproduced with permission from Ref. 40. Copyright 2020 MDPI journals.)*

3.1 Inorganic (silica/metallic) MRs

Catalytic inorganic MR technology is the most studied; the traditional class of catalytic MR was also demonstrated with possible pros and cons of inorganic membranes.[20,41–43] Generally, there are five types of inorganic membranes prescribed by Armor (1998).[44] It includes, (1) precipitated oxides; (2) zeolites; (3) carbon; (4) dense oxides; and (5) dense metals. Most of the applications of these inorganic MRs were performed at high-temperature (>500°C), and widely studied reactions are selective hydrogenation, steam reforming, water gas shift reactions, dehydrogenation (alkanes to simple olefins), oxydehydrogenation, and catalytic decomposition of H_2S, HI, and H_2O. They anticipate the

Table 7.1 Bottlenecks of catalytic MRs.

Types of catalytic MRs	Bottlenecks
Inorganic (Silica/metallic) Catalytic MRs	• Low surface area • Limited product selectivity and absorption capacity • Recovery of the catalysts is poor • Week thermal, chemical, and mechanical stabilities under extensive cycling • Agglomeration- a tendency to aggregate • Lack of reaction specificity in complex systems • Limited selectivity to organics and heavy metals • Little activity under visible light (TiO_2, ZnO) • Fast recombination of photogenerated charge carriers (large bandgap)
Organic (Polymeric) Catalytic MRs	• Postrecovery and reusability issues • Aggregation of carbon materials • Reasonable cost • Hydrophobicity • Toxicity—possible health risk • Poor recovery of materials/reuse after treatment • lack of chemical stability and mechanical strength (corrosion issue) • High-temperature stability issue • Incompatibility and reuse.

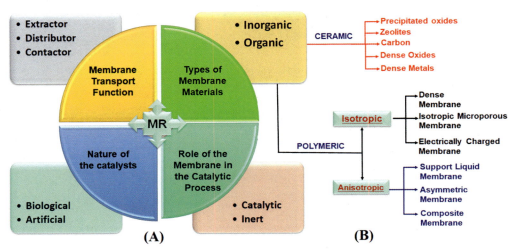

Figure 7.3 (A) Different classifications criteria of the membrane reactor (MR) ; and (b) representative types of catalytic membranes. *((A) Reproduced with permission from Ref. 33. Copyright 2014 Wiley Online Library.)*

significant number of apparent opportunities within refineries and industrial catalysis. However, Pd or Pd alloy—based MRs have shown remarkable developments for hydrogen purification and production as well.[45] Despite the challenges of all upscaling aspects of inorganic MR, mainly because of their inadequate membrane materials, they are unable to commercialize due to severities in process operation. Inorganic membranes are expensive and have a limited lifetime. Moreover, long-term and/or continuous operation of inorganic MRs causes inhomogeneity. Also, it competes for the adsorption process, which may lead to poisoning with sulfide ions and fouling the carbon deposits. Intrinsic properties such as the permeability of a membrane can be improved by reducing its size of thickness, and reaction temperature. However, the low thickness can increase the risk that creates defects in the layers of inorganic membrane, and low-temperature reactions won't be feasible for inorganic MRs.

Along with these drawbacks of inorganic membranes, the MRs and desirable reaction processes are also essential and need to consider their issues. Shortly, the key points are (1) delamination of the surface layer of the membrane and inadequate sealing are consequences thereof; (2) heat and mass transfer between the reactor and catalyst; and (3) cost-economic study of the overall process—the cost of the inorganic membrane must be lower than the desired product. Thus, the manufacturing of inorganic membranes for the development of MR and the temperature aspect of ceramic material is not mature enough. These missing developments in this area are unable to make commercialization of ceramic MR technology at large scale.

3.2 Organic (polymeric) MRs

Organic membranes, popularly known as polymeric membranes, are the most widely utilized membranes in industrial processes because of their ease of processability at a large scale and low manufacturing cost compared with inorganic membranes. Additional properties and basic advantages of polymeric membranes are also overviewed in Section 7.4 of this chapter. The typical morphology of polymeric membranes was classified into isotropic and anisotropic membranes. Subsequently, isotropic membranes mainly include nonporous dense membranes, microporous membranes, and electrically charged membranes. However, anisotropic membranes are distributed into integrated asymmetric membranes, composite membranes, and supported liquid membranes. Fig. 7.3B depicts a representative classification of these membranes. The reactions carried in polymeric MRs are also effective at low temperatures (<300°C), and thus polymeric membrane is a viable alternative to traditional ceramic membranes.

The combined uses of SMNs with polymeric MRs entail specific important and feasible environmental catalytic conversions.[18,34,46,47] Short details of some particular examples are covered under SMNs based polymeric membrane-assisted catalysis (Section 7.6). Indeed, incorporation methodology (Section 7.5), where SMNs fabricated in a

Table 7.2 Characteristic details of different polymeric materials and their application processes.

Polymer name	Acronym	Membrane fabrication method	Tg (°C)	Application process
Cellulose diacetate	CA	Casting	80	RF, UF, MF, GS
Cellulose nitrate	CN	Casting	53	UF
Polyethylene	PE	Melt extrusion	−120	PV, GS
Polyvinylalcohol	PVA	Casting	85	PV
Polyvinylchloride	PVC	Casting	87	MF
Poly(vinyl fluoride)	PVF	Casting	52	UF, MF, ED
Poly(vinylidene chloride)	PVDC	Casting	−17	UF, MF, ED
Poly(chlorotrifluoroethylene)	PCFE	Melt extrusion	87	GS. ED
Poly(acrylamide)	PAA	Melt extrusion	165	UF, RO
Poly(acrylic acid)	PAA	Casting	101	
Poly(methyl acrylate)	PMA	—	10	ED, RO
Poly(propylene)			−10	
Poly(caprolactam)	Nylon 6	Polycondensation	51	MF, ED, GS
Poly(1,2-propylene carbonate)	PPC		37	MF, UF
Poly(1,4-butadiene)			−105	
Poly(caprolactone)	Nylon 66	Electrospinning	−66	MF, ED
Poly(ethylene terephthalate)		Electrospinning	72	PV
Polyglycolic acid	PGL		39	
Polylactic acid	PLA	Electrospinning	59	
Poly(ether ketone ketone)	PEKK	Melt/Casting	164	GS
Poly(ethylene glycol),	PEG	Melt extrusion	−67	
Poly(ethyleneketone)	PEK	Melt extrusion	15	PV
Polystyrene	PS	Electrospinning	100	MF
Poly(methacrylonitrile)	PMAN	Casting	115	PV

ED, electrodialysis; *GS*, gas separation; *MD*, membrane distillation; *MF*, microfiltration; *NF*, nanofiltration; *PV*, pervaporation; *RO*, reverse osmosis; *UF*, ultrafiltration.

polymeric membrane; herein, the choice of polymeric membrane and its functionality can regulate the selective conversions and influence the overall performance of catalysts. Moreover, a more comprehensive selection of polymeric membranes is available (listed in Table 7.2) to select the appropriate one. As a result, the following basic overview of polymeric membranes makes polymeric MRs technology attractive.

4. Basic overview of polymeric MRs

The polymeric membrane is the foremost catalytic membrane known among all MRs for dynamic environmental processes. The polymeric membrane will stay at the heart of MR because of its competitive performance, superior processability, low cost, and abundance.[12,20,43,48,49] The state of the polymeric membrane is manifested through its potential role, advanced structural chemistry, surface morphology, and processing performance paradigm. The important mechanical, chemical, thermal, electrical, and permeation properties provide tuning opportunities to catalytic polymeric MRs. By considering permeation, porous and dense membranes can be distinguished, where polymers choose differently and vary their criteria. Polymer choice is not that important for porous membranes (also known for micro/ultrafiltration processes).[19]

In contrast, chemical, thermal, and electrical properties have a significant impact on developing the selection criteria of polymers. Although, the revolution of catalytic polymeric MRs and their processes was complex for understanding potential promises of polymeric chemistry and structures. In biological applications, polymeric membranes must offer a low binding affinity for separated molecules and endures the required harsh cleaning conditions. Therefore, polymer compatibility is essential for polymer membrane-assisted catalytic applications.[50] It is mainly dependent on three aspects: a former suitable membrane given its molecular chain rigidity, chain interactions, and a functional group that maintains the stereoregularity and polarity of polymer structures. Second, the chosen fabrication methodologies (preparation methods of polymers). Third, amorphous and semicrystalline structures of polymers, which have different glass transition temperatures (Tg), affect the membrane performance. In the case of dense, nonporous membranes, the choice of a polymer directly influences the performance of polymeric MR and changes the value of Tg (Table 7.2). Hence, the crystallinity and change in Tg are also important parameters.

Furthermore, a polymeric MR's low-cost criterion must be constructed from a reasonably priced polymer. As a large number of polymers are available, the selection of polymeric membranes is not a trivial task. A polymer must have appropriate characteristics for the desired application. Various techniques are employed to manufacture membrane polymers, which mainly depend on used materials and selected structure of membranes. This technique includes coating, phase inversions, sintering, stretching, and track-etching. The most commonly employed polymers in membrane synthesis are cellulose acetate (CA), nitrocellulose (CN), cellulose esters (CE), polysulfone (PSU), polyethersulfone (PES), polyacrylonitrile (PAN), polyamide, polyimide, polyethylene (PE), polypropylene (PP), polytetrafluoroethylene (PTFE), polyvinylidene fluoride (PVDF) and polyvinyl chloride (PVC) (Table 7.2).

5. Incorporation of SMNs into polymeric membranes

Incorporating SMNs into polymeric membranes is an emerging research trend that enhances the significant performance of catalytic MR. Although, the SMNs has synergistic effects, when it is incorporated with different types of polymeric materials. They usually improve the performances of MR. However, it might be changed or deteriorated because of the earlier-specified challenges under Section 7.2. Thus, careful attention has to be made to choose adequate fabrication methods and the composition of SMNs incorporated into the polymeric membrane.

5.1 Methods of SMNs incorporation into polymeric MRs

Typically, nanomaterials incorporation into polymer membrane is mainly identified by providing two modification approaches: (1) internal/bulk modification through (mixed-matrix membranes) and (2) external/surface modification. Bulk-modified membranes include a phase inversion process, where NPs dispersed homogeneously onto the polymeric solution.[51] However, surface-modified catalytic membranes are obtained by deposition of NPs onto the surface of polymer membranes. Certainly, premodification (direct) and/or postmodification (indirect) methods are considered to achieve polymeric membrane functionalization. The incorporation of SMNs into the monomer or polymer solution before or during the synthesis process is termed a premodification method. This strategy is advantageous and valuable for homogeneous coverage during one-pot synthesis, controls the required amount of ligand in SMNs, and allows various functional groups.[52] This direct approach mainly includes blending with hydrophilic polymer materials (PVA, PEG).[53] Yet the direct incorporation ability of functional polymers may not be that much compatible during synthesis as they may have a direct reaction with SMNs. Therefore, the postmodification method was employed extensively by introducing a large variety of functional groups. The modification is carried out by grafting approaches and generally applied once the membrane has been formed (after synthesis). This postmodification strategy alters the SMNs composition and surface morphology of polymer membranes; however, the bulk structure modification of essential membrane can be well-maintained. More importantly, preferred conditions of this indirect incorporation methods must be identified with improved permeability and selectivity of catalytic MRs. Under the postmodification technique, which is usually applied to improve the overall performance of polymeric MR, the following major methods have been established. It mainly contains the physical and chemical processes; among them, the most frequent and considerable modifications are (1) direct irradiation (plasma and UV), (2) grafting (chemical and light-induced), and (3) layer-by-layer coating.

5.1.1 Irradiation-based modifications

Radiation-induced surface modification of polymers membrane is evolving continuously, and it is mainly promoted by plasma treatment and ultraviolet (UV) rays.[54]

5.1.1.1 Plasma treatment

This is a low-temperature treatment method utilized under direct/indirect modification, where ions, electrons, radicals, and photons species of plasma energize the surface properties of polymeric membranes.[55] Various functional groups can be introduced onto the surface of the polymeric membrane by selecting an appropriate plasma source and operating parameters. Such functional groups (—OH, =CO and —COOH) are obtained by incorporation of O_2-plasma, while amine, amide, imine, and nitrile surface functional groups are incorporated by N_2-plasma.[56] These functional groups influence the properties of polymer membranes, which include surface energy, wettability, and roughness. For example, plasma-treated polyacrylonitrile (PAN) employed for ultrafiltration (UF) membranes and efficiently hydrophilized without detrimental effects on barrier properties; this had been interpreted to be caused by parallel oxidation and stabilization (via cyclization) of the membrane polymer.[57]

5.1.1.2 UV-irradiation

Photoirradiation using UV light is another technique that can be used for photodeposition of SMNs into polymer membranes. UV-irradiation improves the surface properties of polymeric materials. Photoirradiation is undergoing a renaissance by adopting a tunable light spectrum to perform the light-induced surface modification. However, it is dependent on the applied wavelength of light (λ), which is the basic requirement to drive the surface chemical reactions.[58] The bonding energy of the surface photon is expressed by,

$$E = h\nu = hc/\lambda \quad (7.1)$$

where h is Planck's constant, ν is the frequency, c is the speed of a photon, and λ is the wavelength. The bonding energy of C—C and C—O bonds has a synergistic impact and absorbs UV radiation to incorporate SMNs into polymeric membranes. For instance, polyacrylic acid (AA) was incorporated on polysulfone (PSU) membrane by UV-irradiation.[54]

5.1.2 Grafting-based modifications

Grafting is a facile process where monomers are covalently bonded and polymerized as side chains. Graft-copolymerization-based modification is an appealing technique that chemically bonds a variety of functional groups to polymer membranes.[52] These chemical attachments offer long-term stability and physical surface coatings to a grafted polymer membranes. Two main approaches, namely grafting to and/or grafting from have received considerable attention so far; it also involves covalent bonding.[7] However, Graft-copolymerization is induced by chemical/electrochemical treatment, photoirradiation, and high-energy radiation techniques, and those are also specified below.

5.1.2.1 Chemical/electrochemical initiated grafting

Chemical/electrochemical initiated grafting can be persuaded by two major paths free radical and ionic. In this chemical-induced graft-copolymerization process, SMNs are getting attached to the polymer surface by electrostatic and weak van der Waals forces.

In this approach, the initiator plays a crucial role that determines the grafting path. For instance, significant advances have been made on the emergence development of surface-initiated controlled radical polymerization (SI-CRP) for catalysis, electronics, nanomaterial synthesis, and biosensing applications.[59]

Free radical grafting: In this case, free radicals can be produced during the initiation step of chemical pretreatment processes such as ozonation, diazotization, and xanthation of polymer chains. The obtained free radicals are transferred to the reaction substrate with monomer, which may incorporate the SMNs onto polymer membranes. In general, free radical generation is a direct/indirect method. Recently, polymeric chains of poly (methyl methacrylate) i.e., PMMA grafted successfully from cork particles, wherein situ atom transfer radical polymerization (ATRP) approach is considered.[60] However, a detailed summary of ATRP is reviewed by Matyjaszewski and Xia.[61]

Ionic grafting: Ionic means the cationic or anionic mode of grafting is there; cationic graft-copolymerization has significant advances. Particular emphasis is placed on ionic initiators; it includes organometallic compounds, alkali metals in suspension (Lewis base), and sodium naphthalene. During cationic grafting, alkyl aluminum (R_3Al) used as coinitiator in conjunction halide (ACL) interacts and form carbonium ions along the polymer membrane, that leads the graft-copolymerization process. The cationic reaction mechanism proceeds through (ACl + $R_3Al \rightarrow A^+R_3AlCl$; $A^+ + M \rightarrow AM^+M \rightarrow$ graft copolymer).[52]

5.1.2.2 Photoirradiation-induced grafting

Photoirradiation-induced grafting is becoming a popular strategy due to the advantages of photon light, such as its availability, low cost, and ability to perform under mild circumstances. It is applied for the surface functionalization of polymer membranes. In this photoirradiation method, initiator presence is not important. However, the radicals can be created first on the surface of the polymer membrane by the photoirradiation process. Indeed, UV-induced grafting is more advantageous because of its low wavelength and the bonding energy concept.[62] Herein, the medium has significant importance; for instance, irradiation is carried out in the air, it may form peroxide on the surface of the polymer. Photoirradiation-induced grafting is also associated with generated free radicals and ions upon, those can be used further to incorporate SMNs into polymer membranes.

5.1.2.3 High energy radiation (plasma) induced grafting

The plasma polymerization technique has received considerable attention by itself. Furthermore, plasma-induced grafting can be used more significantly for NPs incorporation to polymer membranes. It is because of the conditions of plasma achieved through slow discharge offer as like ionizing radiation. Electron-induced excitation, ionization, and dissociations are the main processes of plasma. Therefore, the accelerated electrons obtained from the plasma source have sufficient energy to induce the chemical bonding of the polymeric structure. It is used to form macromolecule radicals, which subsequently initiate the graft-copolymerization of SMNs based membranes. Electrons, ions, and γ-rays are examples of high-energy sources that can be utilized to activate the surface of polymer membranes prior to grafting via radical polymerization onto the membrane surface.[62] In this regard, Fisher and coworkers investigated the modification of porous poly(ether sulfone) (PES) membranes with plasma-induced grafting.[56,63]

5.1.3 Surface coating-based modifications

Surface coating is a simple and inexpensive approach that refers to the postmodification method, where polymeric membranes with incorporated SMNs can be obtained either by wet or dry coating strategy. The surface coating creates several layers over existing membrane polymers, mainly dependent on adhesion properties and intermolecular interactions between coated layer and underlying membranes.[5] The functional/modified coatings are responsible for changing the surface properties of the substrate, such as wettability, adhesion, and corrosion resistance.

5.1.3.1 Gas-phase coatings

Physical/chemical vapor deposition (PVD/CVD) is used in this method, which is also known as the dry coating strategy. This method is a multistep process that includes heating, evaporation, sputtering, and deposition, allowing thin film coating.[64] In the background of designing polymer surfaces via vapor deposition,[65] recent reports are there on catalytic hybrid metal-polymer membranes fabrication, and devices based on CVD polymers open a new paradigm for surface modification technology.[66,67]

5.1.3.2 Wet-phase coatings

Wet-phase coating is a well-known classical method for fabricating polymeric asymmetric porous membranes, and it mainly includes dip-coating or spin-coating strategies.[68] Organic solvent-based coating prevailed aging and additive loading that also decreases the surface energy with time and tends to lose adhesion because of hydrophobic recovery. In water-based coating, resist the effects of aging, and subsequent retreatment is necessary for organic solvent coating. Notably, a recent study signifies that increased intermolecular interaction of the coated films has no significant in surface energy.[69]

5.2 Characterization methods of SMNs based polymer membranes

To characterize the nanostructured material composition and physical properties, a variety of approaches are required. In most cases, the physical properties of nanoscale materials can be identified by more than one technique. So, depending on the strengths and limitations of each method, the selection of the most suitable way is a prime consideration.[70] Essential characterization of polymeric SMNs is based on particle size, surface morphology, zeta potential, polydispersity index, surface area, and composition.

Advanced microscopic techniques, including atomic force microscopy (AFM), transmission electron microscopy (TEM), and scanning electron microscopy (SEM) are used to understand the surface morphology. Besides, these techniques can also identify changes in the physical properties of particles after surface modification. Surface hydrophobicity and change in surface composition can be examined by X-ray spectroscopic measurements. Average particle diameter, size distribution, and surface charge can be identified by light scattering spectroscopy. The significant details of these specific methods are described in the following sections.

5.2.1 Advanced microscopic techniques

5.2.1.1 Cryo-transmission electron microscopy (cryo-TEM/cryo-EM)

Recent single-particle electron microscopy (cryo-EM) is the most powerful technique applied at cryogenic temperatures of samples, which is used to visualize the structure of nanomaterials.[71,72] To achieve a frozen-hydrated state, a liquid sample solution is useful to a grid-mesh and plunge-frozen in liquid ethane and/or a mixture with propane. An aqueous sample solution is applied to a grid-mesh and plunge-frozen in liquid ethane or a mixture of liquid ethane and propane. Cryo-EM is known for a futuristic foundational technique in materials science.[73] Albeit the development of this technique was initiated in 1970, recent advancement in high resolved detector and software algorithms allows determining the molecular structure at near-atomic resolution. Complex structures of NPs based polymer membrane can also be identified by cryo-TEM. In 2016, cryo-EM was named as the "Method of the Year" by *Nature Methods* and it also received a Nobel prize in chemistry-2017.[74,75] It is now regarded as an alternative to XRD and NMR spectroscopy for the imaging of macromolecular structures as a result of the recent attention it has garnered. For instance, Lieberwirth et al. highlighted the prospects and pitfalls of cryo-TEM by paying particular emphasis on the electron microscopical imaging of polymeric NPs (Fig. 7.4). Notably, the microscopic structures of the hard polymer, such as polystyrene nanoparticles (PS-NP) have been discussed.[76]

5.2.1.2 Atomic force microscopy (AFM)

The AFM measures 3D characteristics features of nanoparticles with subnanometer resolution. AFM generates 2D/3D images based on the surface force. SMNs and polymer membrane characterization using AFM have many significant advantages over light scattering spectroscopy, TEM/SEM, and optical methods. The unique advantages of AFM for SMNs characterization include (1) typical NPs characterization resolution range is 0.5 nm to micron scale; (2) mixture of NPs can be distributed and visualize below 30 nm; (3) change is the shape of NPs can be characterized; (4) direct visualization of liquid NP is possible; (5) physical properties (e.g., magnetic and contact angle) can be measured; (6) size of NPs can be identified; and (7) the most crucial 2D/3D surface topography can visualize. Fig. 7.5 shows the 2D view images of AFM properties, where polystyrene nanostructures obtained from the dewetting of grafted RAFT polystyrene after 75 min incubation time.[77]

SMN-based catalytic membranes for environmental catalysis 185

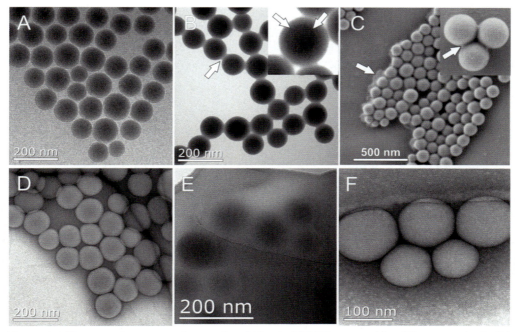

Figure 7.4 Electron microscopy micrographs of identical PS-NP prepared by different methods: (A) cryo-TEM, (B) TEM of a drop cast sample, (C) SEM of a drop cast sample, (D) TEM of a uranyl acetate negative stained sample embedded in trehalose, (E) TEM of a sample prepared by embedding into EMI-BF4 ionic liquid, and (F) embedded in EMI-BF4 ionic liquid stained with uranyl acetate. *(Reproduced with permission from Ref. 76. Copyright 2016 Wiley Online Library.)*

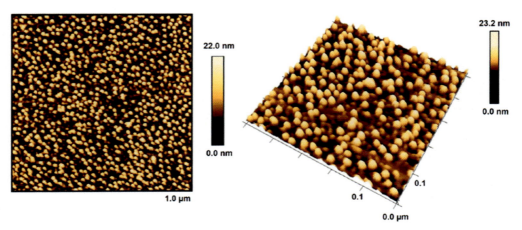

Figure 7.5 AFM images in 2D view (left) and 3D view (right) of preformed polystyrene nanostructures from the dewetting of grafted RAFT polystyrene with of $M_n = 64$ kg/mol on gold surfaces after 75 min incubation times. *(Reproduced with permission from Ref. 77. Copyright 2020 MDPI journals.)*

5.2.1.3 Transmission electron microscopy (TEM) and scanning electron microscopy (SEM)

Surface morphological insights in response to imaging, diffraction, and spectroscopic information about the shape and size of NPs can be ascertained. The basic principle of TEM has a high-resolution limit and is slightly different from that of SEM. Typically SEM has a focused electron beam that can be scanned over a surface and create an image. The obtained signals are often used to assess surface topography and composition. Thus, SEM simply allows morphological examination with direct visualization of the surface. The critical requirement in TEM is to prepare ultrathin samples to enable electron transmittance, TEM samples preparation is crucial and time-consuming as well. For example, TEM and SEM images of PS-NP are represented in Fig. 7.4B–D, where they have used a drop-casting method for sample preparations.[76] Fig. 7.6 shows the nanostructured

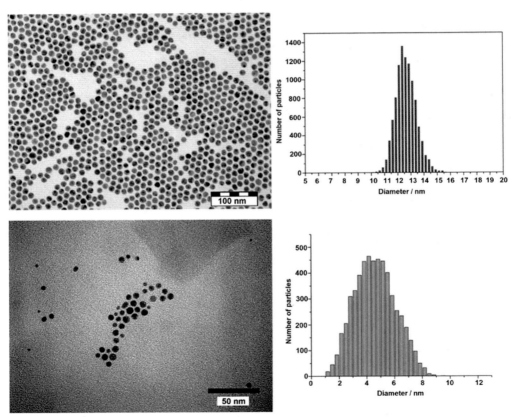

Figure 7.6 Representative transmission electron micrographs of citrate-stabilized gold nanoparticles (top-left) and AuNPs from Brust-Schiffrin synthesis (bottom-left). The corresponding size distribution of the AuNPs is shown in the figures on the right side. *(Reproduced with permission from Ref. 77. Copyright 2020 MDPI journals.)*

properties of polystyrene samples derived via (RAFT) polymerization that grafted directly to gold (Au) NPs via thiocarbonylthio-end groups, as well as the precise arrangement of these Au-NPs and their size distribution.[77]

5.2.2 X-ray spectroscopic techniques
5.2.2.1 X-ray diffraction (XRD)
One of the most prominent techniques is X-ray diffraction (XRD), used extensively for NPs characterization of solid materials. Typical XRD measures the crystalline structure, nature of the phase, lattice parameters, and crystalline grain size. Regarding XRD, certain ample advantages are there; one is commonly performed with powder samples. The composition of the NPs can be measured by comparing the two theta positions and peak intensity with the reference XRD patterns and the possible phases obtained from International Center for Diffraction Data (ICDD) database. However, the XRD technique may not be suitable for some amorphous materials and the NPs having a size below 3 nm.

5.2.2.2 X-ray photon correlation spectroscopy (XPS or XPCS)
One of the most extensively used techniques for analyzing the surface chemical contents of nanoscale materials is X-ray photoelectron spectroscopy (XPS). The basic principle of XPS is underlying the photoelectric effect.[70] Besides, surface hydrophobicity includes chromatographic interactions, adsorption processes, and contact angle measurements that, with change in surface composition, can also be examined by XPS. Among all X-ray spectroscopic measurements, XPS is unique and can identify the specific functional groups on the surface of nanomaterials. D. D. Sarma and coworkers have published a review that describes the essence of XPS as a unique tool. They provided quantitative, reliable, and compelling descriptions of highly complex heterostructures. For instance, they have given an accurate description and prerequisite to understanding the internal structure of nanocrystals with their physical properties.[78]

Despite the potential use of the above microscopic techniques, scanning transmission electron microscope (STEM) combined with HAADF, EDX has a significant impact on surface morphology study, crystal structure, elemental composition. These methods also visualize the atomic structure of heterointerfaces. Attenuated Total Reflection-Fourier Transform Infrared spectroscopy (ATR-FTIR) FTIR spectroscopy has also been used as a vital tool to understand the surface modification of nanoparticles.

6. SMNs based polymeric membrane-assisted catalysis
SMNs based polymeric MRs is extensively used in a wide range of applications, including separation, water desalination, biomembrane reactor, and environmental industry, etc. It is mainly based on the structure and function of the polymer membrane. However, the

solubility and diffusivity of polymer membrane is the basis of separation applications such as pervaporation (PV) and gas separation. These are known traditional applications; appropriate modification with nanomaterials is essential to innovate the MRs applications. Indeed, the precise selection of polymer membrane allowed complete retention to the catalyst and obtained product, which makes the overall process more effective for desired applications. The most common NPs based polymer membrane catalytic reactions can be performed in gas or liquid phases, and by their catalysis reaction type, NPs based polymer MRs applications are specified below;

6.1 Pervaporation for esterification

PV-assisted esterification is the most studied reaction so far.[79] It has the most relevant methodologies for the creation of esters in an equilibrium-limited reaction. The reaction mechanism is well-developed, thermodynamically feasible, and the process has industrious relevance as well. The context of esterification reaction is accompanied by the rise in the cost of product purification and distillation. PV has interesting energy-efficient and separation properties and can be operated at low temperatures as well. This can provide a better match to the optimal condition of esterification reactions. With the help of dense polymer membranes having hydrophilic nature can enhance the selective permeation of water over organics, while hydrophobic membrane permeates the organics over water. Such a complex situation can be overcome with SMNs based polymer membranes having high pore volume/SA and selective polymer membrane. For instance, hydrophilic polymer membranes [e.g., poly (vinyl alcohol), polyamide (PA), and polyimide (PI), etc.] are large in numbers in comparison to hydrophobic polymers [e.g., polydimethylsiloxane (PDMS) and poly (ether-block-amide) (PEBA).[80]

6.2 Hydrogenation

In general, hydrogenation reaction performed at high pressure, which increases the H_2 solubility in organic solvents to carry out the reaction. However, the liquid phase hydrogenation of organic compounds can also be carried in catalytic polymer MR, which evades the high-pressure conditions. Palladium nanoparticle (Pd NP) plays a significant role in hydrogenation reactions.[81] However, Pd NP based on polymeric-flat-sheet and hollow fibers membrane synthesized via UV-irradiation of poly (acrylic acid) at the surface of PES. The obtained polymeric materials with immobilized Pd NPs are effective for hydrogenation reactions and successfully applied in continuous mode as well. For example, in the first configuration, reactions performed at 60°C for short residence times (<1 min) and low H_2 pressure (1.4–2.2 bar), and relatively low Pd loading (3.2 mol%) were used (Fig. 7.7). In the case of ketones and aldehydes, hydrogenation reactions are

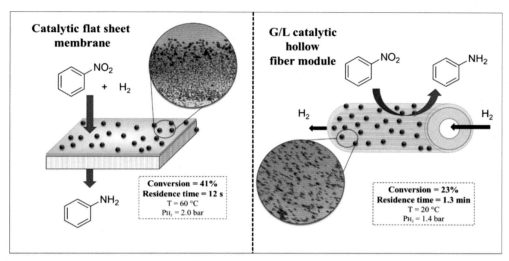

Figure 7.7 Flat sheet membranes (Left) polymeric G/L catalytic hollow fiber contactors (Right) for hydrogenation reactions. *(Reproduced with permission from Ref. 82. Copyright 2020 Elsevier.)*

not efficient in continuous flow-through configuration. However, this study reveals, flat sheet membranes were active under batch mode conditions, and it was stable at harsh applied conditions (20 bar H_2, 16 h). Moreover, the G/L catalytic hollow fiber contactor has shown promising results, which were obtained for the hydrogenation of nitro-arenes. It was also proposed that the scope of the application for both flow modes can be enlarged by monitoring increasing temperature and/or residence time by decreasing liquid flow rate or by increasing H_2 pressure; tuning the design of the contactor module with higher packing density.[82]

Dang et al.[83] investigated the catalytic reduction of 4-nitrophenol (4-NP) by using polymer NPs with polyamidoamine (PAMAM) dendrimer-Ag shell. The cross-linked polystyrene (PS) microspheres were obtained by dispersion copolymerization of styrene with acrylic acid and cross-linked monomer of 1,2- divinylbenzene. The average particle size of PS microsphere is 450 nm, where narrow size distribution was used to immobilize the Ag shell dendrimer. Ag NPs formed directly inside the PAMAM shell through the reduction with $NaBH_4$. The obtained PS@PAMAM-Ag NPs were packed in stainless steel (SS) column for the reduction of 4-NP (Fig. 7.8).

6.3 CO₂ sequestration: hydration of CO₂

The potential application of nickel (Ni) NPs stabilized by adsorbing polymers for enhanced carbon sequestration (mitigation of atmospheric CO_2), has been demonstrated

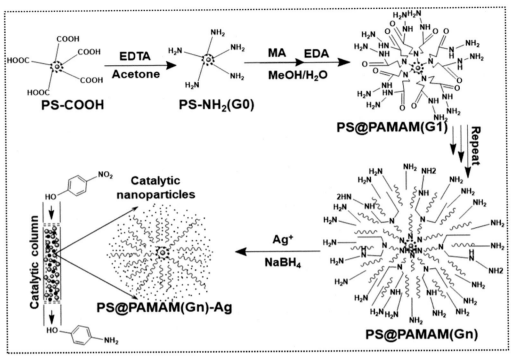

Figure 7.8 Synthesis procedures from PS-COOH polymer microspheres to PS@PAMAM-Ag nanoparticles and their application in the catalytic column for reduction of 4-NP. *(Reproduced with permission from Ref. 83. Copyright 2020 Elsevier.)*

in a recent study by Seo et al.,[81] where they have accelerated carbon dioxide (CO_2) dissolution into saline aquifers. The characteristic catalytic activity of Ni NP based on polymer membrane was monitored with changes in the diameter of CO_2 microbubbles. With the increase of ionic strength, an electrostatic repulsive force reduces in pristine Ni NPs, and that decreases the catalytic potential. This study examines the dispersive behavior of Ni NPs by applying cationic [dextran (DEX), nonionic poly(vinyl pyrrolidone) (PVP)] and anionic polymers [anionic carboxymethylcellulose (CMC)] in their optimal concentration, which improved the catalytic capabilities of CO_2 dissolution in unfavorable conditions as well. Fig. 7.9 elaborates the overall conceptual illustration for the hydration of CO_2.

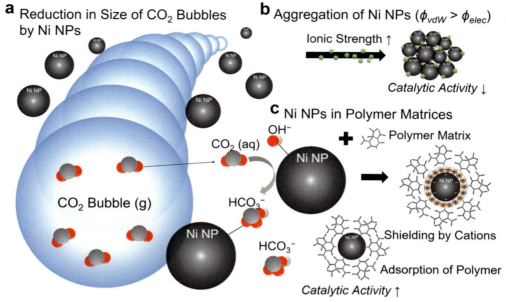

Figure 7.9 Conceptual illustration for the hydration of CO_2: (A) accelerated CO_2 dissolution by Ni NPs catalytic potential; (B) the decreased catalytic potential of Ni NPs by the aggregation behavior of NPs in high ionic strengths; and (C) stabilized Ni NPs by polymers in brine at high salinity levels. *(Reproduced with permission from Ref. 84. Copyright 2018 Nature.)*

7. Summary and future perspectives

SMN-based catalytic polymer membranes are, without doubt, the most studied and applied research of MRs. Herein, we have reviewed the recent advances of SMNs and catalytic polymeric MR and the associated opportunities and challenges. SMN-based polymer membranes have a potential role and promising performance in catalysis. We have provided the possible advancement in the incorporation methodologies that can improve the molecular interactions and results in good mechanical, optical, and electrical properties for desired applications. Advanced characterization methods are also presented to better understanding the structural changes in the developed hybrid polymer membranes with NPs. Particular emphasis has been given to polymeric MR, and its catalytic applications. However, we believe that more critical focus is being needed to understand the past achievements of NP-based polymer membranes. In the future, we may be able to move beyond nanomaterials, particular atomic site catalysts on a porous membrane could greatly influence environmental catalysis.

Abbreviations

4-NP 4-nitrophenol
AFM atomic force microscopy
CA cellulose acetate
CE cellulose esters
CN nitrocellulose
cryo-TEM cryo-transmission electron microscopy
CVD chemical vapor deposition
ED electrodialysis
GS gas separation
ICDD international center for diffraction data
MD membrane distillation
MF microfiltration
MRs membrane reactors
NF nanofiltration
NPs nanoparticle
PAMAM polyamidoamine
PAN polyacrylonitrile
PCDF polyvinylidene fluoride
PE polyamide, polyimide, polyethylene
PES polyethersulfone
PP polypropylene
PSU polysulfone
PTFE polytetrafluoroethylene
PV pervaporation
PVC polyvinylchloride
PVD physical vapor deposition
RO reverse osmosis
SEM scanning electron microscopy
SI-CRP surface-initiated controlled radical polymerization
SMNs surface-modified nanomaterials
TEM transmission electron microscopy
UF ultrafiltration
UV ultraviolet
XPS X-ray photoelectron spectroscopy
XRD X-ray diffraction

References

1. Li Z, Ji S, Liu Y, Cao X, Tian S, Chen Y, et al. Well-defined materials for heterogeneous catalysis: from nanoparticles to isolated single-atom sites. *Chem Rev* 2020;**120**(2):623–82. https://doi.org/10.1021/acs.chemrev.9b00311.
2. Hilal AFIN, Wright JJCJ. *Nanofiber membranes for medical.* Environmental, and Energy Applications; 2020.
3. Hussain CM. *Functionalized nanomaterials for industrial applications.* 2020.
4. Darwish M, Mohammadi A. Functionalized nanomaterial for environmental techniques. *Nanotechnol Environ Sci* 2018;**1–2**:315–50. https://doi.org/10.1002/9783527808854.ch10.

5. Makvandi P, Iftekhar S, Pizzetti F, Zarepour A, Zare EN, Ashrafizadeh M, et al. Functionalization of polymers and nanomaterials for water treatment, food packaging, textile and biomedical applications: a review. *Environ Chem Lett* 2021;**19**:583–611. https://doi.org/10.1007/s10311-020-01089-4.
6. Gopakumar DA, Pai AR, Pasquini D, Ben LS, S AKHP, Thomas S. *Chapter 1 - nanomaterials-state of art, new challenges, and opportunities*. Elsevier Inc.; 2019. https://doi.org/10.1016/B978-0-12-813926-4.00001-X.
7. Kango S, Kalia S, Celli A, Njuguna J, Habibi Y, Kumar R. Surface modification of inorganic nanoparticles for development of organic-inorganic nanocomposites - a review. *Prog Polym Sci* 2013;**38**(8):1232–61. https://doi.org/10.1016/j.progpolymsci.2013.02.003.
8. Guerra FD, Attia MF, Whitehead DC, Alexis F. Nanotechnology for environmental remediation: materials and applications. *Molecules* 2018;**23**:1–23. https://doi.org/10.3390/molecules23071760.
9. Zhu W, Chen Z, Pan Y, Dai R, Wu Y, Zhuang Z, et al. Functionalization of hollow nanomaterials for catalytic applications: nanoreactor construction. *Adv Mater* 2019;**31**:1–30. https://doi.org/10.1002/adma.201800426.
10. Khin MM, Nair AS, Babu VJ. A review on nanomaterials for environmental remediation. *Energy Environ Sci* 2012;**5**:8075–109. https://doi.org/10.1039/c2ee21818f.
11. Asiri IAM. *Applications of Nanotechnology for Green Synthesis. Nanotechnology in the Life Sciences*. Springer; 2020. http://link.springer.com/10.1007/978-3-030-44176-0.
12. Sousa JM, Madeira LM, Santos JC. *Polymeric membrane reactors*. 2008.
13. Noreña LE, Wang JA. *Advanced catalytic materials - photocatalysis and other current trends*. ExLi4EvA; 2016.
14. Hai FI, Fattah KP, Saroj DP, Moreira MT. *Membrane reactors for bioethanol production and processing. Membrane Reactors for Energy Applications and Basic Chemical Production*. Woodhead; 2015. p. 313–43.
15. Seidel-morgenstern A. *Membrane reactors*. 1996. https://doi.org/10.1016/0926-860x(96)80044-5.
16. Basile A, De Falco M, Centi G, Iaquaniello G. *Membrane reactor engineering applications for a greener process industry*. 2016.
17. Basile A, Gallucci F. *Membranes for membrane reactors: preparation, optimization and selection*. 2011. https://doi.org/10.1002/9780470977569.
18. Algieri C, Coppola G, Mukherjee D, Shammas MI, Calabro V, Curcio S, Chakraborty S. Catalytic membrane reactors: the industrial applications perspective. *Catalysts* 2021;**11**. https://doi.org/10.3390/catal11060691.
19. Warsinger DM, Chakraborty S, Tow EW, Plumlee MH, Bellona C, Loutatidou S, et al. A review of polymeric membranes and processes for potable water reuse. *Prog Polym Sci* 2018;**81**:209–37. https://doi.org/10.1016/j.progpolymsci.2018.01.004.
20. Vankelecom IFJ. Polymeric membranes in catalytic reactors. *Chem Rev* 2002;**102**:3779–810. https://doi.org/10.1021/cr0103468.
21. Koros WJ, Ma YH, Shimidzu T. Terminology for membranes and membrane processes (IUPAC Recommendations 1996). *Pure Appl Chem* 1996;**68**:1479–89. https://doi.org/10.1351/pac199668071479.
22. Astruc D. Introduction: nanoparticles in catalysis. *Chem Rev* 2020;**120**:461–3. https://doi.org/10.1021/acs.chemrev.8b00696.
23. Baig N, Kammakakam I, Falath W. Nanomaterials: a review of synthesis methods, properties, recent progress, and challenges,. *Mater Adv* 2021;**2**:1821–71. https://doi.org/10.1039/d0ma00807a.
24. Sharma N, Ojha H, Bharadwaj A, Pathak P. Preparation and catalytic applications of nanomaterials : a review. *RSC Adv* 2015;**5**:53381–403. https://doi.org/10.1039/c5ra06778b.
25. Kudr J, Id YH, Richtera L, Adam V, Zitka O. Magnetic nanoparticles : from design and synthesis to real world applications. *Nanomaterials* 2017;**7**. https://doi.org/10.3390/nano7090243.
26. Hill A, Torrente-Murciano L. Low temperature H2 production from ammonia using ruthenium-based catalysts: Synergetic effect of promoter and support. *Appl. Catal. B: Environ.* 2015;**172-173**:129–35. https://doi.org/10.1016/j.apcatb.2015.02.011.
27. Auffan M, Rose J, Bottero JY, Lowry GV, Jolivet JP, Wiesner MR. Towards a definition of inorganic nanoparticles from an environmental, health and safety perspective. *Nat Nanotechnol* 2009;**4**:634–41. https://doi.org/10.1038/nnano.2009.242.
28. Likodimos V. Advanced photocatalytic materials. *Materials (Basel)* 2020;**13**:13–5. https://doi.org/10.3390/ma13040821.

29. Pervez MN, Balakrishnan M, Hasan SW, Choo KH, Zhao Y, Cai Y, et al. A critical review on nanomaterials membrane bioreactor (NMS-MBR) for wastewater treatment. *Npj Clean Water* 2020;**3**. https://doi.org/10.1038/s41545-020-00090-2.
30. Chaturvedi S, Dave PN, Shah NK. Applications of nano-catalyst in new era. *J Saudi Chem Soc* 2012;**16**:307−25. https://doi.org/10.1016/j.jscs.2011.01.015.
31. Centi G, Perathoner S. Creating and mastering nano-objects to design advanced catalytic materials. *Coord Chem Rev* 2011;**255**:1480−98. https://doi.org/10.1016/j.ccr.2011.01.021.
32. Wieszczycka K, Staszak K, Woźniak-Budych MJ, Litowczenko J, Maciejewska BM, Jurga S. Surface functionalization — the way for advanced applications of smart materials. *Coord Chem Rev* 2021;**436**. https://doi.org/10.1016/j.ccr.2021.213846.
33. Fontananova E, Drioli E. Membrane reactors: advanced systems for intensified chemical processes. *Chem Ing Tech* 2014;**86**:2039−50. https://doi.org/10.1002/cite.201400123.
34. Abdallah H. A review on catalytic membranes production and applications. *Bull Chem React Eng Catal* 2017;**12**:136−56. https://doi.org/10.9767/bcrec.12.2.462.136-156.
35. Zhang F. Grand challenges for nanoscience and nanotechnology in energy and health. *Front Chem* 2017;**5**:1−6. https://doi.org/10.3389/fchem.2017.00080.
36. Khan I, Saeed K, Khan I. Nanoparticles: properties, applications and toxicities, arab. *J Chem* 2019;**12**:908−31. https://doi.org/10.1016/j.arabjc.2017.05.011.
37. Drioli E, Macedonio F, Tocci E. Membrane Science and membrane Engineering for a sustainable industrial development. *Separ Purif Technol* 2021;**275**. https://doi.org/10.1016/j.seppur.2021.119196.
38. Gawande MB, Goswami A, Felpin FX, Asefa T, Huang X, Silva R, et al. Cu and Cu-based nanoparticles: synthesis and applications in catalysis. *Chem Rev* 2016;**116**:3722−811. https://doi.org/10.1021/acs.chemrev.5b00482.
39. Wang Y, Vogel A, Sachs M, Sprick RS, Wilbraham L, Moniz SJA, et al. Current understanding and challenges of solar-driven hydrogen generation using polymeric photocatalysts. *Nat Energy* 2019;**4**:746−60. https://doi.org/10.1038/s41560-019-0456-5.
40. Ray SS, Bakshi HS, Dangayach R, Singh R, Deb CK, Ganesapillai M, et al. Recent developments in nanomaterials-modified membranes for improved membrane distillation performance. *Membranes (Basel)* 2020;**10(7)**(140):1−29. https://doi.org/10.3390/membranes10070140.
41. Wang K, Amin K, An Z, Cai Z, Chen H, Chen H, et al. Advanced functional polymer materials. *Mater Chem Front* 2020;**4**:1803−915. https://doi.org/10.1039/d0qm00025f.
42. Bhattacharjee S. Polymeric nanoparticles. *Princ Nanomedicine* 2019:195−240. https://doi.org/10.1201/9780429031236-8.
43. Ulbricht M. Advanced functional polymer membranes. *Polymer* 2006;**47**:2217−62. https://doi.org/10.1016/j.polymer.2006.01.084.
44. Armor JN. Applications of catalytic inorganic membrane reactors to refinery products. *J Membr Sci* 1998;**147**:217−33. https://doi.org/10.1016/S0376-7388(98)00124-0.
45. Gallucci F, Medrano J, Fernandez E, Melendez J, van Sint Annaland M, Pacheco A. Advances on high temperature Pd-based membranes and membrane reactors for hydrogen purification and production. *J Membr Sci Res* 2017;**3**:142−56. https://doi.org/10.22079/jmsr.2017.23644.
46. Wang J, Gu H. Novel metal nanomaterials and their catalytic applications. *Molecules* 2015;**20**:17070−92. https://doi.org/10.3390/molecules200917070.
47. Dong X, Jin W, Xu N, Li K. Dense ceramic catalytic membranes and membrane reactors for energy and environmental applications. *Chem Commun* 2011;**47**:10886−902. https://doi.org/10.1039/c1cc13001c.
48. Ng LY, Mohammad AW, Leo CP, Hilal N. Polymeric membranes incorporated with metal/metal oxide nanoparticles: a comprehensive review. *Desalination* 2013;**308**:15−33. https://doi.org/10.1016/j.desal.2010.11.033.
49. Lin H, Ding Y. Polymeric membranes: chemistry, physics, and applications. *J Polym Sci* 2020;**58**:2433−4. https://doi.org/10.1002/pol.20200622.
50. Mülhaupt R. Catalytic polymerization and post polymerization catalysis fifty years after the discovery of Ziegler's catalysts. *Macromol Chem Phys* 2003;**204**:289−327. https://doi.org/10.1002/macp.200290085.

51. Yalcinkaya F, Boyraz E, Maryska J, Kucerova K. A review on membrane technology and chemical surface modification for the oily wastewater treatment. *Materials (Basel)* 2020;**13**. https://doi.org/10.3390/ma13020493.
52. Bhattacharya A, Misra BN. Grafting: a versatile means to modify polymers: techniques, factors and applications. *Prog Polym Sci* 2004;**29**:767−814. https://doi.org/10.1016/j.progpolymsci.2004.05.002.
53. Zahid M, Rashid A, Akram S, Rehan ZA, Razzaq W. A comprehensive review on polymeric nanocomposite membranes for water treatment. *J Membr Sci Technol* 2018;**08**:1−20. https://doi.org/10.4172/2155-9589.1000179.
54. Weibel DE, Michels AF, Horowitz F, da Silva Cavalheiro R, da Silva Mota GV. Ultraviolet-induced surface modification of polyurethane films in the presence of oxygen or acrylic acid vapours. *Thin Solid Films* 2009;**517**:5489−95. https://doi.org/10.1016/j.tsf.2009.03.204.
55. Xiao K, Chen L, Chen R, Heil T, Lemus SDC, Fan F, et al. Artificial light-driven ion pump for photoelectric energy conversion. *Nat Commun* 2019;**10**:1−7. https://doi.org/10.1038/s41467-018-08029-5.
56. Kull KR, Steen ML, Fisher ER. Surface modification with nitrogen-containing plasmas to produce hydrophilic, low-fouling membranes. *J Membr Sci* 2005;**246**:203−15. https://doi.org/10.1016/j.memsci.2004.08.019.
57. Li X, Basko M, Du Prez F, Vankelecom IFJ. Multifunctional membranes for solvent resistant nanofiltration and pervaporation applications bases on segmented polymer network. *J Phys Chem B* 2008;**112**:16539−45. https://doi.org/10.1021/jp805117z.
58. Pinson J, Thiry D. *Surface modification of polymers: methods and applications*. 2019. https://doi.org/10.1002/9783527819249.
59. Zoppe JO, Ataman NC, Mocny P, Wang J, Moraes J, Klok H-A. Surface-initiated controlled radical polymerization: state-of-the-art, opportunities, and challenges in surface and interface engineering with polymer brushes. *Chem Rev* 2017;**117**:1105−318. https://doi.org/10.1021/acs.chemrev.6b00314.
60. Barros-timmons A. Grafting poly(methyl methacrylate) (PMMA) from cork via atom transfer radical polymerization (ATRP) towards higher quality of three-dimensional (3D) printed PMMA/Cork-g-PMMA materials. *Polymers (Basel)* 2020;**12**.
61. Matyjaszewski K, Xia J. Atom transfer radical polymerization. *Chem Rev* 2001;**101**:2921−90. https://doi.org/10.1021/cr940534g.
62. Clough RL. High-energy radiation and polymers: a review of commercial processes and emerging applications. *Nucl Instrum Methods Phys Res Sect B Beam Interact Mater Atoms* 2001;**185**:8−33. https://doi.org/10.1016/S0168-583X(01)00966-1.
63. Wavhal DS, Fisher ER. Modification of porous poly(ether sulfone) membranes by low-temperature CO2-plasma treatment. *J Polym Sci Part B Polym Phys* 2002;**40**:2473−88. https://doi.org/10.1002/polb.10308.
64. Fotovvati B, Namdari N, Dehghanghadikolaei A. On coating techniques for surface protection: a review. *J. Manuf. Mater. Process.* 2019;**3**. https://doi.org/10.3390/jmmp3010028.
65. Asatekin A, Barr MC, Baxamusa SH, Lau KKS, Tenhaeff W, Xu J, et al. Designing polymer surfaces via vapor deposition. *Mater Today* 2010;**13**:26−33. https://doi.org/10.1016/S1369-7021(10)70081-X.
66. Coclite AM, Howden RM, Borrelli DC, Petruczok CD, Yang R, Yagüe JL, et al. 25th anniversary article: CVD polymers: a new paradigm for surface modification and device fabrication. *Adv Mater* 2013;**25**:5392−423. https://doi.org/10.1002/adma.201301878.
67. Wang M, Wang X, Moni P, Liu A, Kim DH, Jo WJ, et al. CVD polymers for devices and device fabrication. *Adv Mater* 2017;**29**:1604606. https://doi.org/10.1002/adma.201604606.
68. Scriven LE. Physics and applications of DIP coating and spin coating. *MRS Online Proc Libr* 1988;**121**:717−29. https://doi.org/10.1557/PROC-121-717.
69. Nemani SK, Annavarapu RK, Mohammadian B, Raiyan A, Heil J, Haque A, et al. Surface modification of polymers : methods and applications. *Adv Mater Interfaces* 2018:1−26. https://doi.org/10.1002/admi.201801247.
70. Mourdikoudis S, Pallares RM, Thanh NTK. Characterization techniques for nanoparticles: comparison and complementarity upon studying nanoparticle properties. *Nanoscale* 2018;**10**:12871−934. https://doi.org/10.1039/C8NR02278J.

71. De Yoreo JJ, Sommerdijk NAJM. Investigating materials formation with liquid-phase and cryogenic TEM. *Nat Rev Mater* 2016;**1**:1–18. https://doi.org/10.1038/natrevmats.2016.35.
72. Li Y, Huang W, Li Y, Chiu W, Cui Y. Opportunities for cryogenic electron microscopy in materials science and nanoscience. *ACS Nano* 2020;**14**:9263–76. https://doi.org/10.1021/acsnano.0c05020.
73. Patterson JP, Xu Y, Moradi M, Sommerdijk NAJM, Friedrich H. CryoTEM as an advanced analytical tool for materials chemists. *Acc Chem Res* 2017;**50**:1495–501. https://doi.org/10.1021/acs.accounts.7b00107.
74. Cressey D, Callaway E. Cryo-electron microscopy wins chemistry Nobel. *Nature* 2017;**550**:167. https://doi.org/10.1038/nature.2017.22738.
75. Doerr A. Cryo-electron tomography. *Nat Methods* 2017;**14**(2017). https://doi.org/10.1038/nmeth.4115.
76. Renz P, Kokkinopoulou M, Landfester K, Lieberwirth I. Imaging of polymeric nanoparticles : hard challenge for soft objects. *Macromol Chem Phys* 2016;**217**:1879–85.
77. Hendrich K, Peng W, Vana P. Controlled arrangement of gold nanoparticles on planar surfaces via constrained dewetting of surface-grafted RAFT polymer. *Polymers (Basel)* 2020;**12**:1–12.
78. Sarma DD, Santra PK, Mukherjee S, Nag A. X-Ray photoelectron spectroscopy: a unique tool to determine the internal heterostructure of nanoparticles. *Chem Mater* 2013;**25**:1222–32. https://doi.org/10.1021/cm303567d.
79. Castro-Munoz R, La Iglesia ÓD, Fíla V, Téllez C, Coronas J. Pervaporation-assisted esterification reactions by means of mixed matrix membranes. *Ind Eng Chem Res* 2018;**57**:15998–6011. https://doi.org/10.1021/acs.iecr.8b01564.
80. Korkmaz S, Salt Y, Dincer S. Esterification of acetic acid and isobutanol in a pervaporation membrane reactor using different membranes. *Ind Eng Chem Res* 2011;**50**:11657–66. https://doi.org/10.1021/ie200086h.
81. Chen A, Ostrom C. Palladium-based nanomaterials: synthesis and electrochemical applications. *Chem Rev* 2015;**115**:11999–2044. https://doi.org/10.1021/acs.chemrev.5b00324.
82. López-Viveros M, Favier I, Gómez M, Lahitte JF, Remigy JC. Remarkable catalytic activity of polymeric membranes containing gel-trapped palladium nanoparticles for hydrogenation reactions. *Catal Today* 2021;**364**:263–9. https://doi.org/10.1016/j.cattod.2020.04.027.
83. Dang G, Shi Y, Fu Z, Yang W. Polymer nanoparticles with dendrimer-Ag shell and its application in catalysis. *Particuology* 2013;**11**:346–52. https://doi.org/10.1016/j.partic.2011.06.012.
84. Seo S, Perez GA, Tewari K, Comas X, Kim M. Catalytic activity of nickel nanoparticles stabilized by adsorbing polymers for enhanced carbon sequestration. *Sci Rep* 2018;**8**:1–11. https://doi.org/10.1038/s41598-018-29605-1.

CHAPTER 8

Semiconductor catalysts based on surface-modified nanomaterials (SMNs) for sensors

E. Kuna, P. Pieta, R. Nowakowski and I.S. Pieta
Institute of Physical Chemistry Polish Academy of Sciences, Warsaw, Poland

1. Introduction

Given the extensive body of sensor technologies, there are numerous classifications of sensor-based devices. The most straightforward categorization of sensors is based on the input and output energy (i.e., thermal, electrical, magnetic, radian, or chemical energy) that causes a physical stimulus and corresponding response of devices.[1] A large class of chemical sensors based on semiconductor materials can measure diverse parameters in gaseous and aqueous environments.[2] In such a system, chemical information may be caused by a chemical interaction with the target molecule/analyte or by a physical property (e.g., changes in absorbance, refractive index, conductivity, or temperature) of the system investigated.[2] In some cases, it is not possible to identify whether a sensor operates on chemical or physical bases, such as for a gas sensor, where the signal originates from the adsorption process.[2,3] The sensors are preferably designed to work upon precise conditions and for selected analytes, where the sensing/detection mechanism is mainly determined by the intrinsic properties of the materials that the sensor consists of.[3,4] Therefore, the performance of the semiconductor-based sensors is strongly affected by the morphology and structure of semiconductors.

Different physical, chemical, optical, and electrical properties of semiconductor materials result from their electronic band structure. The electronic band structure is characterized as a series of energetically closed spaced energy levels (called the valence band [VB]) and a series of more distant, energetically similar energy levels localizing at high energy (called the conduction band [CB]).[5] The behavior of semiconductor materials depends then on the energy gap between the VB and CB. The energy band-gap represents the minimum amount of energy required to promote an electron to the conduction band. The electron excitation is caused by a small applied potential difference or photonic energy results in charge carriers' formation.[5] Therefore, the band-gap, charge carrier mobility, and the band edge position affects the electrical, optical, and catalytic semiconductors' properties that govern sensors' responses. The catalytic activity of semiconductors is defined by the upper edge of the VB band, which correlates with the energy that determines the oxidizing ability, and the energy level at the bottom of the CB band, which is correlated with the energy of electrons for reduction processes (Fig. 8.1).[5,7] Knowing the electronic band position is possible to select a suitable semiconductor catalyst for a given chemical reaction involved in the sensing.

The semiconductor catalysts, especially metal oxides, are commonly applied in the gas sensors. Gas sensors can operate by measuring the gas concentrations as a function of the temperature based on the heat liberated during the chemical reaction or, in most cases, by the electrical conductivity changes caused by adsorption/reaction on the solid surface.[8,9] Therefore, the adsorption characteristic of semiconductor materials, like the adsorption selectivity, adsorption capacity, and kinetics of the adsorption processes, is crucial for their role in gas-sensing devices refers to the sensitivity, selectivity, and response/recovery time of sensors.[3,4,10] Significant efforts have been made to reduce a high cross-sensitivity to gasses and improve the low stability of semiconductor catalysts caused by stoichiometry changes and coalescence of the crystal nanostructure at a high operating temperature.[3,4] In this regard, the synthesis of semiconductor materials with better sensitivity, selectivity and long-term stability is required.[3,4,10]

The various synthetic methodologies allow controlling the crystal structure that determines the electronic band and the size and morphology of the semiconductor materials. The possibility of synthesizing semiconductor catalysts with well-defined composition, size, and shape provides different physical and chemical properties, especially when the materials' size falls into the nanoscale.[7,11] The nanostructures of semiconductor catalysts have a high surface-to-volume ratio. Hence, a large specific surface that increased adsorption capacity boosting catalytic activity, which is crucial for gas sensors' performance. Furthermore, surface catalytic properties can be improved by surface functionalization, i.e., grafting/coating nanoparticles with organic/inorganic materials and/or

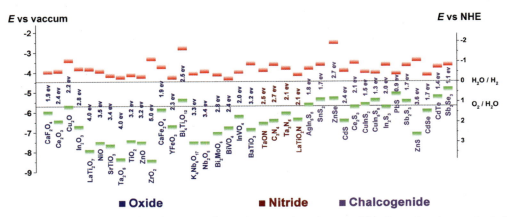

Figure 8.1 The electronic band position for various semiconductors. *(This figure has been adapted from Ref. 6 with the permission of John Wiley and Sons, copyright 2015.)*

doping with metal nanoparticles.[11–15] Surface modification may cause changes in the energy level of semiconductor material and then tailor its electronic (e.g., conductivity, redox properties) and optical (e.g., absorbance, fluorescence, luminescence, light scattering) features that are important for sensing and detection.[15,16]

We summarize the application of different semiconductor catalysts, such as metal oxides, metal chalcogenides, or ternary and quaternary semiconductors (i.e., metallates, oxysulfides, oxyhalides, and oxynitrides) in sensing. These nanomaterials' performance depends critically on their structures that require further development in preparation techniques to receive the desired structures and morphologies.[17] Therefore, the preparation methods and further surface functionalization of zero-dimensional (0D), one-dimensional (1D), two-dimensional (2D), and three-dimensional (3D) nanomaterials are discussed as a versatile tool to improve their materials characteristics for developing ideal sensor devices. Table 8.1 provides a brief overview of materials presenting 0D–3D dimensions and their potential applicative advantages and disadvantages.

2. Zero-dimensional (0D) nanomaterials

Nanoparticles (NPs) constitute the most common representation of nondimension (0D) semiconductor materials. These 0D nanomaterials consist of single- or multichemical elements and have an amorphous, crystalline, or polycrystalline nanostructure. NPs exhibit various shapes and forms like quantum dots, hollow spheres, core-shell nanospheres and nanocluster, and own excellent optical and electronic properties, which will be discussed below regarding their sensing application (Fig. 8.2).

2.1 Quantum dots

Quantum dots (QDs) are well-known nanoscale semiconductor crystals with the size between 2 and 10 nm. They are composed of the atom from groups II–VI of the periodic table, forming typically binary II–VI (e.g., CdSe, ZnS), III–V (e.g., InP, GaAs), or ternary I–III–VI (e.g., $CuInS_2$, AgInSe) semiconductors.[20] Depending on the composition, size, and shell thickness, QDs exhibit different tunable optical properties. In contrast to traditional materials applied for sensor development (such as organic dyes, transition-metal complexes, or carbon materials), QDs are considered as suitable candidates for sensing due to their long luminescence life-times, high quantum yields, reduced photobleaching phenomena, narrow emission bands, and broad absorption windows.[21,22]

The unique properties of QDs arise from their nanostructure and can be explained based on the quantum confinement effects. If the particle size is minimalized to the same order as the exciton radius (i.e., magnitude of the de Broglie wavelength), the atom's energy levels become discrete.[22–24] This phenomenon ultimately leads to different photophysical properties of the QDs compared to the bulk materials. The

Table 8.1 Summary of nanomaterials with various dimension (0D, 1D, 2D, and 3D) regarding potential utility in detecting.

Dimension	Typical structures	Advantages	Disadvantages
0D		Region- and stereo-selective functionalization Atomic control of structure property relationship Large area-to-volume ratio Single molecular electronic-based sensing Tunable size and shape	Low conductivity Difficulty with device integration Limited stability of devices Potential toxicity
1D		High surface-to-volume ratio High aspect ratio Excellent stability High density of reactive sites Good thermal stability Compatible with device miniaturization	Required chemical modification to enhance Selectivity Difficulty in establishing reliable electrical contacts Difficult purification Limited structure and precision control
2D		Wide tunability of conductivity Large surface-to-volume ratio Thickness depending electronic properties Good optical transparency Excellent mechanical flexibility Good functionalization ability Potential for good processibility Compatible with ultrathin silicon channel technology	Lack of mass production of material with large area, high and uniform quality. Lack of facile, effective, and reliable strategies for device integration Limited stability of dome forms at ambient conditions
3D		Good mechanical strength Good thermal stability Easy to interface with solid state device Good design ability to improve selectivity Strong analyte binding	Low surface area Difficulty with miniaturization Slow dynamic of analyte transport

Adapted from Ref. 18 with the permission of Royal Society of Chemistry, copyright 2009.

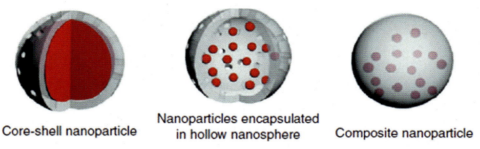

Figure 8.2 Exemplar zero-dimensional (0D) heterogeneous nanostructures. *(This figure has been adapted from Ref. 19 with the permission of Springer Nature, copyright 2016.)*

confinement degree influences the distribution of states' density, which is related to the number of possible electron–hole transitions at given photon energy. For QDs, a pileup of the available transitions' distribution guides to narrow and intense emission peaks desired in optoelectronic devices (Fig. 8.3).[22]

The synthesis and surface modification of QDs affect their size-dependent optical properties as well as physicochemical stability. Several synthetic approaches (including

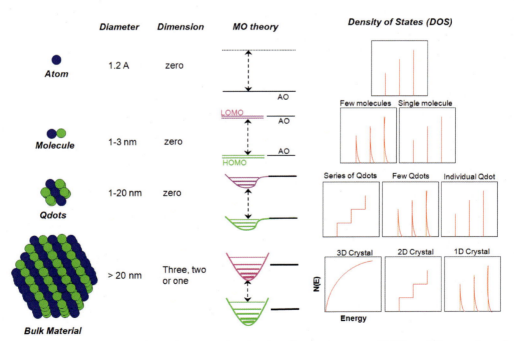

Figure 8.3 Schematic correlation between the density of quantum states (DOS) and the number of atoms in materials (where *AO*, atomic orbital; *HOMO*, highest occupied MO; *LUMO*, lowest unoccupied MO; *MO*, molecular orbital). *(This figure has been adapted from Ref. 24.)*

hot-injection, solvothermal, heating-up methods, or thermolysis) can be applied depending on the potential QDs' application.[20,24,25] The shape, size, and composition can be controlled by selecting suitable reaction conditions, i.e., reaction temperature, time, and the precursor ratio. Control of these parameters allows obtaining a monodispersed size of QDs.[20] In some cases, the hydrophobic nature of QDs excludes them from some application. Therefore, surface functionalization providing by the ligand exchanges or encapsulation, including coating with polymer or various functional groups as $-NH_2$, $-COOH$, $-SH$, and $-silica$ is required.[20,24] Such surface modification is crucial regarding QDs' stability, solubility, and the optical property, i.e., absorption, fluorescence, chemiluminescence, electrochemiluminescence, phosphorescence, or photoinduced electron transfer property.[26]

The sensing mechanism can be based on the aforementioned features and be applied for analyte detection in different sensor devices such as luminescence, electrochemical sensors, optical, electrochemical, and photoelectrochemical biosensors. In such devices, the transduction mechanism is based mainly on the fluorescence/photoluminescence quenching, fluorescence recovery, electrochemical enhancement/recovery, or photoelectrochemical enhancement.[26] Therefore, QDs can be successfully applied to detect any substance that caused changes in their optical properties, including metal ions, small molecules (e.g., amino acids), and biomolecules such as DNA, proteins, and saccharide (Fig. 8.4).[21]

Following the current sensing technology trends, special attention has been given to carbon quantum dots (CQDs) due to their low toxicity, stable photoluminescence, and stability.[27] Compared to the classical semiconductor QDs, CQDs can be synthesized without the need to use heavy metals, approaching top-down or bottom-up synthetic strategy.[21,25] CQDs' photoluminescence arises from π-conjugated domains and surface defects and can be easily modified by surface functionalization or heteroatom doping (Fig. 8.5).[25,27]

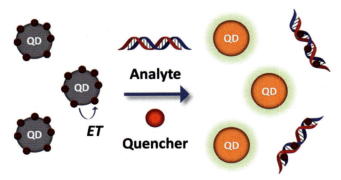

Figure 8.4 Detection of DNA using the photoluminescence turn off-on mechanism. *(This figure has been adapted from Ref. 21 with the permission of RSC Publishing, copyright 2011.)*

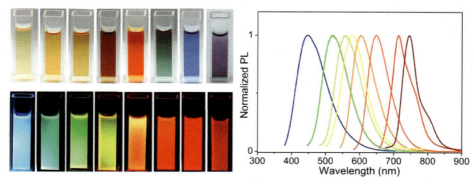

Figure 8.5 QDs in daylight (upper) and UV light (bottom), respectively, and normalized fluorescence emission spectra of all the above samples under excitation of 365 nm. *(This figure has been adapted from Ref. 28 with the permission of The Royal Society of Chemistry, copyright 2011.)*

2.2 Core-shell nanoparticles, hollow spheres, and nanocluster

Core-shell nanoparticles (CSNPs), hollow spheres, and nanoclusters is the next group of 0D nanomaterials with an outstanding sensing characteristic that originates from their unique structure.[29] Core-shell structured NPs consist of two or more materials, wherein one of them forms the central core, and the other develops a shell located around. The inner material of NPs can be covered partially or entirely by a single or multiplies shells' layers, providing different types of architecture, including hollow core-shell, core-multishell, and core-porous-shell NPs (see Fig. 8.6).[31,32] The core-shell materials' distinctive features are determined by the size, composition, and structural order, that is, the number of covering layers and the shell to core ratio. The selection of suitable core and shell materials allows modifying optical, electrical, catalytic properties and improving the thermal stability of CSNPs.[29,31,32]

A wide range of organic (i.e., polymer-based) and inorganic (i.e., silica-based, metal-based, or MOF-based) core-shell materials can be manufactured using top-down and bottom-up techniques. Different synthetic methodologies are applied to form the core and the shell separately.[30–32] In general, previously synthesized NPs' core is further modified by the coating with organic/inorganic components such as polymers, biomolecules, silica, metals, and nonmetals.[33,34] The functionalization procedure determines the final application of core-shell NPs. As in gas sensing application, where the coating of metal-core by the metal oxide-shell (e.g., Au@Cu_2O, Pd@SnO_2) leads to higher sensitivity, better selectivity, and reduced response/recovery time of metal-core@metal oxide-shell compared to pure metal oxide gas sensors.[30] In such a system, the change in the potential energy barrier at the core-shell interface can improve charge carriers' separation and transfer, boosting its sensing performance. Furthermore, the porous nature of the metal oxide shell facilitates gas diffusion to the metal-core and further catalytic

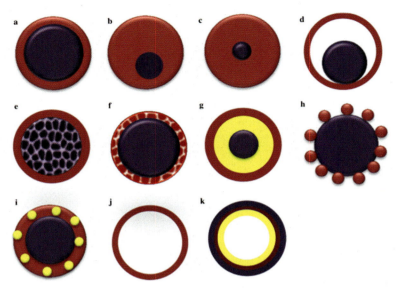

Figure 8.6 Different structures of 0D nanoparticles: (A) core@shell, (B) trapped core in shell, (C) encapsulated core-shell, (D) yolk-shell, (E) porous core@shell, (F) core porous shell, (G) core-multishell, (H) core-discontinuous shell, (I) core-composite shell, (J) hollow shell, and (K) multishell. (This figure has been adapted from Ref. 30 with the permission of Springer Nature, copyright 2015.)

reaction on its surface.[30,31] Finally, increased long-term stability of core-shell based sensor is caused by the presence of a metal oxide shell that works as a shield and protects the core from oxidation and poisoning.[30,31] Apart from the gas sensor, CSNPs found an application in electrochemical sensors as promising electrode modifiers, which promote faster electron transfer between the surfaces' electrode and analytes (e.g., Pd@Pt/multiwalled carbon nanotubes electrode).[31] Moreover, the core-shell NPs, such as AgPt@Ag NPs, were applied to enhance the inelastically scattered Raman signal in optical sensors based on surface-enhanced Raman spectroscopy (SERS) or shell-isolated nanoparticle-enhanced Raman spectroscopy (SHINERS).[31,35] The noble metal CSNPs effectively collect the light due to the resonant oscillation of free surface electrons upon the light excitation, boosting the signal.[35,36] Furthermore, these plasmonic nanostructures can be isolated by the catalytically inactive dielectric shell, preventing the sensors from chemical and thermal degradation.[31]

Besides promising CSNPs, the hollow colloidal spheres appear as a new chemical sensor model.[37] Hollow spheres are a type of unique functional materials possessing an extensive fragment of empty spaces inside intact shells and well-defined morphology.[38] Due to their specific structure, the hollow spheres have a larger surface area and a lower density than their solid analogues.[37,38] Three main strategies have been applied for their synthesis, including hard-templating, soft-templating, and self-templating.[38] The hollow spheres' final shape and size depend mainly on the template size applied for the fabrication. Whereases the shell thickness is primarily governed by the coating process of the desired material on the templates. The hollow spheres are formed after template detaching.[38–40] Such obtained nanostructures provide a sizable surface-to-volume ratio that boosts their adsorption properties. Therefore, the hollow spheres have better adsorptivity than conventional sensing materials based on tightly packed nanoparticles.[37,41] The increased active surface area between sensor material and analyte/gas ensures more active sites and facilitates gas diffusion and mass transport. Furthermore, the possibility of enriching this porous surface area with a noble metal, or metal oxides improves the sensing properties due to the formation of conjugated electron depletion layers, increasing catalytic activity.[37,40,42] Therefore, the hollow spheres have been found as promising materials for gas sensors, e.g., volatile organic compounds (VOCs) and heavy metal ion (HMI) sensors.[40] The gas sensor based on Au–SnO_2,[37,43] Ni–SnO_2,[44] WO_3–C[42] etc., exhibit an enhance the detection limit and selectivity toward a specific target gas such as, e.g., acetone, n-butanol, ethanol, or chloroform.

The last group of 0D nanomaterials consists of nanoclusters (NCs), mainly metallic NCs. NCs are composed of a few to up to 100 atoms, with size between 1 and 10 nm and narrow size distribution.[45] Most nanoclusters form a structure with a size less than 2 nm, which is comparable to the de Broglie wavelength of electrons near the Fermi energy. The dynamic of these electrons gives rise to discrete electronic transitions and ensures outstanding molecular properties.[46] The influence of the quantum confinement effect (described in Section 2.1 for QDs) drives to excellent photoluminescence, electrochemiluminescence, electromagnetism, redox behavior, or molecular chirality.[46,47] Different optical properties can be obtained for the nanoclusters with a size larger than 3 nm, in which, unlike to their smaller counterparts, the surface plasmon resonance (SPR) occurs. Therefore, NCs' physicochemical properties are size-dependent and can be controlled by the selected synthesis method, including top-down syntheses, which are mainly involved in the synthesis of small nanoclusters ≤2 or bottom-up syntheses that are mainly involved in synthesis larger nanoclusters ≥3 nm.[46,48] The high metal-binding affinity of nanoclusters toward various organic groups (including sulfhydryl, hydroxyl, carboxyl, amines, nucleotides stabilizing groups and thiol, tryptophan, and tyrosine reducing groups), allows to tune their electronic structure further and then optical response.[49] Ligand incorporation strategy may boost fluorescence properties (e.g., by increasing the life-time, the quantum yield of nanoclusters) and provides good biocompatibility.[50] Therefore, metallic nanoparticles are commonly applied in chemo-/and bio-sensing, e.g., for detection heavy metal ions,[46,48,51] small organic molecules,[46,49,52] proteins/peptides,[49,53] or DNA/RNA.[49,54] In such a system, the sensing mechanism is based mainly on the fluorescence quenching/recovery, observing differences in its intensity and emission wavelength in the presence of analytes. The dynamic (if caused by collisional quenching) or static (if caused by the complex formation) molecular interaction can cause the observed fluorescence changes. In static quenching, the complex formed between NCs and quencher is nonfluorescent, as can occurs due to ligand exchange reaction on NCs surface.[55] In opposite, the increase in the fluorescence can be observed due to complex formation between NCs and analytes, e.g., for Au NCs in the presence of cyanide anions (CN⁻).[56] Considering the electrical and catalytic properties of metal/metal oxides NCs, they can be successfully applied in gas sensors. According to the literature, they are several tin,[57] zinc,[58] copper,[59] palladium[60] oxides conductometric-based sensors for hydrogen, oxygen, nitrogen oxide or carbon dioxide detection, that exhibit better performances than different nanoclusters regarding their sensitivity and stability.[61] As in the previous cases (i.e., CSs, HSs NPs), the excellent gas sensing properties are assigned to their catalytic properties, including excellent adsorption affinity, selectivity, and stability, arising from their composition, size, and morphology.[61]

3. One-dimensional (1D) nanomaterials

One-dimensional (1D) nanostructures refer to materials like nanorods, nanowire, nanotubes, and nanobelts/nanoribbons, with at least one dimension between 1 and 100 nm where the length is larger than the width (Fig. 8.7).[63,64] Like 0D nanomaterials, 1D nanostructures can form amorphous, single, or polycrystalline structures. Their dimensionality limits the electron movement in two directions.[65] Therefore, the 1D nanomaterials have unique electrical and optical properties that differ from their respective bulk counterparts.[62,65] For the same reason, 1D nanomaterials have found numerous potential applications, such as either functional materials in electronics, optoelectronics, catalysis, or building blocks for composite materials.[62,65] In particular, the 1D semiconducting nanostructures play a crucial role in chemical and biological sensing applications.[62,65]

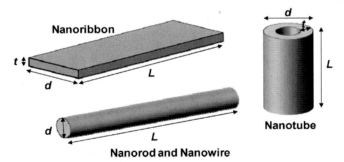

Figure 8.7 Scheme of one-dimensional (1D) nanostructures: nanorod, nanowire, nanoribbon, and nanotube. *(This figure has been adapted from Ref. 62 with the permission of American Chemical Society, copyright 2011.)*

3.1 Synthesis of 1D nanostructures and sensor fabrication

The suitable large-scale fabrication method of 1D nanostructures should provide a uniform control regarding the shape and size of a complex structure, tiny dimension, good spatial alignment over a large area, even pattern density, and high-throughput assembly.[66] There are several potential methodologies to synthesize one (1D) dimensional nanostructures with controlled shape and size, including template-based and nontemplate-based methods. The conventional approach based on the top-down strategy, usually involves different lithography techniques.[62,66] Although those techniques can lead to the synthesis of nanostructures with uniform shapes and suitable spatial alignment, they are limited due to the etching process involved in the fabrication. Etching may cause roughness on sidewalls and/or on the interface, deteriorating device performance.[65–67] Nowadays, the top-down strategy is more often replaced by the bottom-up strategy. In the bottom-up strategy are involved many techniques, including the solution-phase growth methods (i.e., solid-liquid-solid method, solvothermal and hydrothermal growth methods); the vapor-phase growth method (i.e., a vapor-liquid-solid method such as pulsed laser deposition [PLD], chemical vapor deposition [CVD], metal-organic chemical vapor deposition [MOCVD], and physical or vapor deposition [PVD]); the electrochemical deposition method, and many other well-reviewed by several authors.[66,67] The bottom-up strategy allows to overcome physical limits and fabricate nanostructures with good material homogeneity, high crystallinity, and a smooth surface faceting.[66,67] Nevertheless, the selection of suitable synthesis methods depending on the final sensor application.

The reproducible detection system requires a well-controlled performance of the sensing material and the corresponding sensor device. Although, different principles are used for the fabrication of 1D NMs-based sensor, most of them are designed either as conductive sensor devices (monitoring the conductance of a sensor material) or as field-effect transistors (FETs, in which the transverse electric field governs the sensing property) (Fig. 8.8).[67]

3.2 1D NMs-based sensors

1D semiconductor nanomaterials are promising candidates for chemical and biological applications due to two main reasons. First, they possess a great surface-to-volume ratio providing a chemically active surface area. Second, the NPs' radius corresponds to the Debye screening length, leading to the dominant role of surface electrostatics on the carrier conduction through the entire structure.[68] In such a system, the semiconductor catalyst dimension is minimalized to the range of the mean free path of electrons. The transport of electrons becomes unimpeded and causes a significant change in resistance affected by the electronic coupling of adsorbed species.[67] The conduction change providing by the molecular adsorption on the nanomaterial surface constitutes the basis

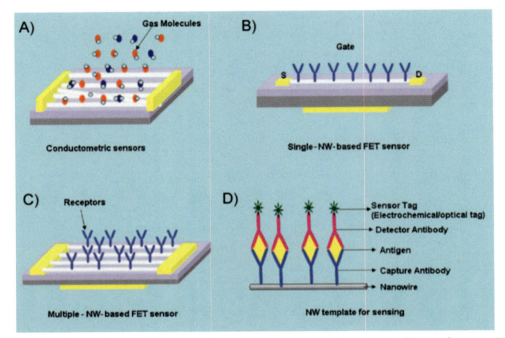

Figure 8.8 Schematic representation of sensor principles based on NWs: (A) conductometric, (B) single-NW-based field-effect transistor, (C) multiple-NW-based field-effect transistor, and (D) NW as template material. *(This figure has been adapted from Ref. 67 with the permission of John Wiley and Sons, copyright 2010.)*

for molecular detection by 1D NMs. Therefore, the 1D nanoparticles are commonly applied in conductometric sensors, e.g., for gas detection.[67] Most of the 1D nanomaterials used for gas sensing are semiconductor metals or metal oxides, such as Pd,[69] ZnO,[70] WO$_3$,[71] SnO$_2$, or TiO$_2$,[72] due to their low-cost, high stability, sensitivity, reliability, and controlled molecular adsorption that can be achieved by material doping or by external fields.[73] Although most of them operate as conductive sensor devices, they can be found in the field-effect transistor configuration.[67,73] FETs devices can work under room temperature compared to resistors sensors, whereas the current modulation by the extra gate electrode increases their sensitivity (Fig. 8.9).[73]

One-dimensional nanomaterials can be applied as template material in biosensing. One-dimensional NMs can form a sizable active site for molecular interaction resulting in specific chemical signatures. Such a configuration enables creating a detection platform for desired markers (e.g., proteins, biomolecules, or enzymes).[67] The controlled immobilization allows for their sensing even at low concentrations. To date, there are several examples of working sensor devices that can be applied for the simultaneous detection of different molecules/proteins.[67] Therefore, 1D NMs were also used as the template for

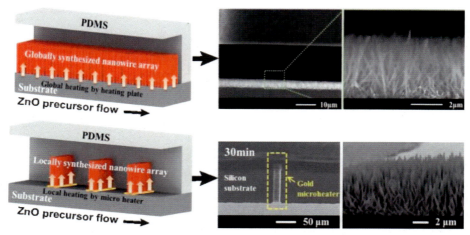

Figure 8.9 The scheme represents a microfluidic-based in situ synthesis of ZnO nanowires as an innovative approach for fabrication sensor devices. *(This figure has been adapted from Ref. 74 with the permission of MDPI.)*

bio-optical sensors by attaching an optical marker.[67] Most of 1D NMs optical sensors are based on the fluorescence detection system, in which the adsorbed species cause the quenching of sensing material. In such a system, a realization of solid contact on sensing NMs is not required in the fabrication process, which is a significant advantage compared to their conductometric counterparts. Besides the metal/metal oxides, 1D carbon NTs can be successfully applied in optical sensors based on their modulated near-IR emission. The detection system measures either changes in the fluorescence or the charge transfer.[67] These electronic properties have triggered various applications of the carbon NTs as microelectrodes in electrochemical sensors. Electrode surface modification with well-aligned 1D NMs results in a high sensitivity of biosensors, e.g., detecting cholesterol.[67,75] The combination of microfluidics and nanomaterials appears as a promising configuration for biosensing applications where high sensitivity and fast response are required. The lab-on-chip system allows reaching a fast detection system using portable, low-power, electronic biosensor chips for medical applications.[67,75]

4. Two-dimensional (2D) nanomaterials

Two-dimensional (2D) nanomaterials usually refer to materials such as graphene, layered double hydroxides (LDH) and many other layered van der Waals solids, like MoS_2, $CaGe_2$, and $CaSi_2$, which present outstanding properties ranging from insulators, semiconductors, metals to superconductors.[18,76–80] Based on chemical composition, this class of layered materials can be divided into two groups, as presented in Fig. 8.10.

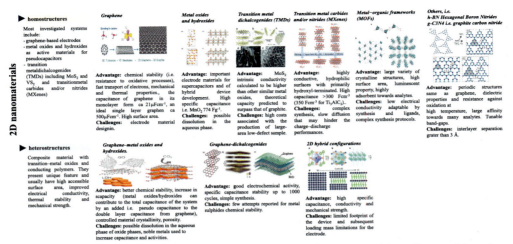

Figure 8.10 The 2D layered materials classified based on their composition. *(Prepared based on Ref. 18,76—81.)*

Two-dimensional nanostructures can be divided into two categories: homo- and heterostructures (Fig. 8.10) according to their chemical composition. Within the first category, usually, following materials group are listed (1) graphene-based metal oxides and hydroxides; (2) transition-metal dichalcogenides (TMDs) including MoS_2 and VS_2; (3) transition-metal carbides, nitrides, and/or carbonitrides (MXenes, with multilayer [ML] or few-layered structure [FL]); (4) metal-organic frameworks (MOFs); and (5) other 2D materials, such as hexagonal boron nitride (h-BN known as "white graphene") and carbon nitrides (especially graphitic carbon nitride g-C_3N_4).[18,77,81]

The composites with the transition-metal oxides and conducting polymers presenting a structure made from one or more homojunction are classified as homostructure composites. They are considered to provide higher energy density than carbon materials. It has been shown that combining such high-energy metal oxides or conductive polymers with graphene-based electrode materials can lead to a high electrochemically accessible area. This combination leads to improvements in electrical conductivity, thermal stability, and mechanical strength of the electrode, while minimizing the restacking of graphene layers.[18,77]

Graphene-metal oxides, hydroxides, dichalcogenides, and other graphene-based composite exhibited larger specific surface area and pore volume than AC-based composites (surface area: 2156 vs. 1547 m2g1, pore size: 2.6 vs. 2.2 nm).

Two-dimensional materials present a few atomic layers' thicknesses with the different two dimensions beyond the nanometric size range.[18,76—78] Among them, graphene is the most investigated 2D material in physics, chemistry, material sciences, biotechnology, and many others compared to other 2D nanomaterials as a conductive material (Fig. 8.11).

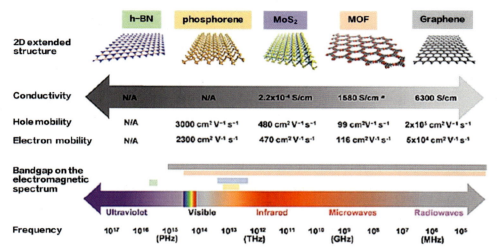

Figure 8.11 Selected properties of different layered material. *(This figure has been adapted from Ref. 18 with the permission of Royal Society of Chemistry, copyright 2009.)*

Among the 2D material, metallic semiconducting as well as isolating properties are observed. Moreover, this type of material presents unique physicochemical properties, i.e., crystalline structure, surface morphology, stability, mechanical strength, electric conductivity, and sorption, which are directly reflected in their electrical, optical, catalytical, and sensing properties. They present a vast possibility of application ranging from electrochemical devices, batteries, optical devices, optoelectronic devices, sensors, up to medical, biotechnological and environmental applications.[18,76–81]

Compared to 0D, 1D, and 3D structures, the 2D materials have the well-developed surface area and high surface-to-volume ratio, which provides a more active species over the surface area for the 2D material-analyte interaction or increased adsorption capacity. This property is unique 2D materials and makes them more suitable for sensor applications, including sensors with low detection limits.

The 2D nanomaterial offers a wide possibility of their modification. Particularly, band-gap and conductivity can be adapted by a tailor-tuning in the selected 2D nanomaterial configuration. Moreover, introducing 0–3D heterostructures into the given 2D structure or composition change or producing defects within the structure, by doping/functionalization or by changing the number of layers would also result in tunable properties change of final composite material (Fig. 8.11).[18]

Most 2D materials possess little or no electrocatalytic activity, making them less attractive for catalytic applications. However, the tuning of selected properties of 2D materials, i.e., the morphological, magnetic, catalytic, and so forth improve their electrical properties and makes them more efficient toward signal transduction due to the molecular binding event.[18,82] Due to their tunable electronic structures, large specific surface areas, and scalable production capabilities, increasing attention is noted toward the 2D abundant material exploration for various sectors, electrochemical, thermochemical, and biomedical. Both the surface chemistry and the bulk properties of 2D nanomaterials can be modified by adopting multiple approaches, i.e., heteroatom doping, defect formation and engineering, strain engineering, ion intercalation, and/or interfacial interaction. Such modification makes them promising for large-scale chemical, environmental, and medical applications.

Advanced technologies and fabrication techniques recently allowed for significant improvement in high-precision physicochemical properties tuning and surface/defects engineering toward expected functionality. This also includes a detailed *in situ/operando* study of 2D materials and a deep understanding of composition-shape-structure-functionality function under no isolated conditions.[81]

A great interest in 2D materials utilization for electro devices and sensors is recently observed and connected with the improvement of 2D materials synthesis and the development of new technologies for NPs and thin surfaces engineering. Compared to other 0–3D materials, 2D materials have better compatibility with metal electrodes because of their large lateral sizes.[18] In contrast to 0, 1, and 3D materials issues related to devices' integration, electrical contact, and device miniaturization can be overpassed. It was shown that 2D materials fulfill the compatibility requirements with thin-film synthesis techniques. Moreover, they have extraordinary compatibility with thin/ultrathin silicon channel technologies, presenting at the same moment many advantages comparing to 0D, 1D, and 3D materials, i.e., mechanical and/or chemical strength, optical transparency. Therefore, this kind of material is of great importance as most suitable for compact electrochemical devices, sensors, including many application fields.[81]

Nikolaou et al. showed an effective immobilization of hybrid organic-inorganic semiconductors (HOIS) based on TiO_2 film, loaded with perovskites.[83] Such electrode was successfully used as a selective electrochemical sensor for CBr_4, with the detectability range of 20 ppb mol/mol. Although many one-dimensional materials (i.e., carbon nanotubes and silicon nanowires) are highly investigated for optical, chemical, electrical, pore, and biological sensing, the most promising devices are based on graphene, graphene-derivatives or functionalized graphene.[84] Among electrochemical sensors, reduced graphene oxide (RGO) take particular attention because (1) it can efficiently transport charges (in contrary to nonconductive graphene oxide (GO); (2) its electrical and chemical properties are highly tunable; (3) it presents structure reach in both defects and chemical groups, which facilitate charge transfer and ensure high electrochemical activity; and (4) chemical moieties on the RGO surface offer the convenience and flexibility for various functionalization to enhance the sensor performance. Apart RGO more recently, h-BN and carbon nitrides receive increasing attention. H-BN shows periodic structures the same as graphene with a different stack ordering, while $g-C_3N_4$ shows layering structure the same as graphene with the carbon and nitrogen atoms present in sp^2 hybridization.[84–86] Diverse structural models for the geometry of $g-C_3N_4$ are explained as triazine and heptazine. Although theoretical calculations predict g-C$_3$N$_4$ consisting of heptazine rings to be the most thermodynamically stable (Fig. 8.12), its formation has not been reported so far.

Depending on the degree of condensation, graphitic carbon nitride resembles to some extent N substituted graphite framework with a very high level of nitrogen doping, consisting of π-conjugated graphitic planes made up from sp^2-hybridized carbon and nitrogen atoms.[85] The surface behavior of g-C$_3$N$_4$ has a broad affinity toward many analytes, which arises from the terminal hydrogen and nitrogen. Moreover, the surface of graphitic carbon nitride can be modified by alkalinization or protonation. It can also be doped with boron, sulfur or metal particles (e.g., Fe, Cu, Ni) to enhance electrical conductivity and photo- and electrocatalytic activity. g-C$_3$N$_4$ is characterized by a bad-gap of 2.5–2.8 eV, implying low electrical conductivity. However, the band-gap can be turned and tailored (i.e., by homogeneously dispersed carbon and nitrogen atoms), leading to better thermal stability and enhanced performance for applications in various detection technologies.

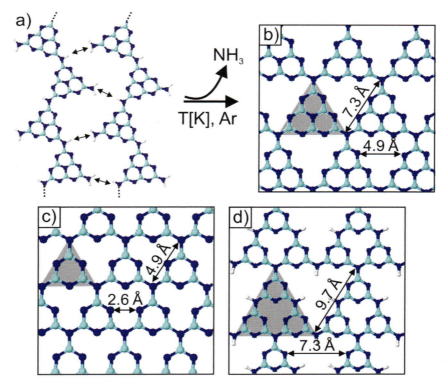

Figure 8.12 Optimized structural elements of graphitic carbon nitride: (A) zig-zag chains of heptazine (tri-s-triazine) units linked by bridging —NH— groups and N—H groups on their edges, (B) condensed heptazine rings, (C) condensed triazine rings, and (D) polytriazine imide. *(With permission Ref. 85 of American Chemical Society, copyright 2020.)*

Due to the low costs, nontoxic properties, and green synthesis protocols, many modified carbon nitrides sensing devices have been reported in the last 10 years, including metal ions, biomolecules, DNA, and immunoglobulins sensing.[87–89] A metal-free graphitic carbon nitride coated onto a glassy carbon electrode has been reported for the herbicide in water and soil samples sensing via the electrochemical technique.[90] This simple method proposed allowed for the trace level concentration detection, revealing biological, agricultural, and environmental relevance of the g-C_3N_4 based materials. Moreover, the excellent fluorescence-quenching phenomenon of g-C_3N_4 nanomaterial brought new insights to utilize them as fluorescent probes for heavy metal ion detection, i.e., Hg^{2+}, Cu^{2+}, Cr^{2+}, Pd^{2+}, Cd^{2+}, among others and iodide ions in aqueous media via "ON-OFF-ON" fluorescence response.[91,92]

5. Tree-dimensional (3D) nanomaterials

Three-dimensional (3D) nanostructures are typically materials as mesoporous carbon and graphene aerogel, which have three dimensions greater than the nanometric size range but still preserve the advantages of the nano-size effect.[80] They can also be obtained by stacking layered structures in the z-direction.[18] The behaviors of 3D nanostructures strongly depend on their chemical composition, sizes, shapes, dimensionality, and morphologies.[81] Consequently, these physicochemical properties are the key factors that condition their ultimate performance, durability, and applications. It is of great importance to obtain 3D nanomaterials with a controlled structure and morphology because 3D nanostructures are widely applied in catalysis, magnetic material science, electrochemistry, and optoelectronic.[93]

3D porous structures provide a well-developed surface area, enhanced electrolyte access, and improved mechanical stability in electrochemistry, i.e., for efficient supercapacitor electrodes. This porous structure can provide an enhanced number of accessible sites for reactants/molecules adsorption, but on the other hand, it would lead to low electrical conductivity. In electrochemical devices, 3D porous structures can be fabricated using metal foam as templates or organizing active materials into 3D nanostructures.[93–96] Metal foams such as nickel foam provide highly porous and conductive substrates for active materials deposition. This would include 1D dimensional material, i.e., carbon nanotubes as well as graphene, conjugated polymers, and metal oxides to build desired electrodes. Compared to, i.e., carbon, such substrate provides overall higher electrical conductivity and better mechanical properties. However, both porosity and specific surface area are low.

In electrochemical sensing, mesoporous carbon (MC) is considered as a promising material because of its large specific surface area and interconnected pores with tunable size.[93] It was shown that the ordered 3D porous structures with controlled pore size and distribution would improve the electrochemical properties. The material with a pore size distribution from range 0.6–3 nm was reported to result in the best performance, depending on different electrolytes.

3D nanostructures, based on pseudocapacitive materials, which offer various types of metal oxide nanostructures, are considered a third class of 3D electrodes. The active material can be produced through multiple techniques, i.e., sol-gel, solvothermal, and other methodologies, including also green chemistry principles. They can also be obtained by stacking 2D materials as well as applying nanotransfer printing (nTP).[95]

Recently mesoporous c-mpg- C_3N_4 was reported as an all-in-one chemosensor to detect trace amounts of metal ions in aqueous solutions.[96] The c-mpg- C_3N_4 was prepared by a nano casting approach using the ordered mesoporous silica KIT-6 as a rigid template. This material was highly selective and sensitive to Cu^{2+} without interference by other metal ions, presenting a fast response (time of 5 min for the 90% of equilibrium PL response). Many 3D hybrid systems were investigated and shown to have increased (1) sorption properties (3D hierarchical RGO foams decorated by various inorganic nanostructures, i.e., ZnO_{nws}/RGO foam/PDMS);[97] (2) activity, stability, and sensitivity toward the electrochemical detection and photodegradation of antibiotic drug (3D flower-like neodymium molybdate $Nd_2Mo_3O_9$),[98] or voltaic organic compounds (ordered Pd–WO_3 doped mesoporous g-CN);[99] and (3) electronics and optoelectronics properties (3D high-density arrays of lead halide perovskite nanowires).[94] These results suggested that 3D nanostructures are a suitable candidate for applications in physics, chemistry, medical (DNA sensing and drug delivery), and environmental science for detecting and degradation of pollutants.

6. Conclusion

The promising physical and chemical properties of 0–3D materials have encouraged many studies on developing highly selective sensors utilizing surface defects, large specific surface area, tunable material modification, and surface functionalization. Many researchers underly the advantage of using 2D materials as they present unique properties. Such materials are studied for various applications, including environmental monitoring, medical (DNS sensing), drug delivery, real-life healthcare, sorbents, and electro and optoelectronic device elements.

Despite these efforts, many of them are still far behind commercialization and will require substantial improvement regarding cost, easier synthesis, greener synthesis protocols, less waste formation, and improved stability and selectivity. In particular, targeting real-life-application in liquid media causes difficulties in developing suitable sensor structures for mobile applications, and further studies are required. Although some semiconductor catalysts based on SNMs have been applied in real-time detection (e.g., metal oxides in gas sensors), there is still room for improvement. Therefore, the development

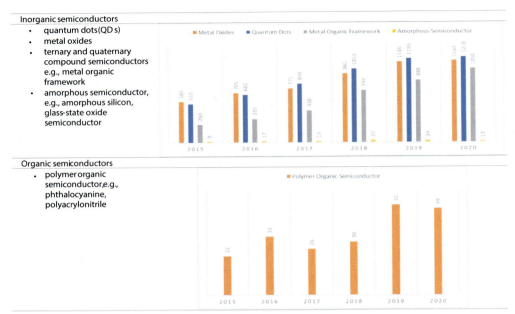

Figure 8.13 The latest developments on the semiconductor catalysts based on surface-modified nanomaterials (SNMs) for sensors during the last five years. The publication search was performed using the Web of Science [v.5.35], Web of Science Core Collection search; keywords: quantum dots; metal oxides; metal-organic framework; amorphous semiconductor; polymer organic semiconductor for sensing.

of semiconductor catalysts for real-time monitoring has to meet the following expectations: (1) semiconductor catalyst has to be stable upon reaction condition; (2) they have to interact only with target molecule/analyte; (3) they should exhibit fast response (regarding changes in physiochemical properties) in the presence of the analyte; and (4) they should show fast recovery times and reproducibility (Fig. 8.13).

Acknowledgments

The present research was financially supported by the European Union's Horizon 2020 research and innovation program under the Marie Skłodowska-Curie co-financed by the Ministry of Science and Higher Education (H2020-MSCA-COFUND-2018 grant agreement No. 847413.).

I.S.P. and P.P. extends their sincere appreciation to NAWA The Polish National Agency for Academic Exchange through Bekker grants PPN/BEK/2019/1/00348 "C1-C4 alkanes to oxygenated fuel electrochemical transformation" and PPN/BEK/2019/1/00345 "Nanostructured carbon-based materials doped with metal nanoparticles as catalytic electrode materials for CO2 electroreduction with the use of surface-plasmon enhancement.

List of resources

LAMTEC GmbH & Co.KG: https://www.lamtec.de/
RIKEN KEIKI Co., Ltd.: https://www.rikenkeiki.com/tech_info/sensor
Figaro Engineering Inc.: https://www.figaro.co.jp/en/
Evikon MCI Ltd: https://www.evikon.eu/en/semiconductor-sensors-c-1224
Interuniversity Microelectronics Center (IMEC): https://www.imec-int.com/en/semiconductor-technology-and-systems/materials-and-components-analysis
High-tech spin-off from the University of Technology Dresden (TUD): https://senorics.com/
Nanoscale Sensors and Materials Research Group at Dundalk Institute of Technology: https://www.dkit.i.e.,/research/research-centres-and-groups/nanoscale-sensors-and-materials-research-group.html
https://www.bisinfotech.com/top-10-sensor-technology-manufacturers-for-2020/

References

[1] Nicholas G. *Expanding the vision of sensor materials*. National Research Council Washington, DC: The National Academies Press.; 1995. p. 1–146. https://doi.org/10.17226/4782.
[2] Hulanicki A, Glab S, Ingman F. Chemical sensors: definitions and classification. *Pure Appl Chem* 1991;**63**(9):1247–50. https://doi.org/10.1351/pac199163091247.
[3] Viter R, Iatsunskyi I. Metal oxide nanostructures in sensing. In: *Micro and nano technologies, nanomaterials design for sensing applications*. Elsevier; 2019. p. 41–91. https://doi.org/10.1016/B978-0-12-814505-0.00002-3.
[4] Zhuiykov S. Electronic devices and functional structures based on nanostructured semiconductors. In: *Nanostructured semiconductor oxides for the next generation of electronics and functional devices*. Woodhead Publishing; 2014. p. 95–138. https://doi.org/10.1533/9781782422242.95.
[5] Hernández-Ramírez A, Medina-Ramírez I. *Photocatalytic semiconductors synthesis, characterization, and environmental applications*. Switzerland: Springer International Publishing; 2015. p. 1–298. https://doi.org/10.1007/978-3-319-10999-2.
[6] Lu Q, Yu Y, Ma Q, Chen B, Zhang H. Composites for photocatalytic and electrocatalytic hydrogen evolution reactions. *Adv Mater* 2016;**28**:1917–33. https://doi.org/10.1002/adma.201503270.
[7] Lia J, Wu N. Semiconductor-based photocatalysts and photoelectron-chemical cells for solar fuel generation: a review. *Catal Sci Technol* 2015;**5**:1360–84. https://doi.org/10.1039/C4CY00974F.
[8] Gentry SJ, Jones TA. The role of catalysis in solid-state gas sensor devices. *Sens Actuators* 1986;**10**:141–63. https://doi.org/10.1016/0250-6874(86)80039-7.
[9] Morrison SR. Selectivity in semiconductor gas sensors. *Sens Actuators* 1987;**12**(4):425–40. https://doi.org/10.1016/0250-6874(87)80061-6.
[10] Korotcenkov G. Gas response control through structural and chemical modification of metal oxide films: state of the art and approaches. *Sensor Actuator B Chem* 2005;**107**:209–32. https://doi.org/10.1016/j.snb.2004.10.006.
[11] Vengatesan MR, Mittal V. Surface modification of nanomaterials for application in polymer nanocomposites: an overview. In: *Surface modification of nanoparticle and natural fiber fillers*; 2015. p. 1–27. https://doi.org/10.1002/9783527670260.ch1.
[12] Yin Y, Talapin D. The chemistry of functional nanomaterials. *Chem Soc Rev* 2013;**42**:2484–7. https://doi.org/10.1039/C3CS90011H.
[13] Yan F, Jiang Y, Sun X, Bai Z, Zhang Y, Zhou X. Surface modification and chemical functionalization of carbon dots: a review. *Microchim Acta* 2018;**185**:1–34. https://doi.org/10.1007/s00604-018-2953-9. 424.

[14] Sarkar D, Xie X, Kang J, Zhang H, Liu W, Navarrete J, Moskovits M, Banerjee K. Functionalisation of transition metal dichalcogenides with metallic nanoparticles: implications for doping and gas-sensing. *Nano Lett* 2015;**15**(5)):2852−62. https://doi.org/10.1021/nl504454u.

[15] Medhi R, Marquez MD, Lee TR. Visible-light-active doped metal oxide nanoparticles: review of their synthesis, properties, and applications. *ACS Appl Nano Mater* 2020;**3**(7):6156−85. https://doi.org/10.1021/acsanm.0c01035.

[16] Vilan A, Cahen D. Chemical modification of semiconductor surfaces for molecular electronic. *Chem Rev* 2017;**117**(5):4624−66. https://doi.org/10.1021/acs.chemrev.6b00746.

[17] Zeng W, Wang H, Li Z. Nanomaterials for sensing applications. *J Nanotechnol* 2016:1−2. https://doi.org/10.1155/2016/2083948.

[18] Tyagi D, Wang H, Huang W, Hu L, Tang Y, Guo Z, Ouyang Z, Zhang H. Recent advances in two-dimensional-material-based sensing technology toward health and environmental monitoring applications. *Nanoscale* 2020;**12**:3535−59. https://doi.org/10.1039/C9NR10178K.

[19] Lukatskaya M, Dunn B, Gogotsi Y. Multidimensional materials and device architectures for future hybrid energy storage. *Nat Commun* 2016;**7**:1−13. https://doi.org/10.1038/ncomms12647.

[20] Tsolekile N, Parani S, Matoetoe MC, Songca SP, Oluwafemi OS. Evolution of ternary I-III-VI QDs: synthesis, characterization, and application. *Nano-Struct Nano-Objects* 2017;**12**:46−56. https://doi.org/10.1016/j.nanoso.2017.08.012.

[21] Cui L, He XP, Chen GR. Recent progress in quantum dot-based sensors. *RSC Adv* 2015;**5**:26644−54. https://doi.org/10.1039/c5ra01950h.

[22] Goicoechea J, Arregui FJ, Matias IR. Quantum dots for sensing. In: *Sensors based on nanostructured materials*. Boston: Springer; 2019. p. 131−81. https://doi.org/10.1007/978-0-387-77753-5_6.

[23] Matea CT, Mocan T, Tabaran F, Pop T, Mosteanu O, Puia C, Iancu C, Mocan L. Quantum dots in imaging, drug delivery and sensor applications. *Int J Nanomed* 2017;**12**:5421−31. https://doi.org/10.2147/ijn.s138624.

[24] Bera D, Qian L, Tseng TK, Holloway PH. Quantum dots and their multimodal applications: a review. *Materials* 2010;**3**:2260−345. https://doi.org/10.3390/ma3042260.

[25] Valappil MO, Pillai VK, Alwarappan S. Spotlighting graphene quantum dots and beyond: synthesis, properties and sensing applications. *Appl Mater Today* 2017;**9**:350−71. https://doi.org/10.1016/j.apmt.2017.09.002.

[26] Li M, Chen T, Gooding JJ, Liu J. Review of carbon and graphene quantum dots for sensing. *ACS Sens* 2009;**4**(7):1732−48. https://doi.org/10.1021/acssensors.9b00514.

[27] Molaei MJ. A review on nanostructured carbon quantum dots and their applications in biotechnology, sensors, and chemiluminescence. *Talanta* 2019;**196**:456−78. https://doi.org/10.1016/j.talanta.2018.12.042.

[28] Ding H, Wei JS, Zhang P, Zhou ZY, Gao QY, Xiong HM. Nanomaterials-based sensors for applications in environmental monitoring. *Small* 2018;**14**:1800612−22. https://doi.org/10.1002/smll.201800612.

[29] Khatami M, Alijani HQ, Nejad MS, Varma RS. Core@shell nanoparticles: greener synthesis using natural plant products. *Appl Sci* 2018;**8**:411−28. https://doi.org/10.3390/app8030411.

[30] Mirzaei A, Janghorban K, Hashemi B, Neri G. Metal-core@metal oxide-shell nanomaterials for gas-sensing applications: a review. *J Nanopart Res* 2015;**17**:1−17. https://doi.org/10.1007/s11051-015-3164-5. 371.

[31] Kalambate PK, Dhanjai, Huang Z, Li Y, Shen Y, Xie M, Huang Y, Srivastava AK. Core@shell nanomaterials-based sensing devices: a review. *Trends Anal Chem* 2019;**115**:147−61. https://doi.org/10.1016/j.trac.2019.04.002.

[32] Schartl W. Current directions in core−shell nanoparticle design. *Nanoscale* 2010;**2**:829−43. https://doi.org/10.1039/c0nr00028k.

[33] Karnati P, Akbar S, Morris PA. Conduction mechanisms in one dimensional core-shell nanostructures for gas sensing: a review. *Sens Actuator B Chem* 2019;**295**:127−43. https://doi.org/10.1016/j.snb.2019.05.049.

[34] Niu M, Pham-Huy C, He H. Core-shell nanoparticles coated with molecularly imprinted polymers: a review. *Microchim Acta* 2016;**183**:2677−95. https://doi.org/10.1007/s00604-016-1930-4.

[35] Hartman T, Geitenbeek RG, Wondergem CS, D Stam WV, Weckhuysen BM. Operando nanoscale sensors in catalysis: all eyes on catalyst particles. *ACS Nano* 2020;**14**:3725—35. https://doi.org/10.1021/acsnano.9b09834.

[36] Li JF, Huang YF, Ding Y, Yang ZL, Li SB, Zhou XS, Fan FR, Zhang W, Zhou ZY, Wu DY. Shell-isolated nanoparticle-enhanced Raman spectroscopy. *Nature* 2010;**464**:392—5. https://doi.org/10.1038/nature08907.

[37] Zhang J, Liu X, Wu S, Xu M, Guo X, Wang S. Au nanoparticle-decorated porous SnO_2 hollow spheres: a new model for a chemical sensor. *J Mater Chem* 2010;**20**:6453—9. https://doi.org/10.1039/c0jm00457j.

[38] Zhong K, Song K, Clays K. Hollow spheres: crucial building blocks for novel nanostructures and nano photonics. *Nanophotonics* 2018;**7**(4):693—713. https://doi.org/10.1515/nanoph-2017-0109.

[39] Hu J, Chen M, Fang X, Wu L. Fabrication and application of inorganic hollow spheres. *Chem Soc Rev* 2011;**40**:5472—91. https://doi.org/10.1039/C1CS15103G.

[40] Li S, Pasc A, Fierro V, Celzard A. Hollow carbon spheres, synthesis and applications - a review. *J Mater Chem A* 2016;**4**:12686—713. https://doi.org/10.1039/C6TA03802F.

[41] Zhang Y, He Z, Wang H, Qi L, Liu G, Zhang X. Front. Applications of hollow nanomaterials in environmental remediation and monitoring: a review. *Environ Sci Eng* 2015;**9**(5):770—83. https://doi.org/10.1007/s11783-015-0811-0.

[42] Shen JY, Zhang L, Ren J, Wang JC, Yao HC, Li ZJ. Highly enhanced acetone sensing performance of porous C-doped WO_3 hollow spheres by carbon spheres as templates. *Sens Actuator B Chem* 2017;**239**:597—607. https://doi.org/10.1016/j.snb.2016.08.069.

[43] Zhang J, Wang S, Xu M, Wang Y, Xia H, Zhang S, Guo X, Wu S. Polypyrrole-coated SnO_2 hollow spheres and their application for ammonia sensor. *J Phys Chem C* 2009;**113**:1662—5. https://doi.org/10.1021/jp8096633.

[44] Liu X, Zhang J, Guo X, Wu S, Wang S. Enhanced sensor response of Ni-doped SnO_2 hollow spheres. *Sens Actuator B Chem* 2011;**152**:162—7. https://doi.org/10.1016/j.snb.2010.12.001.

[45] Goswami N, Li J, Xie J. Functionalisation and application. In: *Frontiers of nanoscience*, vol. 9. Elsevier; 2015. p. 297—345. https://doi.org/10.1016/B978-0-08-100086-1.00011-7.

[46] He Z, Shu T, Su L, Zhang X. Strategies of luminescent gold nanoclusters for chemo-/bio-sensing. *Molecules* 2019;**24**:3045. https://doi.org/10.3390/molecules24173045.

[47] Chakraborty I, Pradeep T. Atomically precise clusters of noble metals: emerging link between atoms. *Nanoparticles Chem Rev* 2017;**117**:8208—71. https://doi.org/10.1021/acs.chemrev.6b00769.

[48] Chen LY, Wang CW, Yuan Z, Chang HT. Fluorescent gold nanoclusters: recent advances in sensing and imaging. *Anal Chem* 2015;**87**(1):216—29. https://doi.org/10.1021/ac503636j.

[49] Wang B, Zhao M, Mehdi M, Wang G, Gao P, Zhang KQ. Biomolecule-assisted synthesis, and functionality of metal nanoclusters for biological sensing: a review. *Mater Chem Front* 2019;**3**:1722—34. https://doi.org/10.1039/c9qm00165d.

[50] Kang X, Zhu M. Tailoring the photoluminescence of atomically precise nanoclusters. *Chem Soc Rev* 2019;**48**:2422—57. https://doi.org/10.1039/c8cs00800k.

[51] Halawa MI, Lai J, Xu G. Gold nanoclusters: synthetic strategies and recent advances in fluorescent sensing. *Mater Today Nano* 2018;**3**:9—27. https://doi.org/10.1016/j.mtnano.2018.11.001.

[52] Yuan X, Tay Y, Dou X, Luo Z, Leong DT, Xie J. Glutathione-protected silver nanoclusters as cysteine-selective fluorometric and colorimetric probe. *Anal Chem* 2013;**85**:1913—9. https://doi.org/10.1021/ac3033678.

[53] Qing Z, He X, Qing T, Wang K, Shi H, He D, Zou Z, Yan L, Xu F, Ye X. Poly(Thymine)-Templated fluorescent copper nanoparticles for ultrasensitive label-free nuclease assay and its inhibitors screening. *Anal Chem* 2013;**85**:12138—43. https://doi.org/10.1021/ac403354c.

[54] Petty JT, Zheng J, Hud NV, Dickson RM. DNA-templated Ag nanocluster formation. *J Am Chem Soc* 2004;**126**:5207—12. https://doi.org/10.1021/ja031931o.

[55] Zhang Y, Li M, Niu QQ, Gao PF, Zhang GM, Dong C, Shuang SM. Gold nanoclusters as fluorescent sensors for selective and sensitive hydrogen sulfide detection. *Talanta* 2017;**171**:143—51. https://doi.org/10.3390/molecules24173045.

[56] Zong C, Zheng LR, He W, Ren X, Jiang C, Lu L. Fluorescent gold nanoclusters: recent advances in sensing and imaging. *Anal Chem* 2014;**86**:1687−92. https://doi.org/10.1021/ac503636j.
[57] Shukla S, Agrawal R, Cho HJ, Seal S. Effect of ultraviolet radiation exposure on room-temperature hydrogen sensitivity of nanocrystalline doped tin oxide sensor incorporated into microelectromechanical systems device. *J Appl Phys* 2005;**97**:1−14. https://doi.org/10.1063/1.1851597.
[58] Zhong J, Muthukumar S, Chenet Y, Chen Y, Lu Y. Ga-doped ZnO single crystal nanotips grown on fused silica by metal-organic chemical vapor deposition. *Appl Phys Lett* 2003;**83**(16):3401−5. https://doi.org/10.1063/1.1621729.
[59] Ayesh AI, Alyafei A, Anjum RS, Mohamed RM, Abuharb MB, Salah B, El-Muraikhi M. Production of sensitive gas sensors using CuO/SnO_2 nanoparticles. *Appl Phys A* 2019;**125**:550−8. https://doi.org/10.1007/s00339-019-2856-6.
[60] Ayesh AI, Mahmoud ST, Ahmad SJ, Haik Y. Novel hydrogen gas sensor based on Pd and SnO_2 nanoclusters. *Mater Lett* 2014;**128**:354−7. https://doi.org/10.1016/j.matlet.2014.04.173.
[61] Ayesh AI. Metal/metal-oxide nanoclusters for gas sensor applications. *J Nanomater* 2016:1−5. https://doi.org/10.1155/2016/2359019.
[62] Kim FS, Ren G, Jenekhe SA. One-dimensional nanostructures of π-conjugated molecular systems: assembly, properties, and applications from photovoltaics. *Sens Nanophoton Nanoelect Chem Mater* 2011;**23**:682−732. https://doi.org/10.1021/cm102772x.
[63] Bashir S, Liu J. Nanomaterials and their application. In: *Advanced nanomaterials and their applications in renewable energy*. Elsevier; 2015. p. 1−50. https://doi.org/10.1016/B978-0-12-801528-5.00001-4.
[64] Gupta VK, Alharbie NS, Agarwal S, Grachev VA. New emerging one dimensional nanostructure materials for gas sensing application: a mini review. *Curr Org Chem* 2019;**15**(2):131−5. https://doi.org/10.2174/1573411014666180319151407.
[65] Weber J, Singhal R, Zekri S, Kumar A. One dimensional nanostructures: fabrication, characterization and applications. *Int Mater Rev* 2018;**53**(4):235−55. https://doi.org/10.1179/174328008X348183.
[66] Liao L, Duan X. More recent advances in one-dimensional metal oxide nanostructures: optical and optoelectronic applications. In: *One-dimensional nanostructures*. John Wiley & Sons, Inc; 2013. p. 359−78. https://doi.org/10.1002/9781118310342.
[67] Ramgir NS, Yang Y, Zacharias M. Nanowire-based sensors. *Small* 2010;**6**:1705−22. https://doi.org/10.1002/smll.201000972.
[68] Fan Z, Ho JC, Takahashi T, Yerushalmi R, Takei K, Ford AC, Chueh YL, Javey A. Toward the development of printable nanowire electronics and sensors. *Adv Mater* 2009;**21**:3730−43. https://doi.org/10.1002/adma.200900860.
[69] Favier F, Walter EC, Zach MP, Benter T, Penner RM. Hydrogen sensors and switches from electro-deposited palladium mesowire arrays. *Science* 2001;**293**:2227−31. https://doi.org/10.1126/science.1063189.
[70] Li QH, Liang YX, Wan Q, Wang TH. Oxygen sensing characteristics of individual ZnO nanowire transistors. *Appl Phys Lett* 2004;**85**:6389−93. https://doi.org/10.1063/1.1840116.
[71] Polleux J, Gurlo A, Barsan N, Weimar U, Antonietti M, Niederberger M. Template-free synthesis and assembly of single-crystalline tungsten oxide nanowires and their gas-sensing properties. *Angew Chem Int Ed* 2006;**45**:261−6. https://doi.org/10.1002/anie.200502823.
[72] Li Z, Wu M, Liu T, Wu C, Jiao Z, Zhao B. Preparation of TiO_2 nanowire gas nanosensor by AFM anode oxidation. *Ultramicroscopy* 2008;**108**:1334−43. https://doi.org/10.1016/j.ultramic.2008.04.059.
[73] Zhao X, Cai B, Tang Q, Tong Y, Liu Y. One-dimensional nanostructure field-effect sensors for gas detection. *Sensors* 2014;**14**:13999−4020. https://doi.org/10.3390/s140813999.
[74] Xing Y, Dittrich PS. One-dimensional nanostructures: microfluidic-based synthesis, alignment and integration towards functional sensing devices. *Sensors* 2018;**18**:1−21. https://doi.org/10.3390/s18010134.
[75] Aravamudhan S, Kumar A, Mohapatra S, Bhansali S. Sensitive estimation of total cholesterol in blood using Au nanowires based micro-fluidic platform. *Biosens Bioelectron* 2007;**22**:2289−94. https://doi.org/10.1016/j.bios.2006.11.027.
[76] Geim AK, Grigorieva IG. Van der Waals heterostructures. *Nature* 2013;**499**:419−25. https://doi.org/10.1038/nature12385.

[77] W Lee C, Suh JM, Jang HW. Chemical sensors based on two-dimensional (2D) materials for selective detection of ions and molecules in liquid. *Front Chem* 2019;**7**:708−29. https://doi.org/10.3389/fchem.2019.00708.

[78] Li T, Miras HN, Song YF. Polyoxometalate (POM)-Layered double hydroxides (LDH) composite materials: design and catalytic applications. *Catalysts* 2017;**7**:260−77. https://doi.org/10.3390/catal7090260.

[79] Tan T, Jiang XT, Wang C, Yao BC, Zhang H. 2D material optoelectronics for information functional device applications: status and challenges. *Adv Sci* 2020;**7**:2000058−83. https://doi.org/10.1002/advs.202000058.

[80] Yu Z, Tetard L, Zhai L, Thomas J. Supercapacitor electrode materials: nanostructures from 0 to 3 dimensions. *Energy Environ Sci* 2015;**8**:702−30. https://doi.org/10.1039/C4EE03229B.

[81] Kuna E, Mrdenovic D, Jönsson-Niedziółka M, Pieta P, Pieta IS. Bimetallic nanocatalysts supported on graphitic carbon nitride for sustainable energy development: the shape-structure-activity relation. *Nanoscale Adv* 2021:1−10. https://doi.org/10.1039/d0na01063d.

[82] Yu Q, Luo Y, Mahmood A, Liu B, Cheng HM. Engineering two-dimensional materials and their heterostructures as high-performance electrocatalysts. *Electrochem Energ Rev* 2019;**2**:373−94. https://doi.org/10.1007/s41918-019-00045-3.

[83] Nikolaou P, Vassilakopoulou A, Papadatos D, Topoglidis E, Koutselas I. A chemical sensor for CBr4 based on quasi-2D and 3D hybrid organic-inorganic perovskites immobilized on TiO_2 films. *Mater Chem Front* 2018;**2**:730−40. https://doi.org/10.1039/C7QM00550D.

[84] Liu Y, Donga X, Chen P. Biological and chemical sensors based on graphene materials. *Chem Soc Rev* 2012;**41**:2283−307. https://doi.org/10.1039/C1CS15270J.

[85] Lewalska-Graczyk A, Pieta P, Garbarino G, Busca G, Hołdynski M, Kalisz G, Sroka-Bartnicka A, Nowakowski R, Naushad M, Gawande MB, Zbořil R, Pieta IS. Graphitic carbon nitride-nickel catalyst: from material characterization to efficient ethanol electrooxidation. *ACS Sustain Chem Eng* 2020;**8**:7244−55. https://doi.org/10.1021/acssuschemeng.0c02267.

[86] Pieta IS, Rathi A, Pieta P, Nowakowski R, Hołdynski M, Pisarek M, Kaminska A, Gawande MB, Zboril R. Electrocatalytic methanol oxidation over Cu, Ni and bimetallic Cu-Ni nanoparticles supported on graphitic carbon nitride. *Appl Catal B Environ* 2019;**244**:272−83. https://doi.org/10.1016/j.apcatb.2018.10.072.

[87] Dong Y, Wang Q, Wu H, Chen Y, Lu CH, Chi Y, Yang HH. Graphitic carbon nitride materials: sensing, imaging and therapy. *Small* 2016;**12**:5376−93. https://doi.org/10.1002/smll.201602056.

[88] Chen L, Song J. Tailored graphitic carbon nitride nanostructures: synthesis, modification, and sensing applications. *Adv Funct Mater* 2017;**27**:1702695−710. https://doi.org/10.1002/adfm.201702695.

[89] Wang A, Wang C, Fu L, Wong-Ng W, Lan Y. Recent advances of graphitic carbon nitride-based structures and applications in catalyst, sensing, imaging, and LEDs. *Nano-Micro Lett* 2017;**9**(47):1−21. https://doi.org/10.1007/s40820-017-0148-2.

[90] Shetti NP, Malode SJ, Vernekar PR, Nayak DS, Shetty NS, Reddy KR, Shukla SS, Aminabhavi TM. Electro-sensing base for herbicide aclonifen at graphitic carbon nitride modified carbon electrode − water and soil sample analysis. *Microchem J* 2019;**149**:103976−84. https://doi.org/10.1016/j.microc.2019.103976.

[91] Ahmad R, Tripathy N, Khosla A, Khan M, Mishra P, Ansari WA, Syed MA, Hahn YB. Review-recent advances in nanostructured graphitic carbon nitride as a sensing material for heavy metal ions. *Electrochem Soc* 2019;**167**:037519−39. https://doi.org/10.1149/2.0192003JES.

[92] Barman S, Sadhukhan M. Facile bulk production of highly blue fluorescent graphitic carbon nitride quantum dots and their application as highly selective and sensitive sensors for the detection of mercuric and iodide ions in aqueous media. *J Mater Chem* 2012;**22**:21832−7. https://doi.org/10.1039/C2JM35501A.

[93] Tiwari JN, Tiwari RN, Kim KS. Zero-dimensional, one-dimensional, two-dimensional and three-dimensional nanostructured materials for advanced electrochemical energy devices. *Prog Mater Sci* 2012;**57**(4):724−803. https://doi.org/10.1016/j.pmatsci.2011.08.003.

[94] Gu L, Tavakoli MM, Zhang D, Zhang Q, Waleed A, Xiao Y, Tsui KH, Lin Y, Liao L, Wang J, Fan Z. 3D arrays of 1024-pixel image sensors based on lead halide perovskite nanowires. *Adv Mater* 2016;**28**: 9713−21. https://doi.org/10.1002/adma.201601603.

[95] Zaumseil JM, Meitl MA, Hsu JWP, Acharya BR, Baldwin KW, Loo YL, Rogers JA. Three-dimensional and multilayer nanostructures formed by nanotransfer printing. *Nano Lett* 2003;**3**(9): 1223−7. https://doi.org/10.1021/nl0344007.

[96] Lee EZ, Jun YS, Hong WH, Thomas A, Jin MM. Cubic mesoporous graphitic carbon(IV) nitride: an all-in-one chemosensor for selective optical sensing of metal ions. *Angew Chem Int Ed* 2010;**122**(50): 9900−4. https://doi.org/10.1002/ange.201004975.

[97] Song C, Yin X, Han M, Li X, Hou Z, Zhang L, Cheng L. Three-dimensional reduced graphene oxide foam modified with ZnO nanowires for enhanced microwave absorption properties. *Carbon* 2017; **116**:50−8. https://doi.org/10.1016/j.carbon.2017.01.077.

[98] Kumar JV, Karthik R, Chen SM, Chen KH, Sakthinathan S, Muthuraj V, Chiu TW. Design of novel 3D flower-like neodymium molybdate: an efficient and challenging catalyst for sensing and destroying pulmonary toxicity antibiotic drug nitrofurantoin. *Chem Eng J* 2018;**346**:11−23. https://doi.org/10.1016/j.cej.2018.03.183.

[99] Malik R, Tomer VK, Dankwort T, Mishra YK, Kienle L. Cubic mesoporous Pd-WO3 loaded graphitic carbon nitride (g-CN) nanohybrids: highly sensitive and temperature dependent VOC sensors. *J Mater Chem A* 2018;**6**:10718−30. https://doi.org/10.1039/C8TA02702A.

CHAPTER 9

Surface-modified carbonaceous nanomaterials for CO$_2$ hydrogenation and fixation

Hushan Chand, Priyanka Choudhary and Venkata Krishnan
School of Basic Sciences and Advanced Materials Research Center, Indian Institute of Technology Mandi, Kamand, Himachal Pradesh, India

1. Introduction

The consistent utilization of fossil-fuels resulted in the excessive emissions of CO$_2$ in Earth's atmosphere. CO$_2$ being a greenhouse gas can give rise to serious problems like global warming and climate change.[1] To overcome these issues CO$_2$ storage, capture, and utilization seem to be a tangible process toward sustainable development.[2] According to a United Nations Environmental Program (UNEP) current report, it has been found that if no strict global measures were not taken against climate change, the Earth's temperatures may rises by 2°C by 2050, and 4°C by 2100 to the existing temperature. To reduce the excessive amount of environmental CO$_2$, various strategies have been adopted. Recently three strategies include (1) control in CO$_2$ emission[3]; (2) CO$_2$ capture and storage[3]; and (3) CO$_2$ conversion to produce value-added chemicals and fuels[4] have been extensively studied. The emission of CO$_2$ in the atmosphere can be reduced by increasing the efficiency of fossil fuels or by replacing fossil fuels carrying a large amount of carbon content (coal) with other fossil fuels carrying less carbon content (oil or natural gas). However, the large reduction in CO$_2$ emissions could only be ascertained by switching off the persistent utilization of fossil fuels and by adopting nonfossil fuels like hydrogen, solar energy, wind energy, hydro energy tidal energy, and so forth. CO$_2$ utilization and storage involves capture of CO$_2$ from industries and other energy-related resources, then compression and insertion to a secure geological formation to ascertain prolonged storage in order to attenuate the threat of global warming and to ultimately get rid of the alarming climate change.[5] However, CO$_2$ storage can result in leakage issues and also CO$_2$ storage could not be ultimate long term solution for CO$_2$ mitigation. Since CO$_2$ contains a carbon atom which could be a very good source of carbon, it provides an opportunity for CO$_2$ to synthesize carbon-containing useful products and feedstocks. Hence utilization of renewable carbon resources will be more useful to provide a long-term solution toward unceasing progress of our community, wherein CO$_2$ could be utilized to synthesize a large class of products as shown in Fig. 9.1.

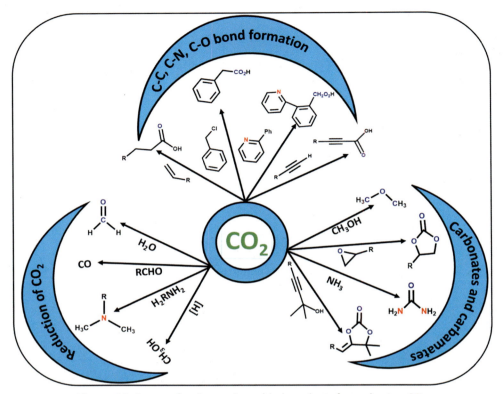

Figure 9.1 Scope of various value-added products formed using CO_2.

Fig. 9.1 describes that CO_2 can be inserted via C—N, C—C, and C—O bond evolution or it can be reduced chemically, photochemically, and electrochemically, or chemically fixed to other substrates forming a plethora of valuable products. As CO_2 is a stable molecule so enabling it in a chemical reaction is a thermodynamically uphill task.[6] Therefore CO_2 accompanying reactions require an input of energy, and its coupling with renewable energy polishes this technique to more encouraging in terms of sustainability and environmentally benign. This input of energy could be minimized if the reaction is enabled in the presence of suitable catalysts. Generally, two types of catalysts were employed in CO_2 conversions, namely heterogeneous and homogeneous catalysts, but we restrict ourselves to heterogeneous catalysts in this chapter, because of their own merits over the homogeneous catalysts. In heterogeneous catalysis, we particularly discuss the use of surface-modified carbonaceous nanomaterials in CO_2 conversion to useful products. The CO_2 conversion reactions can be done via photocatalysis,[7] electrocatalysis,[8] and thermocatalysis,[9] whereas thermocatalysis is the first choice among these as a consequence of its fast kinetics and more flexible combination of active components.

Due to the low efficiency and poor selectivity issues with photocatalysis and electrocatalysis, these are not often the preferred choice of methods. In addition, due to the several advantages pertaining to the use of various nanocarbon materials as catalysts, we restrict our discussion to only them.

1.1 Surface modified carbonaceous nanomaterials

Carbon-based materials like carbon black, carbon nanotubes, fullerenes, activated carbon, graphite, graphitic carbon nitride and graphene, and so forth are well-known carbonaceous nanomaterials used in environmental remediation and energy storage applications.[10] Shan et al.[11] in their review article described the significance of carbon nanotubes, fullerenes, and graphene-based nanomaterials to remove the pollutants from wastewater. They have discussed the simple routes for the synthesis and properties of these exceptional carbonaceous nanomaterials are summarized. They have elucidated that the strong adsorption capacity of these carbonaceous materials is mainly due to the abundant pore size distribution, incredibly high specific surface area, and achievable surface properties, which allow them to interact strongly with organic contaminants, via hydrophobic effects, hydrogen bonding interactions, and electron donor-acceptor interactions. The carbon nanotubes aggregate in the aqueous phase generating a substantial number of interstitial spaces and grooves, which consequently provide high-energy adsorption sites and affording the removal of organic contaminants. For graphene oxide (GO), various oxygen-containing functionalities like $-COOH$, $C=O$, and $-OH$ on the edge would incapacitate their hydrophobic effect in conjunction with nonpolar organics, and reduces their adsorption on its surface. The carbonaceous nanomaterials discussed above are not that effective in the removal of contaminants from wastewater. The agglomeration of these carbonaceous nanomaterials in aqueous solutions is another limitation in the decontamination of water. Agglomerated nanomaterials in aqueous solutions decrease the surface area as well as active sites, which commences decay in efficiency of pollutants removal. To increase their effectiveness, some modifications are required which may be surface modifications or combining these materials with other materials. The various functional groups containing moieties and metal oxides have been used for nanomaterials advancement to amplify the efficiency, selectivity and affinity toward removal of specific pollutants. Besides, the loopholes which narrow their full-scale applications which is needed to be discussed.

Surface-modified carbonaceous nanomaterials have been progressively employed in environmental remediation and energy storage applications.[12] Yang et al.[12a] described the water decontamination abilities of different surface-modified carbonaceous nanomaterials. They summarize various synthesis methods which have been adapted to insert heteroatoms like, nitrogen, oxygen and sulfur, upon the carbonaceous nanomaterials to improve the sorptive behavior and exterior functionalities of heavy metals in aqueous

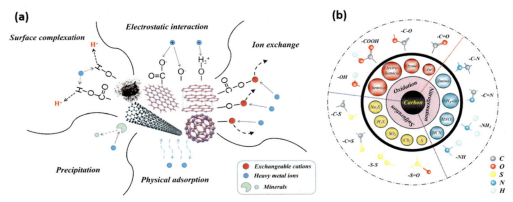

Figure 9.2 (A) Adsorption mechanism of heavy metals on surface-modified carbonaceous nanomaterials, (B) surface moderation techniques to functionalize carbonaceous nanomaterials with various functional groups. *(Reproduced with permission from Ref. 12a.)*

solutions. The study of different functionalities on carbonaceous nanomaterials has been discussed on the adsorption behaviors of heavy metals. Fig. 9.2A shown below represents the mechanism for the adsorption behavior of these heavy metal ions on the carbonaceous nanomaterials. The mechanisms for the adsorption of these heavy metals onto carbonaceous materials involves physical adsorption, surface complexation, electrostatic interaction, ion exchange and precipitation. The particular role of each and every heavy metal ion differ significantly, which depends on the ionic domain of the aq. solution, target metal ion and onto the carbon adsorbent. The compounds having hard Lewis basic functionalities like, deprotonated phenols and carboxylic acids, adsorb through cation exchange whereas, those containing soft Lewis basic functionalities like carbonyl and aromatic structural frameworks adsorb by dipole-dipole interactions.

The surface modification of carbonaceous nanomaterials involves posttreatment with chemical compounds in order to regulate some specific properties of the final products. Therefore, the methods of moderation, kind of raw carbonaceous nanomaterial and the characteristics of these functional groups influence the outcome of a final product prepared. Generally, nitrogenation, oxidation, and sulfuration are mostly used moderation tools to insert nitrogen, oxygen and sulfur on carbonaceous nanomaterial for the creation of these functionalities. Fig. 9.2B is the graphical representation of these moderation techniques, representing the chemical regents employed and generated the desired functional groups on carbonaceous nanomaterials. In a more generalized way, the surface-modified carbonaceous nanomaterials are more encouraging materials for diverse environmental remediation applications, particularly in the removal of contaminants from wastewater.

In this chapter, the use of the surface-modified carbon-based heterogeneous catalysts in CO_2 hydrogenation and fixation is discussed in detail. Under hydrogenation, CO_2

conversion into hydrocarbons, alcohols and other useful products have been discussed. Similarly, under fixation, conversion of CO_2 to CC, cyclic carbamates, urea and other value-added products have been discussed. Finally, some future insights on CO_2 conversion have been discussed.

2. Basic concepts of CO_2 sequestration (hydrogenation and fixation)

The hydrogenation of CO_2 with H_2 or the fixation of CO_2 to other useful chemicals and fuels are two major sources of CO_2 sequestration. The CO_2 hydrogenation[13] with H_2 produced from renewable energy sources[14] brings great enthusiasm for the research community to generate chemicals and fossil fuels,[15] that not only lowers the CO_2 discharge but also simultaneously overcomes the shortage of fuels. The outcome of products upon CO_2 hydrogenation depends on working conditions and reaction systems used. The most common products are dimethyl ether, methanol, methane, higher alcohols and higher hydrocarbons. Two challenges that can pose some difficulties in CO_2 hydrogenations are (1) availability of H_2 source (because pure H_2 gas is not naturally abundant) and (2) isolation of the synthesized products. The first challenge can be overcome by H_2 production through water splitting reactions using photoelectrochemical, photocatalytic, or other chemical processes and water electrolysis using electricity generated by solar or wind energy resources. Whereas the second challenge pertaining to isolation of the synthesized products is still not fully resolved and requires some advancements in this regard.

The CO_2 fixation is another mode for CO_2 consumption. CO_2 has a wide scope in terms of fixation to other substrates. CO_2 could be inserted into epoxides forming CC, which subsequently is a very useful chemical for various purposes.[16] CC were employed as uncooked materials for synthesizing plastics, as polar solvents, in lithium-ion batteries (as electrolytes) and also used in petrochemicals.[9] CO_2 could be inserted into ammonia or amines forming urea or substituted urea which are extensively used as fertilizers.[17] The lucrative transformation of CO_2 into valuable chemicals and fuels through its fixation using some suitable catalysts is of predominant interest in reducing the atmospheric CO_2 concentration. The CO_2 could also be inserted into cyclic aziridines, alcohols, alkynes, and alkenes to form their respective products as shown in Fig. 9.1.

3. Heterogeneous catalyst in CO_2 hydrogenation

The CO_2 hydrogenation into value-added chemicals come up with an alternate root for CO_2 utilization. The thermal stability of CO_2 toward hydrogenation makes it a challenging task. To solve this issue scientists have developed various catalysts, which can effectively hydrogenate CO_2 to hydrocarbons, lower olefins, alcohols, dimethyl carbonates, and other value-added products.

3.1 CO₂ hydrogenation to hydrocarbons

CO$_2$ hydrogenation to hydrocarbons is an effective way of utilizing CO$_2$ for energy storage. These hydrocarbons can be further used as renewable fuels for day-to-day human consumption. It has been observed that nitrogen doped or functionalized materials can effectively catalyze CO$_2$ hydrogenation. In this regard, Wu et al. reported graphene quantum dots doped with nitrogen (NGQDs) for hydrogenation of CO$_2$ to methane.[18] The nitrogen sites present at the edges play a significant role in the thermocatalytic activity of NGQDs. The configuration and defect density of the nitrogen dopants is responsible for the reaction selectivity. The higher N density promotes CO$_2$ conversion selectively to CH$_4$. The increased concentration of pyridinic N sites at the edges leads to lessening the reaction temperature for CO$_2$ hydrogenation and increased selectivity toward CH$_4$. Fig. 9.3A represents N1s XPS spectra of NGQDs with divergent proportions of N species by varying the N sources and Fig. 9.3B shows the different N content obtained from XPS. The pyridinic nitrogen proportions is higher in DMF solution.

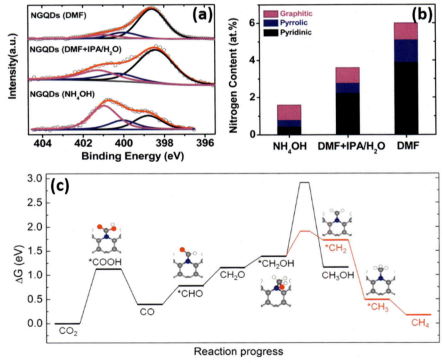

Figure 9.3 (A) XPS spectra of N 1s in three NGQDs catalysts doped by varying N precursor namely NH$_4$OH, DMF diluted by IPA/H$_2$O, and DMF (B) XPS determined specific N content and (C) Gibbs free-energy profiles diagram for CH$_3$OH, CO and CH$_4$ synthesis from CO$_2$ hydrogenation at 683 K and 10 atm. *(Reproduced with permission from Ref. 18.)*

On dilution of DMF solution with 1:1 (IPA/H$_2$O), the total N proportion declines but pyridinic N remains still as the prominent N configuration, while in NH$_4$OH the N proportions further decrease. The trend of hydrogenation of CO$_2$ and CH$_4$ selectivity against pyridinic N proportions shows a linear trend, indicating that the pyridinic N are the prominently active N sites for CO$_2$ conversion. The selectivity toward CH$_4$ over CH$_3$OH can be explained by Fig. 9.3C which represents the different steps involved in the hydrogenation of CO$_2$. The Lewis basic N sites absorb acidic CO$_2$. The DFT calculations represent that hydrogenation of CO into CH$_4$ and CH$_3$OH through intermediate *CHO is thermodynamically more desired intermediate. The synthesis CH$_3$OH and CH$_4$ is defined by hydrogenation of *CH$_2$OH on C or O sites. It was analyzed that the energy barrier for the conversion of *CH$_2$OH to CH$_3$OH is 1.76 eV, which is much higher as compared to 0.65 eV for *CH$_2$OH conversion to *CH$_2$. Therefore the production of CH$_3$OH is unfavorable and the reaction proceeds with CH$_4$ formation with high selectivity.

In another report, Jurca et al. reported N-doped graphene (N-dG) derived from the pyrolysis of chitosan at high temperatures as a defect rich catalyst for CO$_2$ hydrogenation to CH$_4$.[19] The catalytic performance of graphene is remarkably increased by N-doping, which can be attributed to the reason that pyridinic N atoms provide basic sites for CO$_2$, and which was further confirmed by adsorption studies. The role of N-doping and the nature of the active sites were examined by titration of basic and acidic and sites by CO$_2$ and NH$_3$ adsorption-desorption measurements respectively. The measurements suggested that the basicity of N-dG was significantly high in contrary to acidity. These outcomes are well in agreement as reported by Wu and coworkers,[18] which suggests that the active sites for CO$_2$ hydrogenation are pyridinic N at graphene periphery.

Iron nanoparticles were also observed to be a reasonably selective and effective catalyst toward CO$_2$ hydrogenation. Williamson et al. reported Fe nanoparticles impregnated on carbon nanotubes doped with nitrogen (Fe@NCNTs) for hydrogenation of CO$_2$ via combined reversal water gas-shift process (RWGS) as well as Fischer−Tropsch (FT) catalysis.[20] Fe@NCNT catalyst was synthesized by one-step decomposition of ferrocene (FcH) using chemical vapor deposition (CVD) method. Fe@NCNT shows remarkable catalytic performance toward CO$_2$ hydrogenation as compared to Fe@CNT. Fe@CNT results in lower CO$_2$ conversion forming a mixture of CO, olefins and C$_{n > 1}$ hydrocarbon, whereas Fe@NCNT increases the CO$_2$ and CO conversions significantly, shifting the reaction toward the synthesis of paraffin and methane. This observation suggests that N-doping performs a prominent character in manipulating the activity as well as the selectivity of the as-synthesized catalyst. N-doping also increases the reductive nature of iron through electron donation which increases the selectivity toward methane formation. Fe@NCNT enhances the RWGS performance and initial FT transformation into the methane. Simultaneously the synthesis of longer hydrocarbons occurs by reduction of FT intermediates through subsequent FT polymerization. The resulting C−N dipoles created by N-doping help in the attraction of reagents onto the catalyst surface which consequently enhances the catalytic performance by enhancing the concentration of dipoles accomodating RWGS and FT reactants at expense of the formation of long-chain hydrocarbons. These studies suggest that N-doping can effectively improve the performance of hydrogenation catalysts for methane formation rather than the production of hydrocarbons. Developing nitrogen functionalities on carbonaceous material implanted with iron nanoparticles was also used for hydrogenation of CO_2.[21] The catalysts were synthesized through the one-pot hydrothermal method. Various nitrogen reagents were used for functionalization, i.e., pyrrolidine (PYL), ethylenediamine (EDA), diethylformamide (DFM), and pyridine (PYD). Employment of different nitrogen sources influences the structural and morphological changes in carbon materials. The iron embedded carbon material (Fe/C) without N-doping is spherical in morphology. In the presence of PYL nitrogen source, the carbon material displays smooth pellet cementation with reduced dispersibility. In the presence of EDA the carbon support shows rough and nonuniform granules. The incorporation of DFM did not affect the spherical morphology of the carbon material. Nitrogen functionalization onto the carbonaceous materials facilitates the incorporation of defect sites. It was well concluded that the degree of carbonization of Fe precursor, surface area as well as the number of defect sites performs a key role in enhancing catalytic activity toward CO_2 hydrogenation.

3.2 CO$_2$ hydrogenation to alcohols

The hydrogenation CO_2 to alcohols is another synthetic root of CO_2 mitigation. Alcohols can form extensive chemical feedstock. These can be employed as feedstock for fuel cells and combustion engines as they provide an excellent platform for the preparation of diverse useful chemicals as well as fossil-fuels. Different carbon-based materials employed as catalysts for The hydrogenation of CO_2 to higher alcohols.[22] Working in this field Liang et al. synthesized a catalytic system wherein they supported Pd—ZnO on multiwalled carbon nanotubes (MWCNTs) for selective oriented hydrogenation of CO_2 to methanol.[23] MWCNTs act as both catalyst support and reaction promoter. The surface concentration of the Pd^0 increases as the MWCNTs added to reaction mixture, which is a catalytically active site for methanol generation. The turnover frequency for CO_2 hydrogenation using MWCNTs as catalyst was 1.17 and 1.18 times superior to activated carbon (AC) and γ-Al_2O_3. The herringbone type (h-type) MWCNTs possess dangling bonds on catalyst surface which furnishes greater active sites for hydrogen adsorption. This creates a microenvironment and the adsorbed active H species can be easily converted to Pd active sites which increases the pace of hydrogenation reactions. In another report, Wang et al. fabricated CNTs supported Cu—ZrO_2 catalysts for hydrogenation of CO_2 to methanol.[24] The CNTs were converted nitrogen groups functionalities. These nitrogen functionalities increase the dispersion of CuO onto the surface, enhances H_2 and

CO_2 adsorption and reduces the Cu crystal size which in turn enhances the methanol production. 84.0 mg g_{cat}^{-1} h^{-1} was the maximum yield of methanol ascertained for Cu/ZrO$_2$/CNTs catalyst (CZ/CNT-3) with 10.3 wt% Cu loading.

It has been observed that graphene nanosheets can act as excellent support and cocatalysts to metal-based systems for the CO_2 hydrogenation to alcohols. In one such report, Deerattrakul et al. reported reduced graphene oxide (RGO) supported Cu—Zn catalystic system for hydrogenation of CO_2 to methanol.[25] The catalytic systems were synthesized by the inceptive wetness impregnation technique with varying weight percentages (5, 10, 20, and 30 wt%) of Cu—Zn metals. The as-synthesized catalyst was subjected to hydrogenation of CO_2 in a tubular stainless steel fixed-bed reactor. It is observed that upon enhancing the Cu—Zn content of metal loading, undesirable metal aggregation takes place which inhibits the catalytic activity and results in lower yields of methanol. Hence the Cu—Zn/RGO catalyst with 10 wt% loadings was the optimal catalyst for the hydrogenation process. The temperature effect was also studied and it was found that the yield of methanol increases up to a temperature of 250°C, and then starts decreasing. This could be due to the increased agglomeration at a higher temperature which results in lower yield. Witoon and coworkers studied the effect of the presence of different GO weight percentages in CuO—ZnO—ZrO$_2$ catalytic system.[26] The catalytic performance of as prepared catalysts were examined for catalytic hydrogenation of CO_2 to methanol. It is observed that the 0.5—2.5 wt% of GO resulted in higher yields of methanol which can be ascribed to the enhancement of active sites adsorbing the H$_2$ and CO_2. Further, if the content of GO is more than 2.5 wt% there is a decrease in methanol yield. The higher content of GO hinders the precipitation of mixed metal oxides and results in the synthesis of segregated metal oxides which reduces yield of methanol synthesis. The selectivity of methanol synthesis was also found to be higher when metal-GO catalysts were used as compared to the metal free GO catalyst. The higher selectivity could be ascribed to the promotional effect of GO nanosheets. These act as aqueduct between different mixed metal oxides that increases the hydrogen spillover process. This hydrogen spillover reaction took part from the surface of Cu to adsorbed carbon moieties onto the isolated metal oxide surface. The maximum formation of methanol 274.63 g kg_{cat}^{-1} h^{-1} was recorded for 1 wt% GO content i.e., CZZ-1 GO catalyst.

Carbon nanofibers can also be used as support materials. In this regard Din et al. synthesized a wide range of carbon nanofibers supported on bimetallic catalyst copper oxide/zirconia (CuO/ZrO$_2$/CNF) for the hydrogenation of CO_2 to methanol.[27] These catalysts were prepared via deposition precipitation method with varying CuO weight percentages (5, 10, 15, 20, and 25 wt%), as CuO is well-known CO_2 adsorption site. Upon increase in CuO loading above 15 wt% agglomeration takes place within the particles which in turn reduces the catalyst surface area resulting to lower methanol yield. A linear relationship has been found between the Cu surface area and methanol yield. An increase in surface area can provide more atomic hydrogen which is supplied to ZrO_2 basic sites for CO_2 reduction, which results in higher methanol yield.

3.3 CO_2 hydrogenation to value-added products

CO_2 can also be hydrogenated into many other useful products, e.g., as dimethyl ether, acids, olefins, and many more. All these products are of great value in industrial and synthetic chemistry. Wang and coworkers reported carbon-supported metal carbide (M/ZIF-8-C) catalysts doped N and used for selective CO_2 hydrogenation to CO.[28] Four different metals Cu, Co, Ni and Fe were used for catalytic hydrogenation and for the same metal percentage loading the catalytic activity follow the order: Cu/ZIF-8-C < Co/ZIF-8-C < Fe/ZIF-8-C ≈ Ni/ZIF-8-C. Pyridinic nitrogen and carbides help in effective CO_2 adsorption and activate the catalyst surface to mediate the reaction. The metal carbides Ni or Fe dispersed on the surface provide active sites for hydrogenation of CO_2 to CO. The Ni-based catalysts showed remarkable activity, which could be ascribed to high dispersion and exposure of the metal carbide. DFT was used to analyze the mechanism involved in selective CO_2 hydrogenation. These studies suggested that the CO generation occur from CO_2* by the direct breakage of C—O bonds rather than the decay of COOH*. CO* desorbs from the catalyst surface instead of undergoing further reactions. This results in highly selective hydrogenation of CO_2 to CO.

The Pd supported graphitic carbon nitride (Pd/g-C_3N_4) catalysts were also employed for CO_2 conversion to formic acid by hydrogenation.[29] The basic nitrogen sites present of g-C_3N_4 acts as the CO_2 adsorption sites. The yield of formic acid obtained with Pd/g-C_3N_4 catalysts (306.7 $\mu mol \cdot mol_{Pd}^{-1} s^{-1}$) was 12 times higher compared Pd/CNT catalysts. This increased catalytic activity could be due to the higher affinity of CO_2 toward g-C_3N_4. The proposed mechanism suggests that CO_2 activation occur on the surface of Pd/g-C_3N_4 and simultaneous activation of H_2 takes place at Pd surface. Upon decreasing the size of Pd from 7.2 to 3.4 nm the TOF for the hydrogenation reaction gets doubled. Pd particles with smaller size can relatively supplies the more surface area with g-C_3N_4 resulting in a higher TOF.

Zha et al. reported MWCNTs supported CuO—ZnO—Al_2O_3/HZSM-5 catalyst for CO_2 hydrogenation to dimethyl ether (DME).[30] The hydrogenation of CO_2 to DME involves four different reactions as follows:

(1) $CO_2 + 3H_2 \rightarrow CH_3OH + H_2O$ -90.4 kJ mol^{-1}
(2) $CO + 2H_2 \rightarrow CH_3OH$ -49.3 kJ mol^{-1}
(3) $CO + H_2O \rightarrow CO_2 + H_2$ -41.0 kJ mol^{-1}
(4) $2CH_3OH \rightarrow CH_3OCH_3 + H_2O$ -23.4 kJ mol^{-1}

When the MWCNTs were oxidized by a mixture of H_2SO_4 and HNO_3, generation of surface hydroxyl, carbonyl, and carboxylic groups occurs, which were analyzed using FTIR analysis. The FTIR spectra show peaks at 3240 cm^{-1} and 1630 cm^{-1} which were assigned to —OH and C=O groups of the oxidized MWCNTs. These functional groups provide active hydrogen binding sites which makes them hydrophilic in nature, hence improving their suspension and dispersion stability. The catalytic activity for CO_2 hydrogenation was examined employing a fixed-bed reactor. At the optimum conditions of reaction, 20.9% yield of DME was obtained with a selectivity of 45.2%. Based on all these reports, it is evident that CO_2 hydrogenation is a very useful strategy to obtain valuable products.

4. Heterogeneous catalyst in CO_2 fixation

As discussed above, three modes of catalysis, i.e., photocatalysis, electrocatalysis, and thermocatalysis were adopted so far for CO_2 conversion to valuable products. In this section, we have considered the thermal mode of catalysis because of its own merits over photocatalysis, electrocatalysis and other modes.

4.1 CO_2 fixation to cyclic carbonates

The fixation of CO_2 to CC is one of the most common and prevalent approaches for CO_2 conversion. There are various reports on CO_2 fixation to CC but our main focus will be on fixation of CO_2 to CC catalyzed using carbon-based heterogeneous catalysts. Since CO_2 is an inert gas so CO_2 fixation reactions require some input of energy which may be in the form of heat or other energy resources. The CO_2 fixation to CC may be carried out at higher pressure or ambient pressure depending upon the efficiency of the catalyst and other parameters, while the choice of catalyst depends upon the nature of reactants. In CO_2 fixation to CC, the starting materials are CO_2 and epoxides. According to Lewis's acid-base concept, CO_2 is an electrophile, as C of CO_2 is highly oxidized and epoxide acts as a nucleophile, as the oxygen of epoxide has more electron density. Keeping the nature of these reactants in mind, scientists have designed several suitable catalysts in such a way that it can activate both the reactants to undergo the chemical reaction.

Working on this strategy Zhu et al.[6] developed silica-supported boron-doped graphitic carbon nitride (BCN) and used it for catalytic CO_2 fixation to CC. The idea of using BCN can be ascribed to simultaneously occurrence of acid-base sites on BCN which synergetically activates both the reactants. The boron of BCN accounts for Lewis acidity, whereas the nitrogen of BCN accounts for its Lewis basicity. Boron activates the electron-rich epoxides whereas the nitrogen of BCN activates the CO_2 and sustains the reaction in the forward direction. Silica as support provides a significant exterior which assist in the adsorption of the reactants on the catalyst surface and brings them in close proximity to each other. The authors synthesized a series of B_xCN (x = represents boron amount) catalysts with varying boron amounts and tested their catalytic activity for CO_2 fixation to CC, wherein it is observed that $B_{0.1}CN$ shows the best catalytic performance. The synthesis of BCN was confirmed by using PXRD and FTIR techniques as shown in Fig. 9.4. The PXRD pattern shown in Fig. 9.4A represents a peak at $2\theta = 27.4$ degrees, which was assigned to interplanar stacking of BCN nanosheets. The FTIR spectra of BCN shown in Fig. 9.4B contain broadband at 3000–3500 cm^{-1}, which was assigned to –NH or –OH bonds. A small band at 2165 cm^{-1} ascribed to C≡N bond whereas the broadband at 1100–1650 cm^{-1} ascribed to C=N bond. The peaks at 1398, 1110 and 810 cm^{-1} assigned to B–N (B–O), B–O–H, and C–N bonds, respectively. The optimum reaction condition showing the best activity was found to be 1 mL SO,

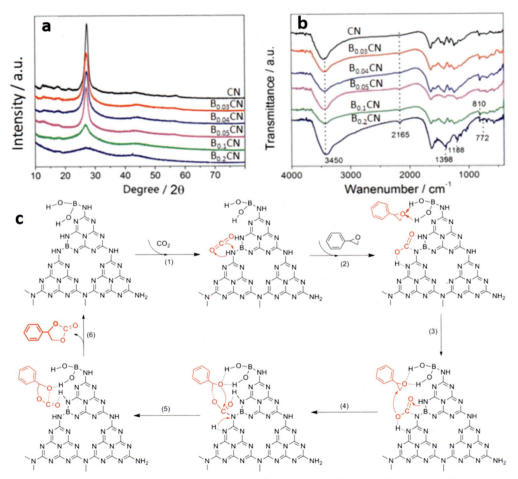

Figure 9.4 (A, B) PXRD pattern and FTIR spectra of B_xCN. (C) mechanism of reaction taking styrene oxide as a reactant. *(Reproduced with permission from Ref. 6.)*

30 mg catalyst, temp = 130°C, P = 3 MPa CO_2, t = 24 h. The mechanism of the reaction, shown in Fig. 9.4C, involves the adsorption of CO_2 on the >NH basic sites forming carbamate as an intermediate in the first step. This step is followed by the evolution of the N—C and the breakage of the N—H bonds. The H atom ruptured in the first step forms a hydrogen bond with the O atom of the CO_2 molecule and also with the neighboring nitrogen atom. The styrene oxide interacts with the hydrogen atoms of —B(OH)$_2$ group through its oxygen atom forming hydrogen bonds in step 2. At the same time another hydrogen bond was formed among O atom of CO_2 and the H atom of —NH which weakens the existing O—H bond which facilitates the nucleophilic attack by CO_2 oxygen atom on activated epoxide (step 3). In step 4, the oxygen anion of carbamate attacks on a C atom of styrene oxide enabling the epoxide ring-opening followed by dissociation of the O—H bond. In step 5, the oxygen of ring-opened epoxide makes a nucleophilic attack on carbon of CO_2 to yield the cyclic carbonate. Finally, the formed cyclic carbonate desorbs from the catalyst surface regenerating the original catalyst again.

Following the same strategy, there are few more reports where graphitic carbon nitride-based materials were explored for CO_2 fixation to cyclic carbonate. The details of those reports are given in Table 9.1.

Graphene-based materials have been also explored for fixation of CO_2 into CC. Lan et al.[32] explored the catalytic activity of graphene oxide for CO_2 fixation to CC. This material shows excellent catalytic activity in the attendance of cocatalyst tetrabutylammonium bromide (Bu$_4$NBr). The optimized conditions were found to be, 28.6 mmol propylene oxide (PO), 2.5 mol% Bu$_4$NBr, 100°C, 2.25 MPa CO_2, 50 mg catalyst, t = 1 h. The influence of water as a solvent on yields of product was examined wherein it was observed that appropriate content of H_2O increases the epoxides conversion astonishingly. The authors have done a detailed study on the surface active sites which enable the CO_2 fixation to CC. The authors have found that it was the —OH functionalities not —COOH on GO surface form hydrogen bonds with epoxides, and work simultaneously with halide anions to facilitate the CO_2 fixation to CC. In order to verify this observation, the authors have recorded the FTIR spectra of GO and PO (model reactant) independently and then graphene oxide dispersed in PO (Fig. 9.5A). The peaks at 1038, 1220, 1608 and 1715 cm^{-1} were assigned C—O, C—O—C, C=C and C=O bonds respectively in GO. The peaks at 3360 and 1365 cm^{-1} were ascribed to the stretching vibration of —OH bond in GO and when graphene was dispersed in propylene oxide these peaks were red-shifted to 3442 and 1381 cm^{-1}, respectively. This shift in wavenumber of —OH bond was attributed to the interaction of propylene oxide with —OH group of graphene oxide. Further confirmation of this observation was made when reduced graphene oxide (r-GO) was employed for PO reformation to propylene carbonate. A series of r-GO was synthesized (TU-r-GO, CA-r-GO, HH-r-GO, TD-r-GO) where TU, CA, HH, and TD stands for thiourea, chloroacetic acid, hydrazine hydrate and thiourea dioxide, respectively which were the reducing agents used in graphene

Table 9.1 Carbon-based nanomaterials for CO_2 hydrogenation.

Entry	Catalyst	Product	Reaction conditions	Selectivity (%)	TOF	Conersion (%)	References
1	Nitrogen doped graphene quantum dots (NGQDs)	CH_4	380°C	55	$1.5\ s^{-1}$	—	18
2	Nitrogen doped defective graphene (N-dG)	CH_4	500°C, 10 bar pressure	99.2	$73.17\ s^{-1}$	52.3	31
3	Iron nanoparticles supported on carbon nanotubes (Fe@NCNT)	CH_4	100°C, 2.0 MPa	91.4	—	60.3	20
4	Iron embedded nitrogen carbon material (Fe/C)	CH_4, C_2–C_4, C_{5+}	300°C, 1 MPa	$CH_4 = 36.8$ C_2–$C_4 = 46.0$ $C_{5+} = 17.2$	—	21.6	21
5	Carbon nanotube-supported Pd–ZnO $Pd_{0.1}Zn_1$/CNTs(h-type)	CH_3OH	503–543 K, 2.0 MPa	99.6	$1.15 \times 10^{-2}\ s^{-1}$	—	23
6	Carbon nanotubes supported copper catalyst Cu_1Zr_1–10% (4.3% Co/CNT)	CH_3OH	513 K, 5.0 MPa	7.30	$2.83 \times 10^3\ s^{-1}$	92.0	22
7	Cu/ZrO$_2$ catalystic systems supported on carbon nanotubes (Cu/ZrO$_2$/CNTs)	CH_3OH	200–300°C, 3.0 MPa	15.8	$1.61 \times 10^{-2}\ s^{-1}$	67.3	24
8	Cu–Zn catalyst supported on reduced graphene oxide (CuZn/rGO)	CH_3OH	270°C, 15 bar	5.1	—	26	25
9	GO supported CuO–ZnO–ZrO$_2$	CH_3OH	200°C	75.9	—	24.8	26
10	Carbon nanofibers based copper/zirconia catalyst (CZC)	CH_3OH	180°C, 30 bar	—	$5.2 \times 10^{-4}\ s^{-1}$	12	27
11	Graphitic carbon nitride supported Pd (Pd/g-C_3N_4)	Formic acid	40°C, 50 bar	—	$1250 \times 10^{-6}\ s^{-1}$	—	29
12	Multiwalled CNT supported CuO–ZnO–Al$_2$O$_3$/HZSM-5 catalyst	Dimethyl ether	262°C, 3.0 MPa	45.2	—	46.2	30

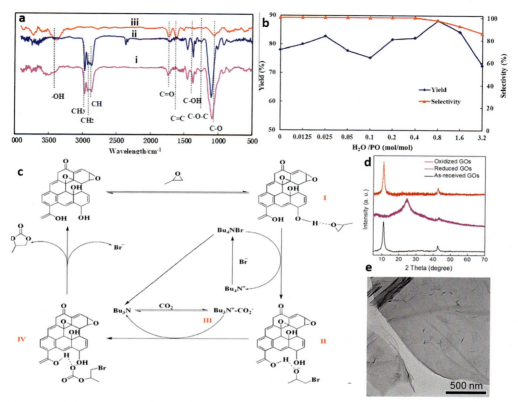

Figure 9.5 (A) FTIR spectra of (i) GO, (ii) propylene oxide and (iii) GO dispersed in propylene oxide. (B) Influence of water incorporation on the activity of Bu₄NBr/GO at optimized reaction conditions described above. (C) Reaction mechanism of propylene carbonate synthesis catalyzed GO. (D) XRD pattern of graphene oxide and r-GO and oxidized graphene oxide. (E) TEM image of GO. ((C, E) Reproduced with permission from Refs. 32 and 33.)

oxide reduction to r-GO. The yields of propylene carbonate synthesis obtained by using GO and r-GO follows the GO (81.3%) > TU-r-GO (57.8%) > CA-r-GO (32.6%) > HH-r-GO (33.2%) > TD-r-GO (14.7%) order. This decrease in the yield of propylene carbonate when catalyzed by r-GO was due to partial deduction of O-functionalities upon reduction which consequently decreases the number of —OH groups on r-GO.

The influence of water solvent on the cycloaddition of PO was also examined as shown in Fig. 9.5B. It was observed that when molar ratio of H_2O/PO was 2.5%, lower as compared to Bu₄NBr/PO the yield of reaction increases remarkably. This increase in yield was attributed to simultaneous interaction of H_2O and Bu₄NBr that affected the activity of Bu₄NBr as predicted by Sun et al.[34] Also the water presence provides extra —OH groups which increases the efficiency of the reaction. If we look at Fig. 9.5B, the H_2O/PO molar ratio is close to 0.1, propylene carbonate yield decreases to 75.2%, while propylene carbonate selectivity remains at 100%. At this amount of water, GO and water compete for propylene oxide and hence there is a decline in propylene carbonate formation. When H_2O/propylene oxide ratio is raised from 0.1 to 0.8, the yield increases from 75.2% to 88.2% with a slight decrease in selectivity from 100% to 97.7%. This surprising behavior was attributed to an increase in acidic sites on GO, which was due to the increased interactions of protons generated in H_2O on GO. On further increasing the H_2O/propylene oxide ratio to 3.2, the selectivity and yield decrease to 85.9% and 72.5% respectively, which could be attributed to the hydrolysis of propylene oxide and decline in concentration of substrate, which collectively causes a decline in substrates contact with active sites. The use of graphite and graphite oxide along with Bu_4NBr was also explored for propylene carbonate synthesis. The yields of propylene carbonate obtained from graphite and graphite oxide were 17.2% and 69.8%, respectively, as tabulated in Table 9.2 (entries 1 and 2). The higher yield in the case of GO was attributed to the rich oxygen functionality on graphite oxide which helps in binding propylene oxide to it through hydrogen bonding.

The mechanism of reaction (Fig. 9.5C) suggests that —OH groups on graphene oxide and the O atom of PO interact through hydrogen bonding in first step forming intermediate **(I)**. The cocatalyst, Bu_4NBr performs two important key roles in this reaction, first is the nucleophilic attack of Br^- on the β-C atom of propylene oxide, and the second was the synthesis of tributylamine cation which consequently interacts with CO_2. The β-C atom of PO was attacked by Br^- leading to the ring-opening of propylene oxide at the same time O anion of the ring-opened epoxy anion forms a hydrogen bond with the —COOH groups of graphene oxide as shown in intermediate **(II)**. The Bu_3N, formed from Bu_4NBr, reacts to CO_2 forming intermediate **(III)** subsequently it interact with **(II)** to form an alkyl carbonate anion **(IV)** and this intermediate was also sustained by hydrogen bonding. In the final step, propylene carbonate was formed by the intramolecular cyclization step and gets desorbed from the catalyst surface to regenerate the original catalyst.

In another work, Zhang et al.[33] used the same graphene oxide for styrene oxide conversion to styrene carbonate. Here in contrast to Lan et al.'s work described above,[32] the cycloadditions were performed at ambient pressure. They performed the reactions in the presence of dimethylformamide (DMF) as a solvent without using any cocatalyst. The synthesis of GO was affirmed by the XRD patterns. The XRD pattern depicted in Fig. 9.5D represents a broader peak indexed at 10.8 degrees, assigned to the (002) characteristics peak of the graphene oxide. In r-GO the intensity of this peak decreases significantly and becomes very weak, whereas in the case of oxidized graphene oxide XRD pattern remains the same. The sheetlike morphology of GO was confirmed by the TEM image shown in Fig. 9.5E. The optimized reaction conditions were found to be

Table 9.2 Carbon-based nanomaterials for CO_2 fixation to cyclic carbonates.

Entry	Catalyst	Catalyst amount (mg)	Cocatalyst	Substrate	Reaction conditions	Selectivity (%)	Yield (%)	Conversion (%)	References
1	Graphitic carbon nitride derived from urea(u-g-C_3N_4)	100	—	Propylene oxide	130°C, 2.0 MPa, 24 h	99.6	98.9		35
2	Graphitic carbon nitride derived from urea(u-g-C_3N_4)	50	—	Epichlorohydrin	130°C, 3.5 MPa, 2 h	99.9	99.1		36
3	Phosphorus doped graphitic carbon nitride (P-g-C_3N_4)	150	TBAB	Propylene oxide	100°C, 2.0 MPa, 3 h	99.9	91.1		37
4	n-butIBr/mp-C_3N_4	200	—	Propylene oxide	140°C, 2.5 MPa, 6 h	99.8	87.7	88	38
5	Zn^{2+}-g-C_3N_4/SBA	100	—	Propylene oxide	150°C, 3.5 MPa, 1.5 h	99	—	94.5	39
6	Mesoporous carbon nitride(MCN)	20	—	Propylene oxide	100°C, 0.5 MPa, 10 h	90	—	34	40
7	Zn—C_3N_4	120	KI	Propylene oxide	130°C, 2 MPa, 5 h	99	99		41
8	C_{60} fullerenol	100	KI	Propylene oxide	120°C, 2 MPa, 2 h	—	95		42
9	Graphite oxide	20	[BMIm]Br (2.5%)	Styrene oxide	80°C, 1 MPa, 6 h	—	99		43
10	Urea derived graphitic carbon nitride(u-C_3N_4)	230	—	Epichlorohydrin	120°C, 2 MPa, 2 h	98.01	96.28	98.31	44

Continued

Table 9.2 Carbon-based nanomaterials for CO$_2$ fixation to cyclic carbonates.—cont'd

Entry	Catalyst	Catalyst amount (mg)	Cocatalyst	Substrate	Reaction conditions	Selectivity (%)	Yield (%)	Conversion (%)	References
11	β-cyclodextrin	100	KI	Propylene oxide	120°C, 6 MPa, 4 h		99		45
12	Lignin	200	KI	Propylene oxide	140°C, 2 MPa, 12 h		93		46
13	Urea-glucose carbonaceous (U-GCM)	200	KI	Propylene oxide	120°C, 2.5 MPa, 2 h	99	—	99.94	47
14	g- C$_3$N$_4$	50	TBAB	Epichlorohydrin	105°C, 1 atm, 20 h	—	—	100	48
15	Urea and thiourea derived sulphonated graphitic carbon nitride(s-C$_3$N$_4$)	50	—	Epichlorohydrin	100°C, 1 MPa, 1 h	99.0	—	93.5	49
16	Oxidized biochar	100	TBAB	Propylene oxide	110°C, 1 MPa, 6 h	—	—	78.5	50
17	Amine-functionalized graphene gxide	30	TBAI	Styrene oxide	70°C, 1 MPa, 12 h	99	94	99	51
18	N-doped active carbons	100	—	Epichlorohydrin	150°C, 1.5 MPa, 15 h	93	—	99	52

5 mmol styrene oxide, 4 mL DMF, 2.5 mg catalyst, temp = 140°C, P = 1.0 atm. CO_2, t = 12 h. For comparison, the other carbonaceous materials like acetylene black, AC and oxidized GO were explored for their catalytic performance for the CO_2 fixation to styrene carbonate and the styrene carbonate yield was found to be 18.7%, 17.1%, and 65.3%, respectively (entries 3, 4, and 5 of Table 9.3). Here similar to the above work the oxidized GO having oxygen functionalities have better catalytic activity than acetylene black and AC. So we can conclude that carbon-based heterogeneous catalysts with rich oxygen functionalities have better catalytic activity in contrast to those which lack oxygen functionalities.

4.2 CO_2 fixation to cyclic carbamates

Aziridines are another class of compounds that can be used for CO_2 fixation into cyclic carbamates as the product. The cyclic carbamates are very useful compounds and have some medicinal applications.[53] For example, 2-oxazolidinones are present in chiral auxiliaries (with 4-substitution) and in super antibiotics, e.g., tedizolid and linezolid while aryl-fused 6-membered rings were found in some HIV-battling antiretrovirals and N-methyl-D-aspartate (NMDA) receptor antagonists.[54] The cyclic carbamates can be synthesized from different starting materials but we limit our discussion to aziridines. Chen et al.[53] synthesized cyclic carbamates using sugarcane bagasse (SCB) a heterogeneous catalyst. Some interesting features of this catalyst include its ease of availability, low cost, chemical stability, biocompatible, ecofriendly and biodegradable.[55] The catalyst was used for cycloaddition of both epoxide and aziridines. The reaction conditions were optimized using propylene oxide as the model reactant. In order to increase the yield of propylene carbonate use of alkali metal halides (KCl, KBr and KI) as cocatalysts was also explored. Out of KCl, KBr and KI, KI exhibited the remarkable performance which could be ascribed to the combined effect of the relatively good leaving ability and strong nucleophilicity of I^- in comparison to Cl^- and Br^-. The optimum reaction condition at which catalyst shows best catalytic activity (92%) was 20 mmol propylene oxide, 0.5 mmol KI, 100°C, 2.0 MPa CO_2, 100 mg SCB, t = 6 h. Fig. 9.6A shown below represents the substrate scope for cyclic carbamate synthesized from different aziridines. When SCB/KI was employed for carbamate synthesis it shows high selectivity, higher catalytic activity and astonishing regioselectivity. Cyclic carbamates are formed from aziridine, two types of products are formed as shown in the chemical reaction shown in Fig. 9.6A. It was observed that 5-aryl-2-oxazolidinones (4) were preferably synthesized under optimized reaction conditions. The yield of products formed from entry two and three was lower than those obtained from entry four. This difference in yield could be attributed to the synthesis of self-oligomers in entries two and three, which show lower yields. However, the product obtained from entry four gave a very low yield and the possible reason could be the presence of a bulky cyclohexyl group on the nitrogen atom of aziridine. Also in contrary to epoxides, the yield obtained from aziridines was relatively low.

Table 9.3 Carbon-based materials from the above two reports[32,52] used in CO_2 fixation.

Entry	Catalyst	Catalyst amount (mg)	Cocatalyst	Substrate	Reaction conditions	Yield (%)	Conversion (%)
1	Graphite	50	Bu$_4$NBr	Propylene oxide	100°C, 2.25 MPa, 1 h	17.2	—
2	Graphite oxide	50	Bu$_4$NBr	Propylene oxide	100°C, 2.25 MPa, 1 h	69.8	—
3	Acetylene black	2.5	—	Styrene oxide	140°C, 1.0 atm., 9 h	—	18.7
4	Activated carbon	2.5	—	Styrene oxide	140°C, 1.0 atm., 6 h	—	17.1
5	Oxidized graphene oxide	2.5	—	Styrene oxide	140°C, 1.0 atm., 6 h	—	65.3

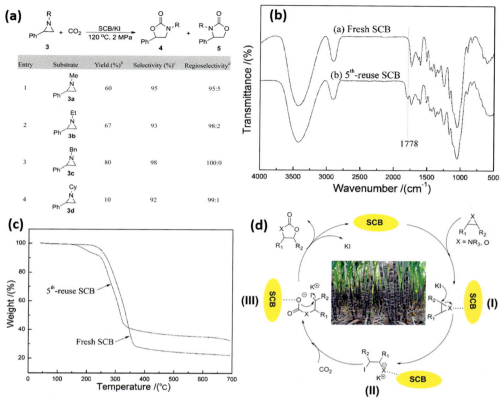

Figure 9.6 (A) Substrate scope of SCB/KI catalyzed cyclic carbamate synthesis, (B, C) FTIR and TGA spectra of SCB before and after the reaction, respectively, (D) mechanism for CO$_2$ fixation to epoxide or aziridine catalyzed by SCB/KI catalytic system. *(Reproduced with permission from Ref. 53.)*

The authors have further studied the reusability and sustainability of the SCB/KI system. The catalyst shows remarkable recyclability where it shows 85% propylene carbonate yield with 99% selectivity after five cycles. The FTIR and thermogravimetric analysis (TGA) of catalyst shown in Fig. 9.6B and C were done before and after the reaction to check the stability of the catalyst. The FTIR spectra of SCB (in Fig. 9.6B) remains almost the same after the reaction except a new peak was observed at 1778 cm^{-1}, which could be attributed to the C=O bond of −COOH groups which may be formed due to carboxylation of −OH functionalities by CO$_2$ in SCB. The TGA plots of SCB (in Fig. 9.6C) show that the catalytic system is stable up to 220°C, which was found to be higher than the temperature of reaction (120°C). The mechanism of reaction (in Fig. 9.6D) involves the interaction of oxygen or nitrogen atom of aziridine or epoxide, respectively with the −OH group of SCB which activates the epoxide or aziridine. Followed by this, I$^-$ of KI attacks the less sterically hindered C atom of the aziridine or epoxide ring forming **(I)** which will commence the ring-opening of aziridine or epoxide to form the intermediate **(II)**. Subsequently, the adsorbed CO_2 molecule on the SCB surface interacts with intermediate **(II)** which then forms intermediate **(III)**. In the last step, intramolecular cyclization takes place leading to the synthesis of the final product with regeneration original catalyst.

4.3 CO₂ fixation to other value-added products

Benzimidazoles and their derivatives formed by CO_2 fixation to o-phenylenediamine (OPD) were found to be most prominent components possessing remarkable biological activity. Benzimidazoles have been used as intermediate in various drugs like mebendazole, rabeprazole, omeprazole and albendazole.[56] Based on such biological applications of benzimidazoles, CO_2 fixation to benzimidazoles is advantageous from both biological as well as CO_2 sequestration points of view. Phatake et al.[57] inspired by such biological applications of benzimidazoles, graphitic carbon nitride derived from urea and supported with copper metal (Cu@U-g-C_3N_4) were used as heterogeneous catalysts for fixation of CO_2 to benzimidazoles. The experiments were performed in propylene carbonate (green solvent) solvent which is used as a mixture of propylene carbonate and water. They synthesized three catalytic systems based on varying copper percentages (Cu@Ug–C_3N_4–1 = 3% Cu, Cu@U-g- C_3N_4-2 = 6% Cu, Cu@U-g-C_3N_4–3 = 10% Cu) and tested their catalytic activity CO_2 fixation to benzimidazoles and Cu@U-g–C_3N_4–2 showed the best catalytic activity. The synthesis of Cu@U-g–C_3N_4–2 has been analyzed by XRD patterns, wherein the plot of Cu@U-g–C_3N_4–2 resembles well with U-g-C_3N_4. The two characteristic peaks at 13.1 and 27.5 degrees were attributed to the (100) in-plane ordering of tri-s-triazine units and (002) plane of interlayer stacking of graphitic layers in U-g-C_3N_4, respectively. When Cu was loaded on U-g-C_3N_4 there was a slight decline in the intensity of the peaks is due to the dilution effect. There were some new peaks at 42.1 degree, 51.2 degrees and 73.5 degrees which were assigned to the dispersion of Cu nanoparticles on the U-g-C_3N_4 nanosheets. These peaks correspond to (111), (200), (220) planes and show the existence of a face-centered cubic structure of Cu. The morphology of Cu@U-g–C_3N_4–2 was obtained using SEM and TEM analysis. The Cu@U-g–C_3N_4–2 has a stacked sheetlike structure that remained intact after using in the reaction. The TEM images of Cu@U-g–C_3N_4–2 before the reaction and after four-step of recyclability were taken which also revealed the stacked sheetlike structure of Cu@U-g–C_3N_4–2. The authors further optimized the various reaction parameters using o-phenylenediamine (1.0 mmol) as a model substrate and dimethylamine borane (DMAB) as a reducing agent. The optimal reaction condition at which Cu@U-g–C_3N_4–2 shows the best catalytic activity was Cu@U-g–C_3N_4–2 (20 mg), the reactant (1.0 mmol), DMAB (3.0 mmol), CO_2 (2.5 MPa), at 100°C for 24 h.

The use of Cu@U-g–C_3N_4–2 was further explored in catalyzing different reactions at the optimized conditions. The substrate scope of o-phenylenediamines conversion to give a long range of benzimidazoles was also explored. The unsubstituted o-phenylenediamines give a better yield than substituted o-phenylenediamines which could be due to less steric bulkiness of unsubstituted o-phenylenediamines. When the o-phenylenediamines are substituted with electron-donating or withdrawing groups, the o-phenylenediamines give moderate yields. Table 9.4 lists some of the companies working in the field of CO_2 mitigation.

5. Summary and perspectives

CO_2 being an environmentally friendly and renewable source of carbon provides a great opportunity for its conversion to fuels and chemicals. The CO_2 hydrogenation and fixation are feasible and sustainable methods in this regard. In order to get rid of the limitations associated with CO_2 conversion and selectivity, scientists have to design various technical directions and specific research strategies, like the design of catalysts, reaction condition optimization, use of some suitable cocatalysts, and exploration of reaction mechanisms. The CO_2 is used as a reactant in industries is only 0.5 wt% megaton per year, which is only a very small amount of the total anthropogenic CO_2 produced in the atmosphere every year. The industrial utilization of CO_2 is not expected to mitigate the enhanced CO_2 concentration with existing technologies. Some more potential processes should be employed to mitigate the CO_2 problem. One possible method could be polymer (polycarbonates and polyurethanes) synthesis, which includes laminates and plastics used in the construction industry. The conversion of CO_2 into liquid carbonaceous transportation fuels is another promising way for CO_2 mitigation. The CO_2 could be used as a raw materials for the formation of various chemical intermediates, fuel additives, chemicals, and environmentally benign fluid in diverse applications can come up with sustainable chemical methods for industrial use, which can consequently lead to CO_2 mitigation in the atmosphere. This will be of great benefit to both humanity and the environment.

Table 9.4 Organizations working on CO_2 mitigation strategies.

Sl. No	Name of the company	Website
1	Carbon recycling international	https://www.carbonrecycling.is/
2	Carbicrete	https://carbicrete.com/
3	Blue planet	http://www.blueplanet-ltd.com
4	Econic	https://econic-technologies.com/
5	Newlight technologies	https://www.newlight.com/
6	Carbon Engineering	https://carbonengineering.com/
7	Shell	https://www.shell.com
8	Global Thermostat	https://globalthermostat.com/
9	Clime works	https://climeworks.com/
10	NRG energy	https://www.nrg.com
11	CO_2 solutions	https://co2solutions.com

References

1. Swart R, Robinson J, Cohen S. Climate change and sustainable development: expanding the options. *Clim Pol* 2003;**3**:S19—40. Suppl. 1.
2. Rahman FA, Aziz MMA, Saidur R, Bakar WAWA, Hainin M, Putrajaya R, Hassan NA. Pollution to solution: capture and sequestration of carbon dioxide (CO2) and its utilization as a renewable energy source for a sustainable future. *Renew Sustain Energy Rev* 2017;**71**:112—26.
3. Steeneveldt R, Berger B, Torp T. CO2 capture and storage: closing the knowing—doing gap. *Chem Eng Res Des* 2006;**84**(9):739—63.
4. Xiaoding X, Moulijn J. Mitigation of CO2 by chemical conversion: plausible chemical reactions and promising products. *Energy Fuel* 1996;**10**(2):305—25.
5. Li L, Zhao N, Wei W, Sun Y. A review of research progress on CO2 capture, storage, and utilization in Chinese Academy of Sciences. *Fuel* 2013;**108**:112—30.
6. Zhu J, Diao T, Wang W, Xu X, Sun X, Carabineiro SA, Zhao Z. Boron doped graphitic carbon nitride with acid-base duality for cycloaddition of carbon dioxide to epoxide under solvent-free condition. *Appl Catal B Environ* 2017;**219**:92—100.
7. Li K, Peng B, Peng T. *ACS Catal* 2016;**6**:7485—527.
8. Gao Y, Li F, Zhou P, Wang Z, Zheng Z, Wang P, Liu Y, Dai Y, Whangbo M-H, Huang B. Enhanced selectivity and activity for electrocatalytic reduction of CO2 to CO on an anodized Zn/carbon/Ag electrode. *J Mater Chem* 2019;**7**(28):16685—9.
9. Biswas T, Mahalingam V. Efficient CO2 fixation under ambient pressure using poly (ionic liquid)-based heterogeneous catalysts. *Sustain Energy Fuels* 2019;**3**(4):935—41.
10. (a) Shan SJ, Zhao Y, Tang H, Cui FY. A mini-review of carbonaceous nanomaterials for removal of contaminants from wastewater. *IOP Conf Ser Earth Environ Sci* 2017;**68**:012003. (b) Ge Y, Shen C, Wang Y, Sun Y-Q, Schimel JP, Gardea-Torresdey JL, Holden PA. Carbonaceous nanomaterials have higher effects on soybean rhizosphere prokaryotic communities during the reproductive growth phase than during vegetative growth. *Environ Sci Technol* 2018;**52**(11):6636—46.
11. Shan S, Zhao Y, Tang H, Cui F. In A mini-review of carbonaceous nanomaterials for removal of contaminants from wastewater. In: *IOP conference series: Earth and environmental science, IOP Publishing*; 2017. p. 012003.
12. (a) Yang X, Wan Y, Zheng Y, He F, Yu Z, Huang J, Wang H, Ok YS, Jiang Y, Gao B. Surface functional groups of carbon-based adsorbents and their roles in the removal of heavy metals from aqueous solutions: a critical review. *Chem Eng J* 2019;**366**:608—21. (b) Sakthivel R, Kubendhiran S, Chen S-M, Ranganathan P, Rwei S-P. Functionalized carbon black nanospheres hybrid with MoS2 nanoclusters for the effective electrocatalytic reduction of chloramphenicol. *Electroanalysis* 2018;**30**(8):1828—36. (c) Reinholds I, Pugajeva I, Bogdanova E, Jaunbergs J, Bartkevics V. Recent applications of carbonaceous nanosorbents for the analysis of mycotoxins in food by liquid chromatography: a short review. *World Mycotoxin J* 2019;**12**(1):31—43. (d) kumar GG, Hashmi S, Karthikeyan C, GhavamiNejad A, Vatankhah-Varnoosfaderani M, Stadler FJ. Graphene oxide/carbon nanotube composite hydrogels—versatile materials for microbial fuel cell applications. *Macromol Rapid Commun* 2014;**35**(21):1861—5.
13. (a) Jessop P. *Handbook of homogeneous hydrogenation*1; 2007. p. 489—511. (b) Jessop PG, Joo F, Tai CC. *Coord Chem Rev* 2004;**248**:2425—42. (c) Wang W, Wang S, Ma X, Gong J. Recent advances in catalytic hydrogenation of carbon dioxide. *Chem Soc Rev* 2011;**40**(7):3703—27. (d) Saeidi S, Amin NAS, Rahimpour MR. Hydrogenation of CO2 to value-added products—a review and potential future developments. *J CO2 Util* 2014;**5**:66—81.
14. (a) Centi G, Perathoner S. CO2-based energy vectors for the storage of solar energy. *Greenh Gases Sci Technol* 2011;**1**(1):21—35. (b) Centi G, Perathoner S. Opportunities and prospects in the chemical recycling of carbon dioxide to fuels. *Catal Today* 2009;**148**(3—4):191—205.
15. (a) Westermann A, Azambre B, Bacariza M, Graça I, Ribeiro M, Lopes J, Henriques C. Insight into CO2 methanation mechanism over NiUSY zeolites: an operando IR study. *Appl Catal B Environ* 2015;**174**:120—5. (b) Shimoda N, Shoji D, Tani K, Fujiwara M, Urasaki K, Kikuchi R, Satokawa S. Role of trace chlorine in Ni/TiO2 catalyst for CO selective methanation in reformate gas. *Appl Catal B Environ* 2015;**174**:486—95. (c) Beaumont SK, Alayoglu S, Specht C,

Michalak WD, Pushkarev VV, Guo J, Kruse N, Somorjai GA. Combining in situ NEXAFS spectroscopy and CO2 methanation kinetics to study Pt and Co nanoparticle catalysts reveals key insights into the role of platinum in promoted cobalt catalysis. *J Am Chem Soc* 2014;**136**(28):9898—901. (d) Porosoff MD, Yan B, Chen JG. Catalytic reduction of CO2 by H2 for synthesis of CO, methanol and hydrocarbons: challenges and opportunities. *Energy Environ Sci* 2016;**9**(1):62—73.

16. (a) Sun J, Fujita S-i, Zhao F, Arai M. Synthesis of styrene carbonate from styrene oxide and carbon dioxide in the presence of zinc bromide and ionic liquid under mild conditions. *Green Chem* 2004; **6**(12):613—6. (b) Sun J, Fujita S-i, Arai M. Development in the green synthesis of cyclic carbonate from carbon dioxide using ionic liquids. *J Organomet Chem* 2005;**690**(15):3490—7. (c) Bayardon J, Holz J, Schaeffner B, Andrushko V, Verevkin S, Preetz A, Boerner A. Propylene carbonate as a solvent for asymmetric hydrogenations. *Angew Chem Int Ed* 2007;**46**(31):5971—4.
17. (a) Kong D-L, He L-N, Wang J-Q. Synthesis of urea derivatives from CO2 and amines catalyzed by polyethylene glycol supported potassium hydroxide without dehydrating agents. *Synlett* 2010; **2010**(08):1276—80. (b) Wu C, Cheng H, Liu R, Wang Q, Hao Y, Yu Y, Zhao F. Synthesis of urea derivatives from amines and CO2 in the absence of catalyst and solvent. *Green Chem* 2010; **12**(10):1811—6.
18. Wu J, Wen C, Zou X, Jimenez J, Sun J, Xia Y, Fonseca Rodrigues M-T, Vinod S, Zhong J, Chopra N. Carbon dioxide hydrogenation over a metal-free carbon-based catalyst. *ACS Catal* 2017;**7**(7): 4497—503.
19. Jurca B, Bucur C, Primo A, Concepción P, Parvulescu VI, García H. N-doped defective graphene from biomass as catalyst for CO2 hydrogenation to methane. *ChemCatChem* 2019;**11**(3):985—90.
20. Williamson DL, Herdes C, Torrente-Murciano L, Jones MD, Mattia D. N-doped Fe@ CNT for combined RWGS/FT CO2 hydrogenation. *ACS Sustain Chem Eng* 2019;**7**(7):7395—402.
21. Guo L, Zhang P, Cui Y, Liu G, Wu J, Yang G, Yoneyama Y, Tsubaki N. One-pot hydrothermal synthesis of nitrogen functionalized carbonaceous material catalysts with embedded iron nanoparticles for CO2 hydrogenation. *ACS Sustain Chem Eng* 2019;**7**(9):8331—9.
22. Zhang H-B, Liang X-L, Dong X, Li H-Y, Lin G-D. Multi-walled carbon nanotubes as a novel promoter of catalysts for CO/CO2 hydrogenation to alcohols. *Catal Surv Asia* 2009;**13**(1):41—58.
23. Liang X-L, Dong X, Lin G-D, Zhang H-B. Carbon nanotube-supported Pd—ZnO catalyst for hydrogenation of CO2 to methanol. *Appl Catal B Environ* 2009;**88**(3—4):315—22.
24. Wang G, Chen L, Sun Y, Wu J, Fu M, Ye D. Carbon dioxide hydrogenation to methanol over Cu/ZrO 2/CNTs: effect of carbon surface chemistry. *RSC Adv* 2015;**5**(56):45320—30.
25. Deerattrakul V, Dittanet P, Sawangphruk M, Kongkachuichay P. CO2 hydrogenation to methanol using Cu-Zn catalyst supported on reduced graphene oxide nanosheets. *J CO2 Util* 2016;**16**:104—13.
26. Witoon T, Numpilai T, Phongamwong T, Donphai W, Boonyuen C, Warakulwit C, Chareonpanich M, Limtrakul J. Enhanced activity, selectivity and stability of a CuO-ZnO-ZrO2 catalyst by adding graphene oxide for CO2 hydrogenation to methanol. *Chem Eng J* 2018;**334**:1781—91.
27. Din IU, Shaharun MS, Naeem A, Tasleem S, Johan MR. Carbon nanofibers based copper/zirconia catalysts for carbon dioxide hydrogenation to methanol: effect of copper concentration. *Chem Eng J* 2018;**334**:619—29.
28. Ji X, Kong N, Wang J, Li W, Xiao Y, Gan ST, Zhang Y, Li Y, Song X, Xiong Q. A novel top-down synthesis of ultrathin 2D boron nanosheets for multimodal imaging-guided cancer therapy. *Adv Mater* 2018;**30**(36):1803031.
29. Park H, Lee JH, Kim EH, Kim KY, Choi YH, Youn DH, Lee JS. A highly active and stable palladium catalyst on a gC3 N4 support for direct formic acid synthesis under neutral conditions. *Chem Commun* 2016;**52**(99):14302—5.
30. Zha F, Tian H, Yan J, Chang Y. Multi-walled carbon nanotubes as catalyst promoter for dimethyl ether synthesis from CO2 hydrogenation. *Appl Surf Sci* 2013;**285**:945—51.
31. Jurca B, Bucur C, Primo A, Concepción P, Parvulescu VI, García Gómez H. N-doped defective graphene from biomass as catalyst for CO2 hydrogenation to methane. *ChemCatChem* 2018.
32. Lan D-H, Yang F-M, Luo S-L, Au C-T, Yin S-F. Water-tolerant graphene oxide as a high-efficiency catalyst for the synthesis of propylene carbonate from propylene oxide and carbon dioxide. *Carbon* 2014;**73**:351—60.

33. Zhang S, Zhang H, Cao F, Ma Y, Qu Y. Catalytic behavior of graphene oxides for converting CO_2 into cyclic carbonates at one atmospheric pressure. *ACS Sustain Chem Eng* 2018;**6**(3):4204−11.
34. Sun J, Cheng W, Fan W, Wang Y, Meng Z, Zhang S. Reusable and efficient polymer-supported task-specific ionic liquid catalyst for cycloaddition of epoxide with CO_2. *Catal Today* 2009;**148**(3−4):361−7.
35. Su Q, Sun J, Wang J, Yang Z, Cheng W, Zhang S. Urea-derived graphitic carbon nitride as an efficient heterogeneous catalyst for CO_2 conversion into cyclic carbonates. *Catal Sci Technol* 2014;**4**(6):1556−62.
36. Huang Z, Li F, Chen B, Yuan G. Cycloaddition of CO_2 and epoxide catalyzed by amino-and hydroxyl-rich graphitic carbon nitride. *Catal Sci Technol* 2016;**6**(9):2942−8.
37. Lan D-H, Wang H-T, Chen L, Au C-T, Yin S-F. Phosphorous-modified bulk graphitic carbon nitride: facile preparation and application as an acid-base bifunctional and efficient catalyst for CO_2 cycloaddition with epoxides. *Carbon* 2016;**100**:81−9.
38. Xu J, Wu F, Jiang Q, Li Y-X. Mesoporous carbon nitride grafted with n-bromobutane: a high-performance heterogeneous catalyst for the solvent-free cycloaddition of CO_2 to propylene carbonate. *Catal Sci Technol* 2015;**5**(1):447−54.
39. Huang Z, Li F, Chen B, Lu T, Yuan Y, Yuan G. Well-dispersed g-C3N4 nanophases in mesoporous silica channels and their catalytic activity for carbon dioxide activation and conversion. *Appl Catal B Environ* 2013;**136**:269−77.
40. Ansari MB, Min B-H, Mo Y-H, Park S-E. CO_2 activation and promotional effect in the oxidation of cyclic olefins over mesoporous carbon nitrides. *Green Chem* 2011;**13**(6):1416−21.
41. Wang X, Liu MS, Yang L, Lan JW, Chen YL, Sun JM. Synthesis of Zn modified carbon nitrides heterogeneous catalyst for the cycloaddition of CO_2 to epoxides. *Chemistry* 2018;**3**(15):4101−9.
42. Sun Y-B, Cao C-Y, Yang S-L, Huang P-P, Wang C-R, Song W-G. C 60 fullerenol as an active and stable catalyst for the synthesis of cyclic carbonates from CO_2 and epoxides. *Chem Commun* 2014;**50**(71):10307−10.
43. Luo R, Zhou X, Fang Y, Ji H. Metal-and solvent-free synthesis of cyclic carbonates from epoxides and CO_2 in the presence of graphite oxide and ionic liquid under mild conditions: a kinetic study. *Carbon* 2015;**82**:1−11.
44. Song X, Wu Y, Pan D, Cai F, Xiao G. Carbon nitride as efficient catalyst for chemical fixation of CO_2 into chloropropene carbonate: promotion effect of Cl in epichlorohydrin. *Mol Catal* 2017;**436**:228−36.
45. Song J, Zhang Z, Han B, Hu S, Li W, Xie Y. Synthesis of cyclic carbonates from epoxides and CO_2 catalyzed by potassium halide in the presence of β-cyclodextrin. *Green Chem* 2008;**10**(12):1337−41.
46. Wu Z, Xie H, Yu X, Liu E. Lignin-based green catalyst for the chemical fixation of carbon dioxide with epoxides to form cyclic carbonates under solvent-free conditions. *ChemCatChem* 2013;**5**(6):1328−33.
47. Mujmule RB, Chung W-J, Kim H. Chemical fixation of carbon dioxide catalyzed via hydroxyl and carboxyl-rich glucose carbonaceous material as a heterogeneous catalyst. *Chem Eng J* 2020:125164.
48. Biswas T, Mahalingam V. g-C3N4 and tetrabutylammonium bromide catalyzed efficient conversion of epoxide to cyclic carbonate under ambient conditions. *New J Chem* 2017;**41**(24):14839−42.
49. Samanta S, Srivastava R. A novel method to introduce acidic and basic bi-functional sites in graphitic carbon nitride for sustainable catalysis: cycloaddition, esterification, and transesterification reactions. *Sustain Energy Fuels* 2017;**1**(6):1390−404.
50. Vidal JL, Andrea VP, MacQuarrie SL, Kerton FM. Oxidized biochar as a simple, renewable catalyst for the production of cyclic carbonates from carbon dioxide and epoxides. *ChemCatChem* 2019;**11**(16):4089−95.
51. Saptal VB, Sasaki T, Harada K, Nishio-Hamane D, Bhanage BM. Hybrid amine-functionalized graphene oxide as a robust bifunctional catalyst for atmospheric pressure fixation of carbon dioxide using cyclic carbonates. *ChemSusChem* 2016;**9**(6):644−50.
52. Samikannu A, Konwar LJ, Mäki-Arvela P, Mikkola J-P. Renewable N-doped active carbons as efficient catalysts for direct synthesis of cyclic carbonates from epoxides and CO_2. *Appl Catal B Environ* 2019;**241**:41−51.

53. Chen W, Zhong L-X, Peng X-W, Sun R-C, Lu F-C. Chemical fixation of carbon dioxide using a green and efficient catalytic system based on sugarcane Bagasse; an agricultural waste. *ACS Sustain Chem Eng* 2015;**3**(1):147—52.
54. (a) Ucar H, Van derpoorten K, Cacciaguerra S, Spampinato S, Stables JP, Depovere P, Isa M, Masereel B, Delarge J, Poupaert JH. Synthesis and anticonvulsant activity of 2 (3 H)-benzoxazolone and 2 (3 H)-benzothiazolone derivatives. *J Med Chem* 1998;**41**(7):1138—45. (b) Brickner SJ. Oxazolidinone antibacterial agents. *Curr Pharmaceut Des* 1996;**2**(2):175—94. (c) Famiglini V, Silvestri R. Focus on chirality of HIV-1 non-nucleoside reverse transcriptase inhibitors. *Molecules* 2016;**21**(2):221.
55. Heinze T, Liebert T, Heublein B, Hornig S. Functional polymers based on dextran. In: *Polysaccharides ii*. Springer; 2006. p. 199—291.
56. (a) Vojčić N, Bregović N, Cindro N, Požar J, Horvat G, Pičuljan K, Meštrović E, Tomišić V. Optimization of omeprazole synthesis: physico-chemical steering towards greener processes. *Chemistry* 2017;**2**(17):4899—905. (b) Keri RS, Hiremathad A, Budagumpi S, Nagaraja BM. Comprehensive review in current developments of benzimidazole-based medicinal chemistry. *Chem Biol Drug Des* 2015;**86**(1):19—65. (c) Zhang J, Wang J-L, Zhou Z-M, Li Z-H, Xue W-Z, Xu D, Hao L-P, Han X-F, Fei F, Liu T. Design, synthesis and biological activity of 6-substituted carbamoyl benzimidazoles as new nonpeptidic angiotensin II AT1 receptor antagonists. *Bioorg Med Chem* 2012;**20**(14):4208—16. (d) Zhou R, Skibo EB. Chemistry of the pyrrolo [1, 2-a] benzimidazole antitumor agents: influence of the 7-substituent on the ability to alkylate DNA and inhibit topoisomerase II. *J Med Chem* 1996;**39**(21):4321—31.
57. Phatake VV, Bhanage BM. Cu@ UgC 3 N 4 catalyzed cyclization of o-phenylenediamines for the synthesis of benzimidazoles by using CO_2 and dimethylamine borane as a hydrogen source. *Catal Lett* 2019;**149**(1):347—59.

CHAPTER 10

Surface-modified nanomaterials for synthesis of pharmaceuticals

Kishore Natte[1,3] and Rajenahally V. Jagadeesh[2]

[1]Chemical and Material Sciences Division, CSIR-Indian Institute of Petroleum (CSIR-IIP), Mohkampur, Dehradun, India; [2]Leibniz Institute for Catalysis, Rostock, Germany; [3]Academy of Scientific and Innovative Research (AcSIR), Ghaziabad, Uttar Pradesh, India

1. Introduction

Transition metals are commonly used as catalysts either in the form of homogeneous complexes or heterogeneous materials in numerous chemical transformations.[1] Due to the increasing demand for the improvements and advancement of industrial chemical procedures, scientific attention has been paid for the development of cost-effective, durable, and convenient catalytic systems.[1] In this respect, a number of homogeneous and solid-based catalysts have been established for the production of all kinds of chemicals. Homogenous catalysts dissolve in reaction media and provide more readily accessible active catalytic sites, which work under mild reaction conditions and exhibit high selectivities.[1,2] However, homogeneous complexes are rather sensitive or require sophisticated ligands and hence they are difficult to apply at drastic conditions.[2] In addition, catalyst recycling and reusability are the major issues with homogeneous catalysts.[2] These homogeneous catalytic processes require additional techniques/works and costs for the catalyst separation and product purification.[2] In this regard, homogeneous catalysts are less feasible for the large scale bulk chemicals productions as well as for the synthesis of pharmaceuticals.[2] In contrast, solid-based catalysts are stable and conveniently reusable, which enable for achieving more cost-effective protocols as well considerably facilitate easy product purification.[3] Hence, they are more preferable to achieve sustainable and cost-effective industrial processes.[3] Among heterogeneous materials, nanocatalysts obtained by the pyrolysis process are of prime importance owing to their excellent activities and selectivities as well as low energy consumptions.[3]

In this book chapter, we discuss number of opportunities offered by supported nanoparticles to promote chemical reactions for the preparation and functionalization of pharmaceuticals and their intermediates. More clearly, we discuss on the preparation of both precious and nonprecious solid catalysts for reduction of nitroarenes and hetero(arenes), N-methylation, and N-alkylation as well as reductive aminations for the production of selected pharmaceutical compounds and their intermediates (Table 10.1). We believe that this book chapter will be helpful to researchers working in the field of nanocatalysis,

Table 10.1 Recent achievements on the preparation of supported nanoparticles and their applications for the synthesis of pharmaceutical-based products.

Catalyst	Reaction conditions	Applications and synthesis of pharmaceuticals-based products
Co-DBCO-TPA@C-800 by the group of Beller[4]	5–7 bar NH$_3$, 40 bar H$_2$, 120°C	Used for the synthesis of general amines and pharmaceutical products.
Ni/Al$_2$O$_3$ by Kempe and coworkers[5]	Aqueous NH$_3$, 10 bar H$_2$, 80°C	Applied for the preparation of amines and pharmaceutical products.
Rh-NP–rGO by Dyson and coworkers[6]	30 bar H$_2$, 100°C, 24–48 h	Applied for the hydrogenation of heterocycles, which allows for the preparation of furan-based selected drugs and bioactive compounds such as Visnagin, Methoxsalen, and Khellin.
Co/Melamine-6@TiO$_2$-800-5 by Beller and coworkers[7]	60 bar H$_2$, 160°C, 48 h	Applied for the preparation of biologically active compounds such as isosolenopsin A and Desoxypipradrol

organic synthesis, and drug discovery. Certainly, this book chapter inspires for the further developments in heterogeneous materials and their applications in chemical synthesis for the manufacture of essential chemicals and drugs.

2. Noble metal-based nanoparticles for the synthesis of pharmaceuticals

Applicability of precious metals-based nanoparticles have received considerable attention in chemical and pharma sectors due to their exceptional catalytic performances under mild conditions.[8] Among these, catalytic hydrogenations are of particular interest, which give access to number of valuable chemicals as well as drugs, pesticides, herbicides, and biomolecules.[9] The ring hydrogenated products of arenes or heteroarenes are frequently appeared in pharmaceuticals, natural products and bioactive compounds.[9,10] In particular, quinolines are suitable starting materials for accessing heteroatom containing cycloaliphatic compounds, which are important motifs found in a number of drug molecules.[11] In this respect, Dyson and coworkers[6] prepared and applied Rh-NP–rGO nanocomposite-catalysts for the selective ring hydrogenation of heterocycles such as quinolines and benzofurans.[6] Interestingly, this Rh-based hydrogenation protocol allows for the preparation of furan-based selected drugs and bioactive compounds such as Visnagin, Methoxsalen, and Khellin in 67%–95% yields (Scheme 10.1).[6]

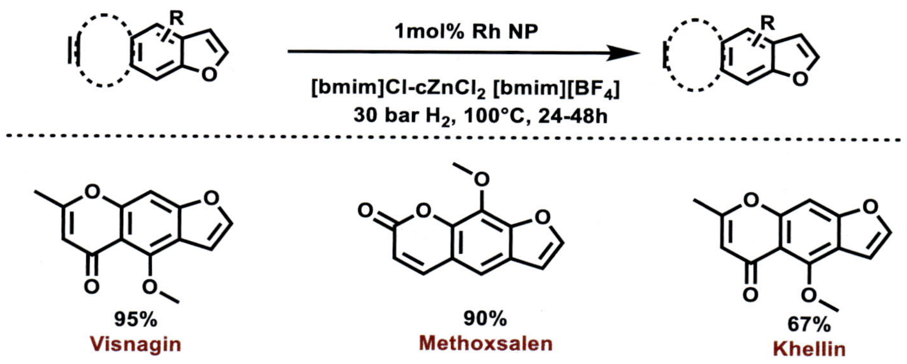

Scheme 10.1 Rh-nanocomposite catalyzed hydrogenation of heterocyclic compounds to access selected drug molecules.

In 2018, Naka and coworkers[12] reported a reusable palladium-loaded TiO₂ photocatalyst that successfully enabled the *N*-methylation of amines including drugs under mild conditions. The authors extended this approach to synthesize pharmaceuticals including deuterium-labeled ones via photocatalytic *N*-methylation or *N*-ethylation approach using deuterated methanol and ethanol. For example, Loxapine, Venlafaxine, and Butenafine drugs were synthesized from amines and methanol as C1 reagent (Scheme 10.2). Also, this Pd/TiO₂ photocatalyst was successfully recyclable up to four times.[12]

Jagadeesh and coworkers[13] prepared Pd-nanoparticles (Pd@SiO₂) by the reduction of palladium(II) acetylacetonate on commercial SiO₂ employing H₂ (Fig. 10.1). The obtained palladium-nanoparticles are found to be recoverable catalysts for the production of different *N*-alkyl amines by applying auto-transfer hydrogen method without external base or additive. The catalyst, Pd@SiO₂ was extensively characterized by ICP, BET, XRD, TEM, XPS, H₂-TPR, and chemisorption. Excellent catalytic performance of Pd@SiO₂ is due to the formation of very small palladium-nanoparticles, enhancement of metallic palladium in the near-surface region and excellent dispersion of active Pd species. Applying this Pd-based methodology, authors have prepared biologically and pharmaceutically relevant amines such as, 4-Benzylpiperidine, Benzylpiperazine, Methylbenzylpiperazine, Piribedil (anti-Parkinson agents), and Cinnarizine (calcium channel blocker) (Scheme 10.3).[13]

Recently, the group of Natte[14] reported common Pd/C catalyst for *N*-methylation of nitroaromatics as well as amines using CH₃OH as one carbon and hydrogen source. In this paper, the authors demonstrated *N*-methylation and *N*-trideuteromethylation of nitro moiety in nimesulide drug molecule using MeOH or CD₃OD as coupling partners and achieved the targeted product in 92% and 89% yields, respectively (Scheme 10.4).

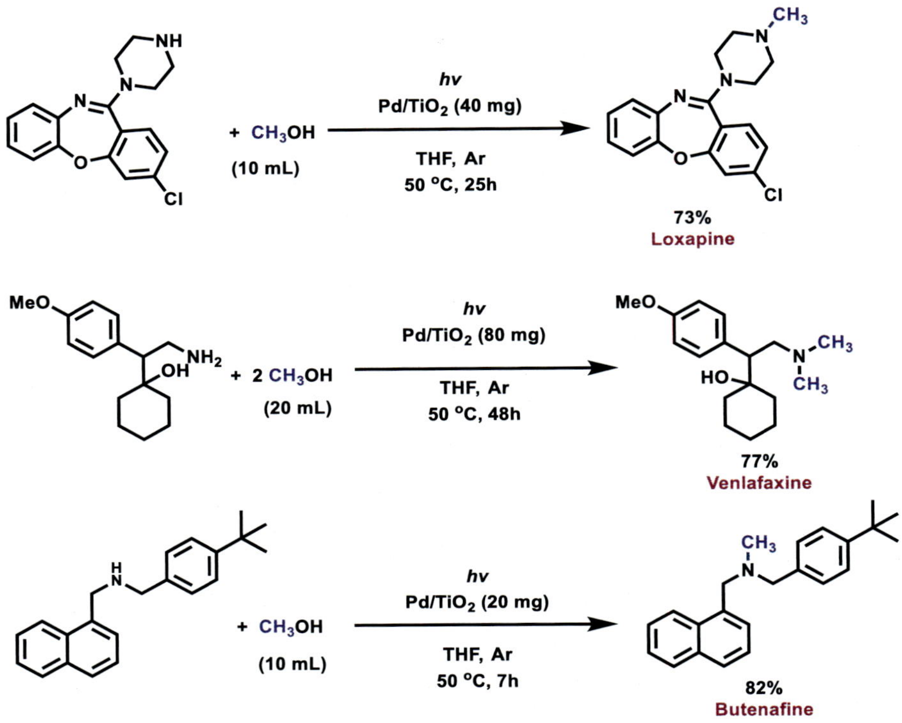

Scheme 10.2 Synthesis of (N), N-dimethylated drug molecules using Pd/TiO$_2$ as a catalyst.

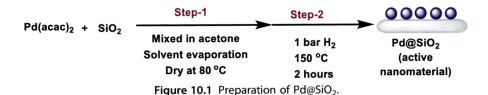

Figure 10.1 Preparation of Pd@SiO$_2$.

3. Nonnoble metal-based nanoparticles for the synthesis of pharmaceuticals

With respect to the present pattern and state-of-the-art-catalysts, the advancement and development of nonprecious metal-based nanocatalysts is one of the significant goal of modern organic synthesis because the base metals are more abundant, inexpensive, and less toxic.[15] In 2015, Beller and coworkers[16] prepared Co$_3$O$_4$—Co/NGr@α-Al$_2$O$_3$ nanocatalysts by the pyrolysis Co-phenanthroline complex α-Al$_2$O$_3$ under argon (Fig. 10.2).

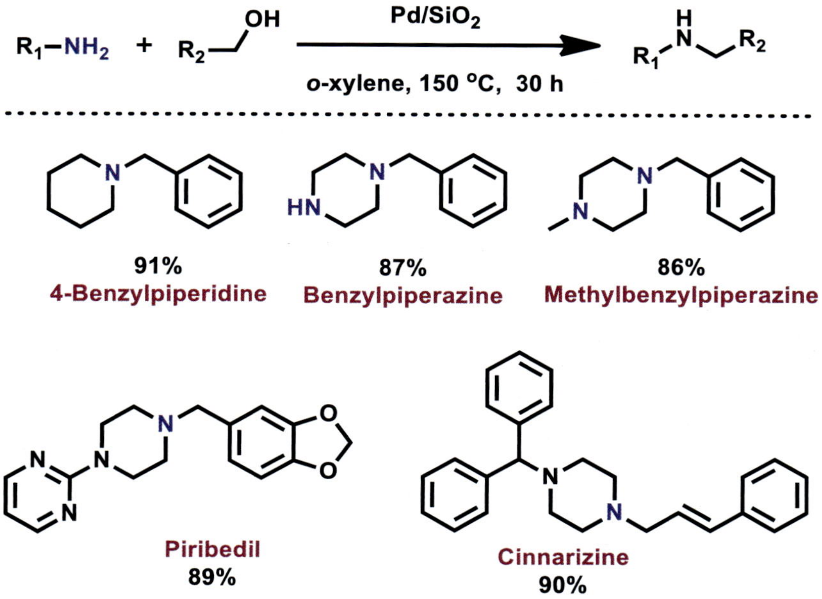

Scheme 10.3 Synthesis of pharmaceutical molecules by using Pd/SiO$_2$ as a catalyst.

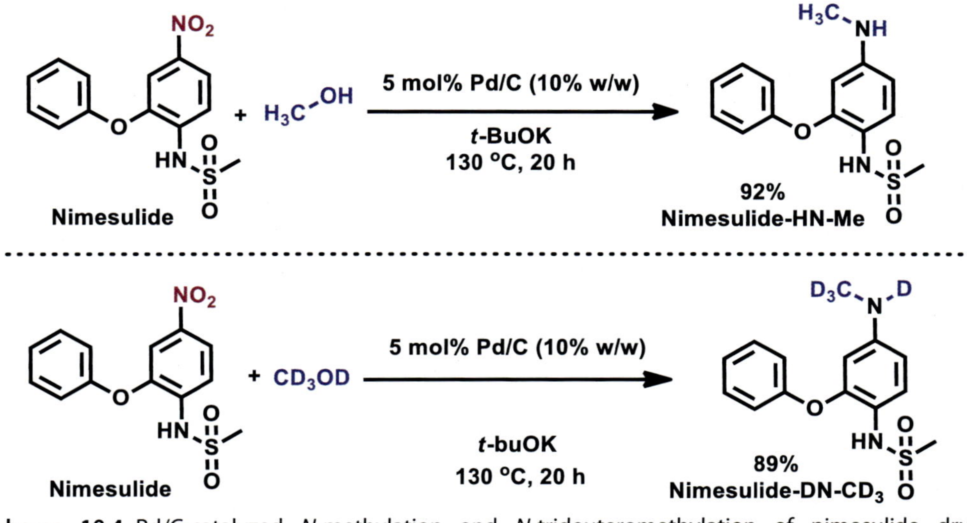

Scheme 10.4 Pd/C-catalyzed N-methylation and N-trideuteromethylation of nimesulide drug molecule.

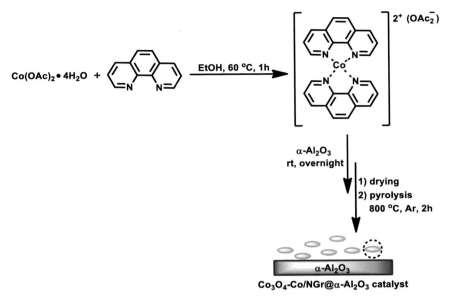

Figure 10.2 Synthesis of Co$_3$O$_4$–Co/NGr@α-Al$_2$O$_3$.

These nanocatalysts exhibited excellent chemoselectivity for the ring hydrogenation of various quinolines and afforded respective 1,2,3,4-tetrahydroquinolines in moderate to good yields. In addition, other N-heterocycles such as acridine and unprotected indoles were also selectively hydrogenated and gave ring hydrogenated products in excellent yields. Next, the authors showcased the applications of 1,2,3,4-tetrahydroquinolines in the preparation of three different pharmaceutically active compounds (Scheme 10.5). As an example, the reduced 6-methoxy-1,2,3,4-tetrahydroquinoline was treated with 3,4,5-trimethoxybenzoyl chloride to afford (6-methoxy-3,4-dihydroquinolin1(2H)-yl)(3,4,5-trimethoxyphenyl)methanone (A) in 92% yield (Scheme 10.5), which is a tubulin polymerization inhibitor. Furthermore, 6-fluoro-2-methyl-1,2,3,4- tetrahydroquinoline was transformed into flumequine in three steps in 78% total yield (B; Scheme 10.5). Lastly, (±)-galipinine was produced from 2-(2-(benzo[d][1,3]-dioxol-5-yl)ethyl)-1,2,3,4-tetrahydroquinoline in 91% yield (C; Scheme 10.5).[16]

Again, the Beller research group[17] for the first time reported an efficient heterogeneous iron-based nanocatalyst for the ring hydrogenation of quinolones and (iso)quinolones using H$_2$ as a reductant.[17] The active nanocatalyst was synthesized by the immobilization and pyrolysis of Fe(II) acetate-N-aryliminopyridines complex on carbon. The Fe-based protocol was applied for the synthesis of biologically active compounds (Scheme 10.6).

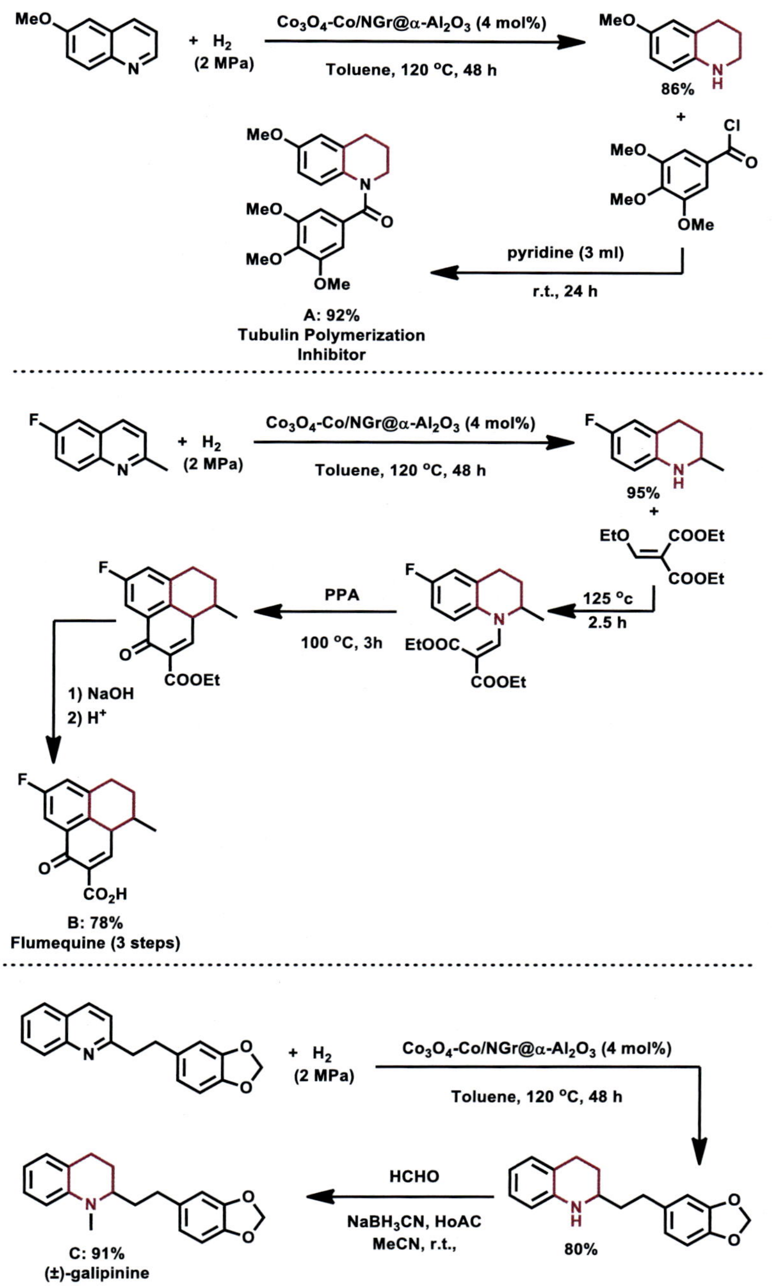

Surface-modified nanomaterials for synthesis of pharmaceuticals   257

Scheme 10.5 Pharmaceutical compounds prepared by using Co_3O_4–Co/NGr@α-Al_2O_3 as a catalyst.

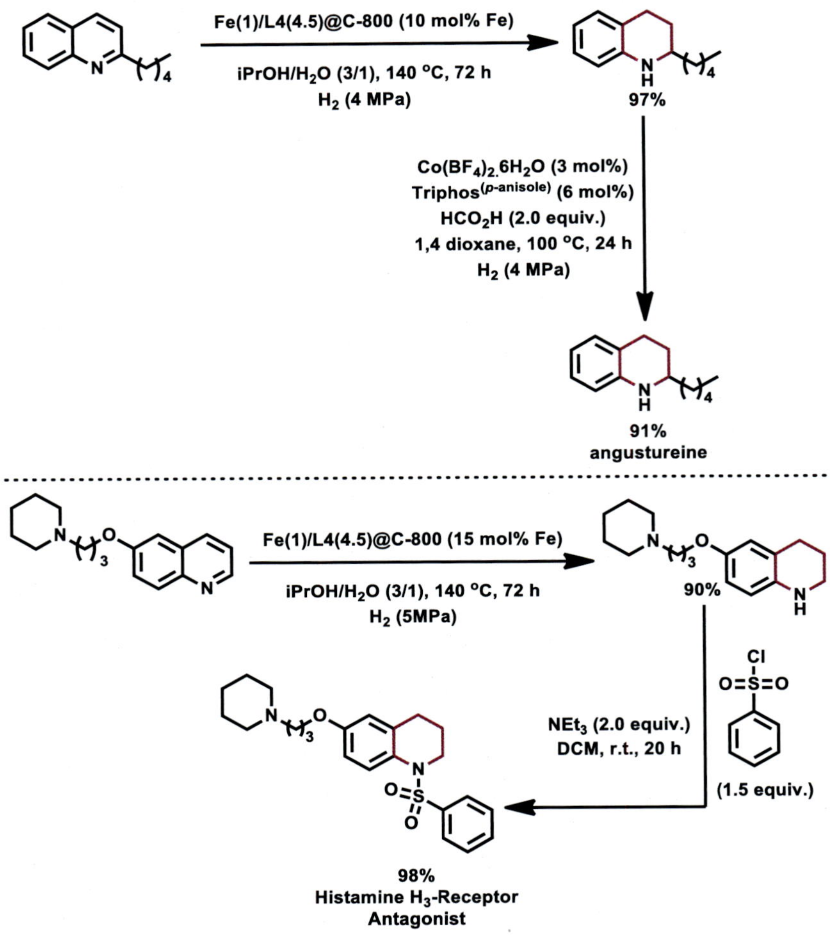

Scheme 10.6 Pharmaceutical compounds prepared by using iron catalysts.

Next, another cobalt-based heterogeneous catalysts, Co/Melamine-6@TiO$_2$-800-5 was reported, which was prepared by the pyrolysis of Co-Melamine in TiO$_2$ support for the hydrogenation of pyridines to pyrrolidines[7] In fact this catalyst constitute as the first and foremost heterogeneous nonprecious metal catalyst for the ring hydrogenation of pyridines. Interestingly, using this cobalt catalyst, the syntheses of bioactive compounds such as isosolenopsin A and Desoxypipradrol in excellent yields (Scheme 10.7) was performed starting from pyridine-based compounds by hydrogenations.[7]

In 2017, Beller and team[4] synthesized MOFs-derived graphitic shell encapsulated cobalt nanoparticles and tested for the reductive amination of aldehydes and ketones to prepare numerous amines. These cobalt nanostructured catalysts (Co-DABCO-TPA@C-800)

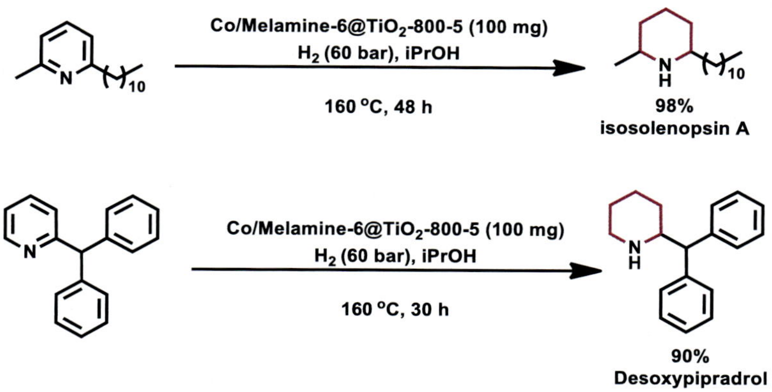

Scheme 10.7 Pharmaceutical compounds prepared by using Co/Melamine-6@TiO$_2$-800-5 catalyst.

were typically prepared by the templated synthesis of cobalt nitrate-DABCO-terephthalic acid MOF on carbon in a controlled manner and subsequent pyrolysis of this templated synthetic material at 800°C under argon (Fig. 10.3).[4]

Co-DABCO-TPA@C-800 nanostructured catalyst was characterized and evaluated using Cs-corrected STEM, electron energy loss spectroscopy, XPS, and XRD spectral analysis. All these characterized results revealed the formation of graphitic shells encapsulated metallic cobalt particles with sizes from 5 to 30 nm. In addition, a small quantity of oxidic cobalt in the surface of core-shell nanoparticles is also noticed. EELS measurements identified the traces of nitrogen which is located in the vicinity of metallic Co particles within Co-DBCO-TPA@C-800 catalyst.[4] This Co-catalyzed amination protocol was applied for the installation of —NH$_2$ motif in structurally diverse molecules. Initially, bioactive amphetamines (Scheme 10.8A), which are effective CNS—stimulating drugs, were prepared in up to 91% yield. Also, this cobalt catalyst was applied for the amination

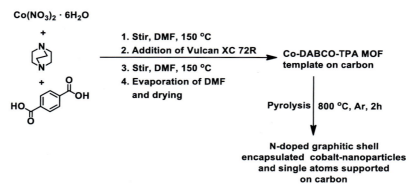

Figure 10.3 Preparation of Co-DABCO-TPA@C-800 by the pyrolysis of Co-MOF on carbon.

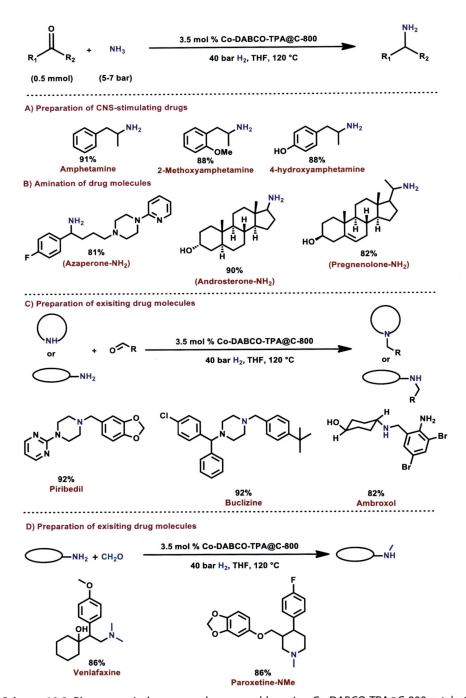

Scheme 10.8 Pharmaceutical compounds prepared by using Co-DABCO-TPA@C-800 catalyst.

of nonsteroidal antiinflammatory (NSAI) drugs and steroid derivatives (Scheme 10.8B). Further, the Co-DABCO-TPA@C-800-based reductive amination approach was also successfully executed for the synthesis of selected commercial pharmaceuticals (Scheme 10.8C). At the end, using this cobalt catalyst selected N-methylated drugs from aldehydes and N, N-dimethylamine were also synthesized (Scheme 10.8D).

Later, Kempe et al.[5] developed Ni/γ-Al$_2$O$_3$ catalyst and tested for the reductive amination using aqueous NH$_3$ and H$_2$ to produce a number of primary amines.[5] These novel Ni-catalysts were in general prepared by the immobilization of homogeneous Ni-salen complex on γ-Al$_2$O$_3$ and pyrolysis under nitrogen atmosphere at 700°C followed by reduction under H$_2$ at 550°C. Ni-nanoparticles were also synthesized on various supports such as Al$_2$O$_3$, SiO$_2$, carbon, CeO$_2$. Characterization by TEM and XPS of Ni/γ-Al$_2$O$_3$ results revealed that this catalyst comprise N-doped carbon encapsulated superparamagnetic Ni-nanoparticles in Ni^0 and Ni^{+2} oxidation state with an average size of 8 nm. By using this Ni/γ-Al$_2$O$_3$ catalyst under mild conditions (10 bar H$_2$, 80°C), few drug molecules were efficiently synthesized in high yields (Scheme 10.9).[5]

Followed by Kempe and team reductive amination,[5] again Beller and Jagadeesh[18] together reported SiO$_2$ supported Ni-nanoparticles as reductive amination catalysts. The obtained Ni-TA@SiO$_2$-800 catalyst promotes the synthesis of primary amines high yields. Outstandingly, this Ni-TA@SiO$_2$-800 catalyst is recoverable and reused up to 10 times. Ni-TA@SiO$_2$-800 nanocatalyst was further characterized by TEM, XPS, and BET. These results revealed that Ni-nanoparticles are formed in spherical shapes with 5—15 nm size. The Ni-TA@SiO$_2$-800 nanocatalyst was applied for the synthesis of drug like molecules in moderate to high yields (Scheme 10.10).[18]

With respect to Fe-based heterogeneous reductive amination catalysts, Kempe and coworkers[19] for the first time reported Fe/(N)SiC catalyst for the reductive N-alkylation of aldehydes and ketones with aqueous NH$_3$ and H$_2$ to afford the respective amines in good yields. This iron catalyst was synthesized by the impregnation and pyrolysis of a

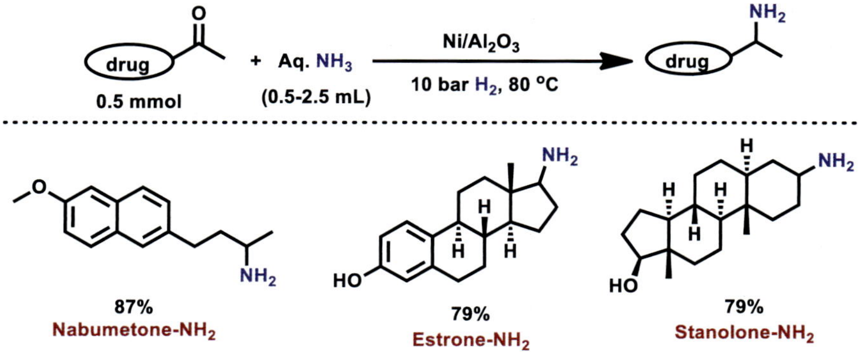

Scheme 10.9 Synthesis of bioactive compounds using Ni/γ-Al$_2$O$_3$.

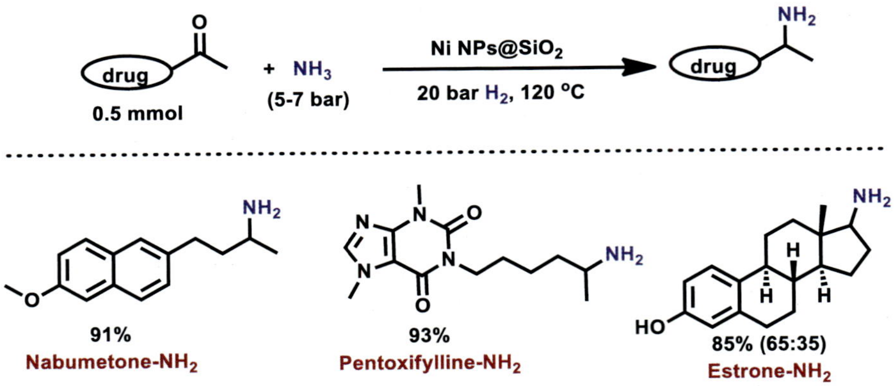

Scheme 10.10 Synthesis of bioactive compounds using Ni-TA@SiO$_2$-800 catalyst.

specific Fe-salen complex on N-doped SiC under a N$_2$ atmosphere at 750°C followed by a reduction at 550°C (Scheme 10.11).[19] By using this Fe/(N)SiC catalyst, the authors synthesized CNS stimulant drug molecules such as 4-Methoxyamphetamine and 2-Methoxyamphetamine.[19]

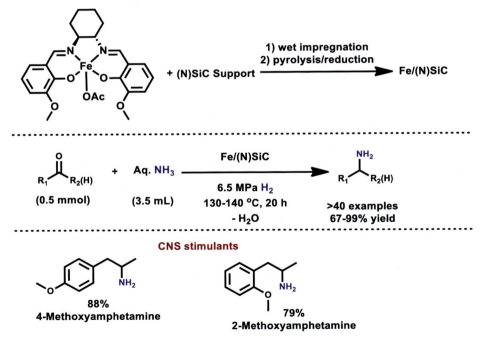

Scheme 10.11 Preparation of Fe catalyst and its application for pharmaceutical compounds via reductive amination.

Surface-modified nanomaterials for synthesis of pharmaceuticals   263

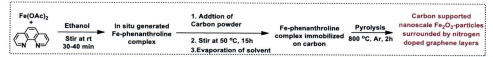

Figure 10.4 Preparation of carbon supported Fe$_2$O$_3$-nanoparticles surrounded by nitrogen doped graphene layers.

Interestingly, nitrogen doped graphene activated nanoscale Fe$_2$O$_3$-particles supported on carbon (Fe$_2$O$_3$/NGr@C) were prepared by the Beller and team,[20] which represent excellent nanocatalysts for the reduction of functionalized nitroarenes as well as chemoselective reduction of —NO$_2$ group in the life science molecules. These Fe$_2$O$_3$-based nanoparticles were prepared by the immobilization and pyrolysis of homogeneous Fe-phenanthroline complex on carbon (Vulcan XC-72R) at 800°C for 2 h under argon (Fig. 10.4). The pyrolysis of specific Fe-phenanthroline complex on carbon furnished the Fe$_2$O$_3$ nanoparticles which are surrounded with nitrogen doped graphene layers, which are the crucial features of these Fe-materials to exhibit higher activity and unprecedented chemoselectivity for the reduction of nitroaromatics to anilines. As a result >80 nitro compounds including pharmaceutical compounds such as Nimodipine, Nilutamide, Nimesulide Flutamide, and Niclosamide in high yields (Scheme 10.12).[20]

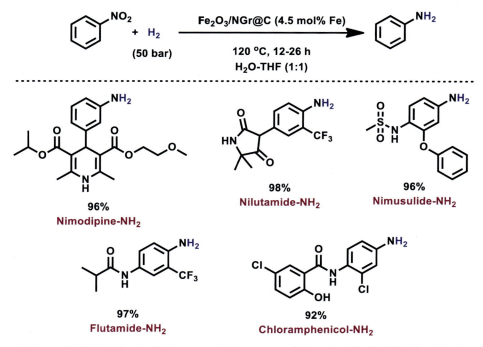

Scheme 10.12 Synthesis of pharmaceutical compounds by using Fe$_2$O$_3$/NGr@C catalyst.

Similar kinds of pharmaceuticals were also prepared by Yang and coworkers by using FeS$_2$/NSC as a heterogeneous catalyst.[21]

Then, few years later, Jagadeesh and Beller et al.[22] applied Fe$_2$O$_3$/NGr@C catalysts for the selective synthesis N-methylated amines. These represent important class of amines and N-methyl motifs presented in number of drugs and other bioactive compounds. Thus, the syntheses of N-methyl amines play crucial roles in drug discovery. Applying Fe$_2$O$_3$/NGr@C catalyst using paraformaldehyde as convenient methylation and hydrogen source, N-methylation of nitro group containing biologically active molecules to the respective N, N-dimethylated analogs were synthesized in good yields. For example, the nitro hydrogenation of calcium channel blockers (e.g., Clinidipine, Nicardipine, and Nimodipine) were converted to the targeted N, N-dimethylamines in up to 76% yields. Interestingly, the authors also synthesized Venlafaxine (antidepressant drug) in 89% yield including imipramine and amitriptyline (Scheme 10.13). Very recently, Yang and coworkers prepared Cu/Al$_2$O$_3$ nanocatalyst and demonstrated the synthesis of N,N-dimethylated pharmaceutical molecules form its corresponding nitro compounds in excellent yields.[22]

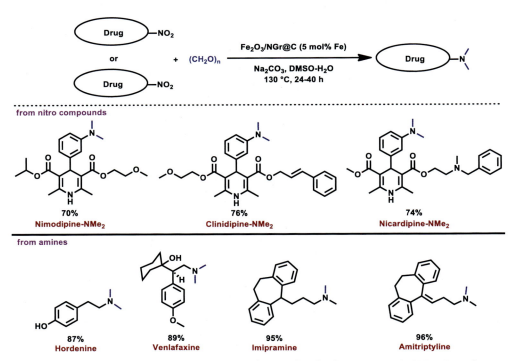

Scheme 10.13 Synthesis of N, N-methylated drug molecules by using Fe$_2$O$_3$/NGr@C catalyst.

4. Conclusions

In summary, the use of nanocatalysts in organic reactions to prepare pharmaceuticals and other life science molecules has received considerable attention in both laboratory and industrial processes. This book chapter offers a short overview in the area of noble and nonnoble metal-based nanoparticles as excellent catalysts for the efficient synthesis of various types of pharmaceutical agents. Initially, the applications of heterogeneous noble catalysts for hydrogenation and N-alkylation/methylation reactions have been reviewed in this chapter. Secondly, nonprecious metal-based nanocatalysts for hydrogenation and reductive amination reactions and their applications in pharmaceutical synthesis have been discussed. The development and improvements of nanocatalysts with enhanced activity and selectivity will have noteworthy implications for organic reactions and drug synthesis. We truly believe that this chapter will be beneficial for researchers and industrialists interested in heterogeneous catalysis for pharmaceutical synthesis.

Important websites (important companies, organizations, and research groups working on synthesis of surface-modified nanomaterials for applications in catalysis):

1. https://www.catalysis.de/en/people/beller-matthias/
2. https://www.catalysis.de/en/people/rajenahally-jagadeesh/
3. https://www.gawandesgroup.com/group-members/manoj-gawande/
4. https://www.profilfelder.uni-bayreuth.de/de/advanced-fields/polymer-kolloidforschung/mitglieder/Kempe_Rhett/index.php
5. https://www.chem.s.u-tokyo.ac.jp/users/synorg/en/index.html
6. http://english.licp.cas.cn/pe/js/sf/
7. https://www.iip.res.in/synthetic-chemistry-and-petrochemicals/dr-kishore-natte/
8. https://catalysts.basf.com/
9. https://catalysts.evonik.com/en/catalyst-brands
10. https://matthey.com/en/products-and-services/pharmaceutical-and-medical/catalysts?q=
11. https://www.exxonmobilchemical.com/en/

References

[1] a) Gawande M, Goswami A, Asefa T, Guo H, Biradar AV, Peng D-L, Zboril R, Varma RS. *Chem Soc Rev* 2015;**44**:7540−90. b) Gawande MB, Brancoa PS, Varma RS. *Chem Soc Rev* 2013;**42**:3371−93. c) Brandsma L, Vasilevsky SF, Verkruijsse HD. *Application of transition metal catalysts in organic synthesis.* Springer-Verlag Berlin Heidelberg; 1999, ISBN 978-3-642-60328-0. d) Tsuji J. *Transition metal reagents and catalysts: innovations in organic synthesis.* John Wiley & Sons, Ltd; 2001, ISBN 9780470854761. e) van Santen RA, Neurock M. *Molecular heterogeneous catalysis: a conceptual and computational approach.* Weinheim: Wiley-VCH Verlag GmbH; 2001, ISBN 978-3-527-29662-0. f) van Leeuwen PWNM. *Homogeneous catalysis.* Springer Netherlands; 2004, ISBN 978-1-4020-2000-1.

[2] a) García-Álvarez J. *Molecules* 2020;**25**:1493. b) Twigg MV. In: Twigg MV, editor. *Mechanisms of inorganic and organometallic reactions*, vol. 8. Boston, MA: Springer US; 1994. p. 363−96. c) Shelke YG, Yashmeen A, Gholap AVA, Gharpure SJ, Kapdi AR. *Chem Asian J* 2018;**13**:2991−3013.

[3] a) Liu L, Corma A. *Chem Rev* 2018;**118**:4981−5079. b) Liu L, Corma A. *Trends in Chemistry* 2020;**2**: 383−400.
[4] Jagadeesh RV, Murugesan K, Alshammari AS, Neumann H, Pohl M-M, Radnik J, Beller M. *Science* 2017:eaan6245.
[5] Hahn G, Kunnas P, de Jonge N, Kempe R. *Nat Catal* 2019;**2**:71−7.
[6] Karakulina A, Gopakumar A, Akçok I, Roulier BL, LaGrange T, Katsyuba SA, Das S, Dyson PJ. *Angew Chem Int Ed* 2016;**55**:292−6.
[7] Chen F, Li W, Sahoo B, Kreyenschulte C, Agostini G, Lund H, Junge K, Beller M. *Angew Chem Int Ed* 2018;**57**:14488−92.
[8] a) Weiqin S, Qing Y, Jian W. *Curr Org Chem* 2011;**15**:3692−705. b) Wu X-F, Anbarasan P, Neumann H, Beller M. *Angew Chem Int Ed* 2010;**48**:9047−50. c) Gawande MB, Goswami A, Felpin F-X, Asefa T, Huang X, Silva R, Zou X, Zboril R, Varma RS. *Chem Rev* 2016;**116**:3722−811.
[9] Wiesenfeldt MP, Nairoukh Z, Dalton T, Glorius F. *Angew Chem Int Ed* 2019;**58**:10460−76.
[10] Huck CJ, Sarlah D. *Chem* 2020;**6**:1589−603.
[11] Sridharan V, Suryavanshi PA, Menéndez JC. *Chem Rev* 2011;**111**:7157−259.
[12] Wang L-M, Jenkinson K, Wheatley AEH, Kuwata K, Saito S, Naka H. *ACS Sustain Chem Eng* 2018;**6**:15419−24.
[13] Alshammari AS, Natte K, Kalevaru NV, Bagabas A, Jagadeesh RV. *J Catal* 2020;**382**:141−9.
[14] Goyal V, Gahtori J, Narani A, Gupta P, Bordoloi A, Natte K. *J Org Chem* 2019;**84**:15389−98.
[15] Wang D, Astruc D. *Chem Soc Rev* 2017;**46**:816−54.
[16] Chen F, Surkus A-E, He L, Pohl M-M, Radnik J, Topf C, Junge K, Beller M. *J Am Chem Soc* 2015;**137**:11718−24.
[17] Sahoo B, Kreyenschulte C, Agostini G, Lund H, Bachmann S, Scalone M, Junge K, Beller M. *Chem Sci* 2018;**9**:8134−41.
[18] Murugesan K, Beller M, Jagadeesh RV. *Angew Chem Int Ed* 2019;**58**:5064−8.
[19] Bäumler C, Bauer C, Kempe R. *ChemSusChem* 2020;**13**:3110−4.
[20] Jagadeesh RV, Surkus A-E, Junge H, Pohl M-M, Radnik J, Rabeah J, Huan H, Schünemann V, Brückner A, Beller M. *Science* 2013;**342**:1073−6.
[21] Duan Y, Dong X, Song T, Wang Z, Xiao J, Yuan Y, Yang Y. *ChemSusChem* 2019;**12**:4636−44.
[22] Natte K, Neumann H, Jagadeesh RV, Beller M. *Nat Commun* 2017;**8**:1344.

CHAPTER 11

Surface-modified nanomaterial-based catalytic materials for modern industry applications

Priti Sharma[1] and Manoj B. Gawande[1,2]

[1]Regional Centre of Advanced Technologies and Materials, Czech Advanced Technology and Research Institute, Palacký University, Olomouc, Czech Republic; [2]Department of Industrial and Engineering Chemistry, Institute of Chemical Technology, Mumbai Marathwada Campus, Jalna, Maharashtra, India

1. Introduction

Nanomaterial and nanoengineering are two parallel faces of future research as a promising scientific pathway.[1,2] Disasters in terms of epidemic, release of greenhouse gases from fossil fuels, deforestation, increasing sea level, and climate change are the greatest threat to humanity and global ecosystems at the same time challenges the widespread scientific evolution of human civilization in the coming years.[3] Due to these fundamental factors, nanotechnology and functionalized nanomaterials are emerging as a key factor in meeting the expanded needs of human civilization in terms of energy, economic stability, and environment protection in the current time situation due to these fundamental factors.[4,5] In recent decades, the discipline of catalysis has played a significant role in the commercialization of a variety of essential products that are exclusively employed as a fundamental component in the developing sectors of energy, organic synthesis, and environmental safety.[6] Since conventional catalysts contribute to environmental issues, low efficiency, and toxicity along with high cost (i.e., metal and metal oxides, metal salts), the worldwide scientific community has explored new sustainable process to replace conventional catalysts with environmentally benign–based green catalysts.[6] For covering the cost and effective applicability in industry, nanotechnology offers exclusively a platform for various application-based scientific research. Functionalized nanomaterials finds a suitable place in agriculture in terms of encapsulation along with entrapment of chemicals used in agriculture such as herbicides, fertilizers,[7] pesticides,[8] and plant growth regulators.[9] On the other hand, in medicinal areas, nanomedicine were prepared with selective targeting technologies with defined specific place of the drug release; such as nanodiagnostics, nanopharmaceuticals, and nanobiomaterials.[10,11] Today, nanopharmaceuticals exclusively covers 70%–80% of the market capture share with widely accepted and approved nanomedicines.[12] Therefore the interaction between nanomaterials engineered field and biomedical engineering have increased multiple folds due to rapid innovations based in nanomedicine and medical science. Currently, nanotechnology opens a wide scope for biomedical sciences and engineering precisely for appropriate nanoparticle synthesis.[13] During the last decade, a wide number of chemical methods via nanotechnology usage have been developed to synthesize functionalized nanoparticles for drug molecular biology, delivery, cancer therapy, and tissue engineering.[14,15] This represents how rapidly nanotechnology in various sectors growing faster and covers the market in commercial sectors.

As we all know, environmentally friendly chemical process engineering with the integrated catalytic efficient protocol are areas of nanomaterials with a huge scope of future research.[16] A huge number of research were published during the last couple of years, precisely keeping a center eye attention over nanomaterial synthesis, its optimization along with precise design and application for catalysis and industrial applications.[17] Due to huge difficulties during separation in synthesis process; an easy facile protocol of heterogenization method were introduced.[18] Although such heterogenization method further enhances the reactivity of supported nanoparticles (NPs) in various cases, but nevertheless their selectivity seeking further investigation and improvement for a large-scale process optimization.

In the present chapter, various type of surfaces modifications techniques were reported for nanomaterials like graphene, silica, TiO_2, and magnetic modified surface and such combination could be further explored for industrially important chemical transformations such as oxidation, Reduction, isomerization, coupling, esterification, and polymerization (Fig. 11.1).[19–27] For making more accessible sites for various industrially important reactions researchers have used various possible surface modification techniques such as acidic functional groups, basic functional groups, miscellaneous species like MOFs, zeolites, fullerenes, and click modification.

Apart from that, mercaptopropyl group containing derivatives, carboxylic acid, phosphoric acid, dopamine derivatives, phosphonate, and amine derivatives were used primarily as anchoring substrates.[28–35]

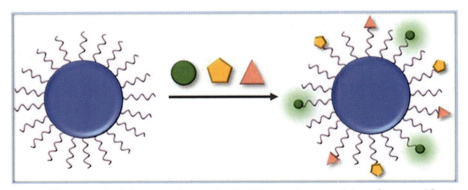

Figure 11.1 Nanoparticle (silica, graphene, titania TiO_2, and magnetic) surface modification via various ligand (NH_2, Cl, SH, COOH, PO_4) and organic complexes, and so forth.

The vast scope with varied challenges of industrial along with its applications in environmental engineering via using nanomaterials are discussed in the present chapter, with a clear emphasis on functionalized nanomaterial which finds suitable applications at the industrial level.

2. Scope of the book chapter

In recent years, nanotechnology has emerged as one of the most encouraging technologies, which provides a unique platform in various industrial sectors such as pharmaceutical, polymer, paints, and so forth for production of various essential chemicals on a commercial scale by using nanomaterials. However, the application of nanomaterials at industrial scale for the production of various fuels and chemicals have emerged as one of the highly promising research platforms due to appreciable yield and optimum cost of products. However, as the name itself is defined, nanomaterials are such small particles that are in the nanometer scale range. Nanomaterials are small particle materials in which a single unit size is in between 1 and 1000 nm. The materials that hold structure at the nanoscale seem to possess a unique and different optical, electronic, and mechanical property. Surface modification strategies of nanomaterials significantly enhance its catalytic activity as well as selectivity in chemical transformation, which is applicable in commercial scale production of any desired chemical to fulfill the society requirements.

Surface parameters of modern age materials are generally incompetent in terms of adhesive property, hydrophobicity, and biocompatibility, and therefore they need be improved before any application or processing protocol such as functional compounds coating. Both the morphological features and chemical nature as well as composition should be well planned in order to get a desirable surface finish.[29,32,35]

In the present chapter we tried to cover the various reported strategies for functionalization of nanomaterials, which could be next generation smart materials with enormous possible applications in industrially important reactions.

3. Active role of surface-modified nanomaterials in industry

Nanomaterials possess a distinctive characteristic that finds versatile applications in industrial sectors such as energy, catalysis, environmental remediation, and medicine technologies. However, nanomaterials are well represented with various advantages, but also suffers with limitations such as limited available reactive sites over surface and insolubility in key solvents. At the same time, there is a huge possibility that sometimes nanomaterials may react with substrate host complexes, solvent mixtures, or other reactive species, which may later the specific reaction proceedings. Such kinds of limitations could be rectified via functionalizing the nanomaterials with targeted specific chemical moieties. The main advantage of functionalization is to fine tune the required characteristics in a

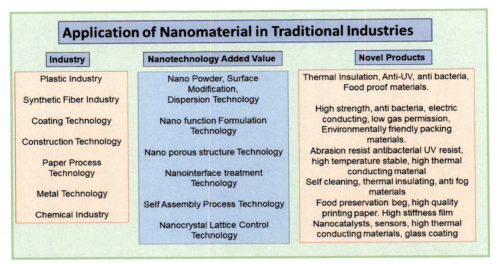

Figure 11.2 Nanotechnology based industrial application in various field.

specific targeted way and maximize usage in certain applications (Fig. 11.2). Specific functionalization of nanomaterials has opened a wide scope for the development of enhanced tools and key protocols for industrial applications such as textile, plastic, paints, and pharmaceuticals (Fig. 11.2).[17,30,32]

Catalysis is well known for playing a vital role in making commercialization of cost-effective chemical synthesis and protocols for material formulation, precisely for energy, organic synthesis, and environmental protection on a wide scale. Catalysis has emerged as a widely acceptable field of research in chemical, pharmaceutical, and petrochemical industries. Therefore, considerable research efforts have been widely explored for formulating the novel catalytic materials with appreciable catalytic efficiency along with good selectivity, stability, and reusability. The huge developments in limited time period in the field of nanotechnology, precisely for the synthesis optimization of NPs, have widely opened gate of catalysis field.

The current chapter gives a brief review over supported nanomaterial-based chemistry, specifically the emphasis on its functionalization over various types of heterogeneous supports such as silica, graphene, TiO$_2$, magnetic, and so forth. Furthermore, their application in industrially important catalytic organic transformation with a clear vision on cost and environmentally friendly approach are discussed.

3.1 Silica-modified nanomaterials

Silica-modified nanomaterials have been deployed for different applications such as industrial catalysis, CO$_2$ capture, separation, dye and chromatographic supports, foams, and emulsion applications.[36] The advanced natured materials development for the upscaling of heterogeneous catalysis-based applications needs controlled optimization in protocols for the synthesis for enhancing active sites as a structural parameter. Mesoporous silica-based nanomaterials have caught the interest of global researchers due to their high surface area, porous design, and metal atoms incorporating possibility over the specific site of mesoporous channels. Particularly in the domain of nanostructured supported materials, which makes them particularly promising supportive materials for the creation of a variety of catalysts. Generally supported nanomaterials have been well divided into active or inert material as per their capacity to take part in catalytic reactions system. Chemically inert supported nanomaterials were classified, which mainly functions for the easy dispersion of metal oxides, metal NPs, or anchoring the organic complexes; at the same time; the active supported materials provide ligand anchoring along with the active site for easy facilitation of the catalytic reaction. However, the mesoporous material can be made by making active site via functionalization the channel surfaces with organic complexes or via immobilizing ligand like metal oxides metal NPs.

Mesoporous surface channels anchoring could be formulated via coating the silica surface with the well-defined active site groups with the various forms of amine, imine sulphonyl, thiol, carboxyl thiol, and click and nitriles groups.[30,31,33–36] The direct postmodification method was comparatively beneficial since it makes available the well-defined exact active site in the nanomaterial in one step along with uniform distribution in the framework. On the other hand, postmodification process can effectively formulate an excellent supported material with effective active site formulation via immobilizing a well-defined functional active site or group over the surface and further searching the catalytic applications.[37] Such functionalized silica nanomaterials could be widely classified into various functional group in use; modification with acidic functional groups (sulphonic, carboxyl), basic functional groups, zeolites, amphoteric nature groups along with enzymes, MOFs, fullerenes, and so forth.[22,28,33–37]

Basset et al.[38] explored a mesoporous silica with well-defined and well-ordered channels which are having one of the following pairs in close vicinity, namely, silylamine, silanol, or bis-silylamine, which is well formulated via controlled exposure of SBA-15 with ammonia. The well-observed complexes [($\equiv$Si−NH−) ($\equiv$Si−O−)]Zr(CH$_2$tBu)$_2$ and [($\equiv$Si−NH−)$_2$]Zr(CH$_2$tBu)$_2$ emerged as a new class of surface hydride complexes (Fig. 11.3, NMR spectrums). These newly synthesized materials exhibit appreciable activities precisely in C−H and C−C bond activation along with the cleavage reactions. An appreciable reactivity is well observed; the silylamido/silyloxo surface ligands presence helps in the formulation of a silylimido−silyloxo ancillary ligand. Such appreciable observation highlights and opens a new area of research along with newly synthesized material, and effectively provide a quite new coordination sites in terms of surface organometallic chemistry.

In continuation of surface modification; Wolosiuk[39] reported in a facile method click chemistry as a tool to react thiocarboxylic acid with vinyltrimethoxysilane under

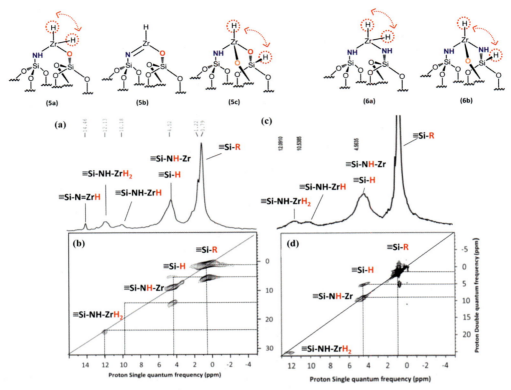

Figure 11.3 (A, C) Displacing the ¹H MAS NMR spectrum of different zirconium hydride form; (B, D) and its DQ rotor-synchronized 2D ¹H MAS NMR spectra. *(Reproduced with permission from Ref. 38. Copyright 2020 Royal Society of Chemistry.)*

photochemical atmosphere, which results into a free carboxylic acid-functionalized anchored silica. Such single-step synthesis protocol route can be eventually followed up for functionalization other various mesoporous silica materials.

In similar context of click chemistry modification extension Sharma et al.[40] demonstrated an efficient postgrafting protocol method for ligand generation along with covalently anchoring to an SBA-15 (heterogeneous material) inner wall via click reaction (Fig. 11.4); click chemistry versatile nature in ligand generation and anchoring of covalent group in one step). In the defined protocol they have demonstrated complex tris(-triphenylphosphine)ruthenium(II) dichloride [RuCl$_2$(PPh$_3$)$_3$] immobilized via anchoring over the click-generated ligand over a wall of SBA-15, which eventually established the protocol for a new SBA-15-Tz-Ru(II)TPP heterogeneous catalyst and evaluated for the multicomponent click cycloaddition reaction precisely in water solvent, and it exhibited surprising unusual excellent selectivity for the 1,4-disubstituted triazole as the main product. Such unusual observations may exceptionally open the door for selective chemical transformations. The newly formed ligand via 1,2,3-triazole formation

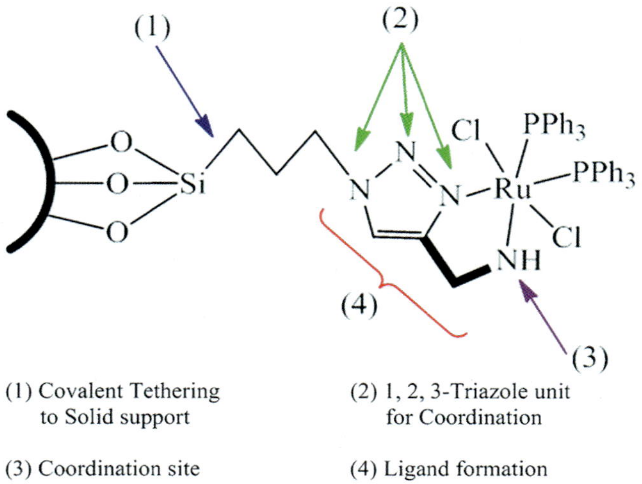

(1) Covalent Tethering to Solid support
(2) 1, 2, 3-Triazole unit for Coordination
(3) Coordination site
(4) Ligand formation

Figure 11.4 Click chemistry versatile nature for ligand generation along with covalent group anchored in single step.

(click reaction products), which provides the various advantage in a single step, namely, covalent tethering to solid support, 1,2,3 triazole unit for coordination, different coordination sites, and ligand formation.

Dendritic fibrous nanosilica: Because silica owns various appreciable physical and chemical properties such as comparable density, lower toxicity, strong biocompatibility, extended surface modification, versatile stable, and cheap, cost-effective raw materials, they are distinguished as a versatile material in research and industry. In the search of improved usage aspects of silica, "dendritic fibrous" is part of the quite recent and pioneering research work from Polshettiwar and his coworkers.[41] Morphological controlled improved nanomaterials exhibit highly improved activity in almost every research field, such as gas absorption, solar-energy generation and storage, catalysis, and sensors which are effectively biomedical compatible applications. Due to the unique morphological nature of fibrous material, it was termed as fibrous nanosilica (KCC-1), where the fibers exist alike thin sheets between 3.5 and 5.0 nm thick instead of sharp needle-shaped rods (Fig. 11.5).

Surface-modified silica NMs utilized as a catalyst for synthesis of several pharmaceutically important organic molecules via click chemistry, cross-coupling reactions (Suzuki, Heck, Sonogashira, Aminations, Stille, Carbonylations, Cyanations, Buchwald–Hartwig reaction), C–H activation, and so forth.[41]

Apart from the aforementioned detailed modification another attention-worthy class for anchoring over the silica nanomaterial is metal-organic frameworks (MOFs), purely inorganic zeolites materials, which is a combination of inorganic and organic hybrid

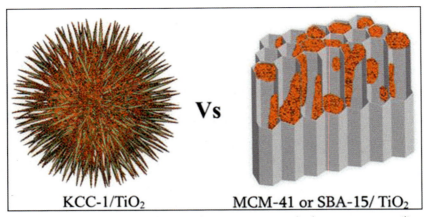

Figure 11.5 Images and schematic differences between KCC-1 and other mesoporous silica material. *(Reproduced with permission from Ref. 41. Copyright 2016 American Chemical Society.)*

materials, eventually came out and have gained attention in the last couple of years.[42] MOFs a crystalline porous materials, which is formulated via making site group anchoring with inorganic units such as metal ions via organic group as linkers, resulted in infinite networks with fixed porosity, high surface area, along with huge pore volume. Such features enrich MOF with up-and-coming usage based applications such as catalysis sector, gas separation, energy generation and storage, magnetism, and luminescence.[43,44]

He and coworker research team[45] have formulated yolk–shell and hollow structured based unique composites (ZnO@SiO$_2$@C/Ni, Zn@SiO$_2$@C/Ni, C@SiO$_2$@C/Ni) via carbonizing the ZIF-8@SiO$_2$@PDA-Ni^{2+} (PDA-Ni^{2+}: polydopamine-Ni^{2+}) at various temperatures range (500, 700, and 900°C) (Fig. 11.6). The synthesis of ZIF-8@SiO$_2$@PDA-Ni^{2+} material with unique structure which could be due to the SiO$_2$

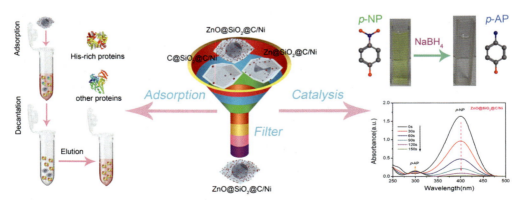

Figure 11.6 Pictorial description the formulation of nickel-based carbonaceous composites and along with carbonization process and application for catalysis and adsorption. *(Reproduced with permission from Ref. 45. Copyright 2019 American Chemical Society.)*

interlayer coating (Fig. 11.6). The ZnO@SiO$_2$@C/Ni composites based material were screened for the nitrophenols reduction to aminophenol and shown an appreciable catalytic result. Furthermore, they also performed the adsorption study using the histidine-rich protein. Aminophenol stands as a reducing agent in various industrial application like dye industry, photograph developer.[46] Based on these current research findings the core—shell/yolk—shell/hollow based structures of nanoparticles have been formulated and due to its peculiar characteristics such as the high tunable nature of porosity, pore functionality, different architectures or compositions in terms of pore, easily accessible open metal sites with hydrophilicity, appreciable structural integrity with specific host-guest interactions, MOFs are supposed to be the new era of advanced materials with various possible tailorable structures for the specific applications. Industrial relevance of such surface-modified MOFs for hydrogen storage, heat-transformation applications such as adsorption chillers and their potential usage in several state-of-the-art applications like water harvesting and energy storage.[47]

3.2 Graphene modified nanomaterials

Among the several carbon-based materials, for example diamond, carbon nanotube, and so forth,[48] graphene stands out as highly advanced carbon-based nanomaterials in today's energy framework,[49] which is composed of two-dimensional (2D) sp^2 carbon lattice honeycomb, wide range of surface area, excellent chemical inert nature, feasible transparent nature, electron mobility channels, appreciable thermal conductivity, and outstanding elasticity.[49] Such composition formulate graphene as a favorable material in terms of various future scientific approach which include sensors, catalysis, transistors, and a precise environmental pollution treatment (Fig. 11.7).[48,49] Graphene oxide (GO) is

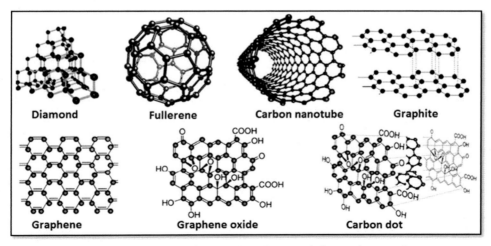

Figure 11.7 Various forms of carbon-based materials for catalytic applications.

a surface oxidized modified form of graphene with various active channels composed of carboxyl epoxide and hydroxyl groups. Such unique functional active sites eventually emerged as the negative charge sites, hydrophilic nature, and faster uniform dispersion of GO in water medium.[50] Such optimized characteristics makes GO a universal potential material for the removal and breaking of various pollutants via adsorption. GO requires a high surface area along with various functional active sites, and GO could be utilized to harvest different NPs. GO is helpful in restricting agglomeration of NPs; therefore it can be a promising candidate for growing various shape nature and properties of NPs.[51] Functionalized GO owns a large surface area and helps for easy dispersion of metal oxide NPs and at the same time helps as a highly conductive support too (Fig. 11.7).[51]

Manna and coworkers[38] have reported the highly flexible and 2D GO nanosheets with different active site of functionality for chemically appropriate sector wise applications. A chemically reactive site enriched with magnetic nature with chemically suitable coated layer was developed over every GO nanosheets individually via following an appropriate chemical route for targeting the 2D magnetically active natured along with superhydrophobic (Fig. 11.8; characterization of magnetically active GO). The synthesized magnetically active graphene oxide (MASHGO: magnetically active superhydrophobic graphene oxide) was highly capable of separating water-oil emulsions (various type of oil). However, apart from the aforementioned characteristics the presence of relevant chemically active composition like high pH, sea-water, and so forth also do not alter the nature of magnetically active superhydrophobic GO. Such impression could be effectively used for further industrial catalytic application.

In the next interesting aspect of graphene-based material, Pyun et al.[52] established the capacity of maximum utilization of the reactivity of GO for application-based various synthetic reactions. The use of GO for "carbocatalysis" opens a gate of possibilities for chemical synthesis in coming age synthesis and research industry.

Therefore, the scope of facile reactivity demonstrated by Pyun et al.[52] can be broadened to other protocols via extensive usage of surface modifications, and enormous edge defects of GO surface (Fig. 11.9). The possible enormous opportunities and the future of "carbocatalysis" will be an interesting new pathway in chemically modified substrates along with materials science.

Furthermore, in another class of nanomaterials, graphite-based nanomaterials can perform multifunctional roles for increasing the strength as mechanical, chemical, and physical nature of future-based materials.[53] However, by dispersing nanomaterials over the cementitious matrix is a crucial point via major usage of their precise geometric and engineering-based chemical nature for the development of future generation-based higher performance cementitious nanocomposites. The suitable surface treatment of nanomaterials along with an interfacial interaction eventually support in fine tuning for the even dispersion of nanomaterials in the water medium of cementitious material

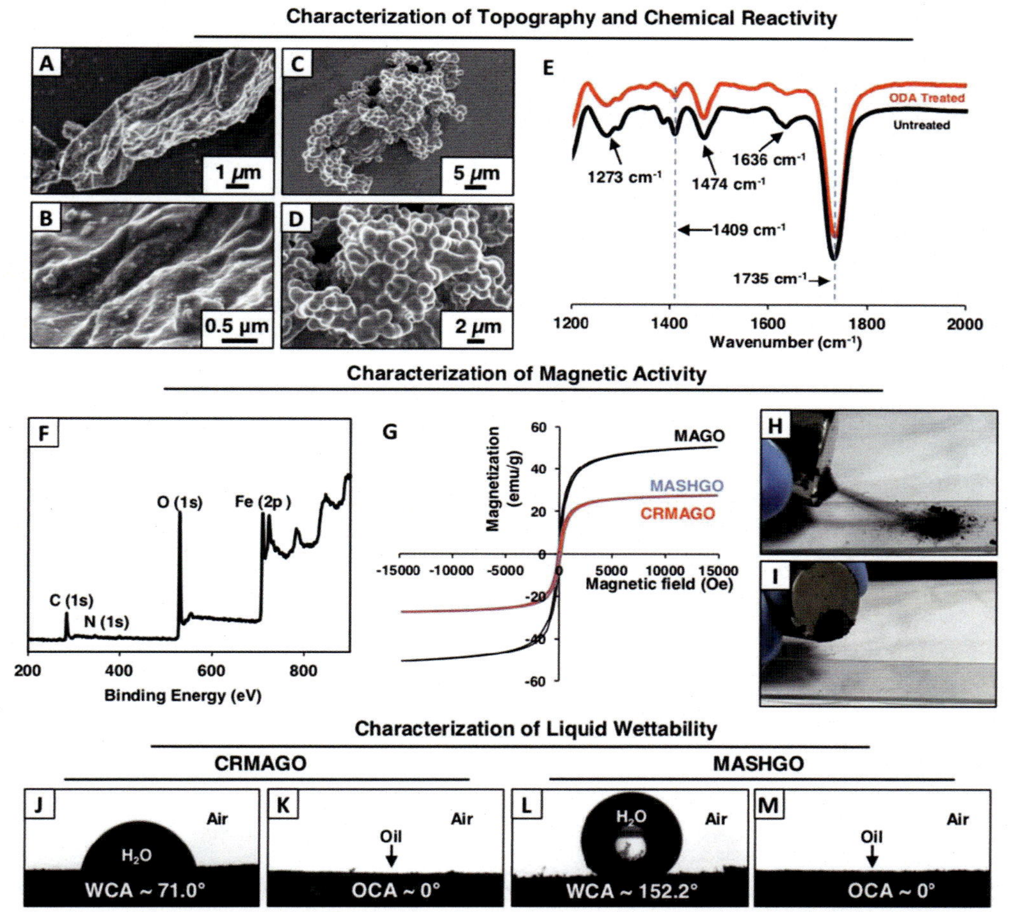

Surface-modified nanomaterial-based catalytic materials for modern industry applications 277

Figure 11.8 Various characterization of magnetically active superhydrophobic graphene oxide (A–D) FESEM images; (E) FTIR spectra; (F) XPS spectra of different analysis; (G) magnetization curves at RT; (H and I) with digital images demonstrating MASHGO (using a neodymium magnet) (0.5 T); and (J–M) static contact angle images displaying the beaded water and oil droplets over different active forms of magnetically active graphene oxide. *(Reproduced with permission from Ref. 38. Copyright 2020 Royal Society of Chemistry.)*

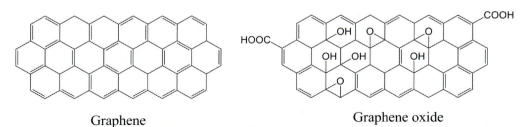

Figure 11.9 GO for carbocatalysis.

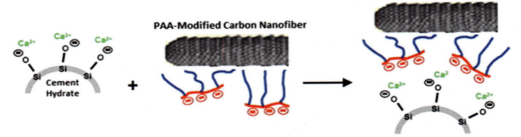

Figure 11.10 Pictorial depiction of interactions of PAA-modified carbon-based nanofiber with cement hydrates. *(Reproduced with permission from Ref. 53. Copyright 2013 Elsevier.)*

matrix. Graphite nanoplatelets and carbon nanofibers provide appropriate needed mechanical and physical property at reduced cost compared with carbon nanotubes in many chemically suitable aspects. The surface modification protocols used to enhance unit of hydrophilic groups over graphite-based nanomaterials surface to make ease their well dispersion in aqueous media or suitable solvent medium (Fig. 11.10). Such protocol consist of (1) polymer covering of oxidized nature of carbon nanofiber; and (2) easily reachable covalent linker functional groups. The eventually generated effects of such surface-modified cementitious nanocomposite were evaluated on the basis of performance. Oxidized graphite nanoplatelet could be helpful in reducing the moisture sorption into appreciable high-performance cementitious paste as nearly 50% effectively (Fig. 11.10).

3.3 Magnetic surface-modified nanomaterials

In another important aspect of modification where an easygoing separation of catalyst from the reaction mixture and along with that recycling protocol are the key step in catalytic technology, and consequently impact the whole economy based over the process.[54] Since it is well known, homogeneous catalysts are efficient, but separation and reuse of these catalysts are significantly difficult with multiple process.[54] Therefore to design a strong and efficient heterogeneous catalyst either via immobilization of homogeneous catalysts or creating the active site inside the active channels heterogeneous support via various modified protocol has attracted worldwide researchers' attention. In this search, immobilization of catalytically active species over the magnetically active surface that eventually that makes easy feasible separation from the reaction mixture system via applying a proper magnetic field (Fig. 11.11).[55] Surface modification with desired functional groups could be easily achieved via precise engineering of magnetic NPs surface properties in a proficient way coating or layered them by suitable material. Such protocol follows the pathway of covalent group anchoring; which seems strong linker connecting the hydroxyl nature groups and the NPs surface using linker agents *viz*; carboxylic acid, phosphonic acid phosphonate, and amine linker family.[56] To design a supported key

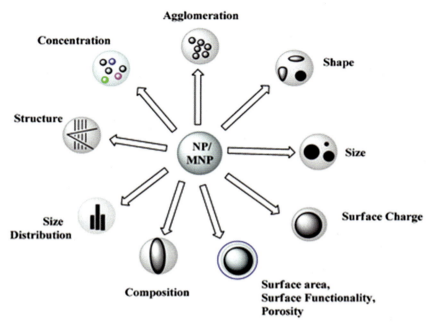

Figure 11.11 Nanomaterials/magnetic nanomaterial characteristics. *(Reproduced with permission from Ref. 55. Copyright 2013 Royal Society of Chemistry.)*

catalyst, functionalization of magnetic nanomaterials is key step, because such immobilization protocols provide various types of molecular catalysts, metal NPs, organocatalysts, and biocatalysts.[54–56]

In the search of supported catalysts, mercaptopropyl anchored ligand could be efficiently used to modify for linking the homogeneous catalysts over magnetic heterogeneous supports.[57] To anchor thiol linker and a nonaromatic C—C bond by following Michael addition was demonstrated to attach (S)-diphenylprolinol trimethylsilyl ether and copper (II)-poly(N-vinylimidazole) catalysts in an easy feasible process.[58] In another significant approach in the anchoring literature, urethane linkage using 3-isocyanatopropyl(triethoxysilane) and 3-methacryloxypropyltrimethoxysilane precursor were also shown as a good candidate for the anchoring.[57,58]

On the other hand, carboxylates-based ligands were also convention ligands to anchor the metal oxide channel surface via easily accessible bidentate metal carboxylate bonds. However; higher stability of the group containing carboxylate-oxide linkage is the key factor for synthesizing an organic ferrofluids formed oleic acid-stabilized magnetic NPs.[26] In this unique combination of stabilization; one side the carboxylate group ties to the active iron oxide surface and on the other hand the hydrophobic natured group ligand supports the steric hindrance resulting into stabilization the NPs in solution. In an appreciable example of anchoring the surface of iron oxide NPs was anchored with

—SH terminal groups via 3-mercaptopropionic acid for the immobilization of a Pd catalyst effectively; which gives effective binding along with recycling benefits too.[59] Few other effective examples are such as; such as linolenic (LLA) and linoleic (LEA) acids or pyridine moieties such as isonicotinic acid (INA), 6-methylpyridine-2-carboxylic acid (MPCA) and 6-(1-piperidinyl)-pyridine-3-carboxylic acid (PPCA), 3-hydroxypicolinic acid (HPA), functional acids having multiple double bonds were also known as a good candidate for the functionalization of magnetic NPs for the ease step of separation and recycling bases research.[60] Whereas at the same time in an effective approach, the complex [Rh(cod) (η6-benzoic acid)]BF$_4$ finds suitable immobilization protocol over the surface of ferrite NPs through the carboxylic acid tethering process.[61] The supported heterogeneous catalyst shows appreciable selective catalytic part in hydroformylation reactions, considered as proportionally better than that of its homogeneous analogue. The [Rh(cod) (μ-S-(CH$_2$)10CO$_2$H)]$_2$ complex could also be easily targeted for immobilized over the surface of ferrite NPs via tethering of carboxylic acid strategy.[61]

Amine ligand modification over magnetic NPs is one the most common reactions for functionalized NPs by organic modification approach. A series of various complex ligands has been formulated via N ligand in the form of imine, amine, and amide bonds using 3-aminopropyl (triethoxysilane) or aminopropyl-functionalized magnetic solid supports for easy separation.[60] On the other hand, various anchored or immobilized Schiff base complexes were also formulated by anchoring 3-aminopropyl (triethoxysilane) along with salicylaldehyde and such modified heterogeneous magnetic catalysts were effectively screened as the catalyst for the various organic transformations such as CO$_2$ fixation, solar cell components, materials science, and corrosion inhibition.[62,63]

In an another appreciable approach, Shylesh et al. demonstrated immobilization process of oxodiperoxomolybdenum complexes by aminopropyl groups anchored onto silica-coated magnetic NPs, and thus prepared magnetically separable catalysts screened for the epoxidation reactions.[64] A similar context; Hirakawa et al. exhibited the anchoring of [CpRu(η3-C$_3$H$_5$) (2-pyridinecarboxylato)]PF6 complex on some ferromagnetic particles, which is effectively screened for deallylation reaction.[65] In the first stage for the preparation of immobilized Ru complex consist of the synthesis of a carboxylic acid chloride via reacting 2-allyl hydrogenpyridine-2,4-dicarboxylate ligand with thionyl chloride active site. Furthermore; in next stage; the direct reacting 5-(triethoxysilyl)pentyl-1-amine functionalization of silica-coated effectively to magnetic support via modified ligand, simultaneously results into formation of the amide bond linkage over the surface.

Similar protocol was exhibited by Xu et al. for the preparation of silica-coated magnetite NPs anchored with 1-benzyl-1,4-dihydronicotinamide, where an amide bond formation took place as linkage of nicotinic anhydride with the enriched amino groups existed over the supported surface further employed in catalytic hydrogenation of α, β-epoxy ketones.[66]

These surface modifications important for recover the catalyst after each cycle of reaction and used for further several cycles of reactions which subsequently reduced the cost of catalysts.[54–56] Nanomagnetite-supported in the field of heterogeneous catalysts are established as a multipurpose materials for heterogeneous catalysis field for fulfillment separation and recycling of costly and rare catalysts at the wide scale of usage. Such kinds of combination widened up the research area precisely for the frequently in use reactions such as aromatic or aliphatic alkylation, feasible oxidation, and reductions, coupling reaction containing C—C or C—N or C—O bond formation, along with asymmetric synthesis protocol along with other named reactions.[54,55] Catalysis precisely carried out using nanomagnetite-supported catalysts is a crucial and emerging sector in research area with various application as it fulfills the green chemistry principles along with selective organic transformations at the same time recyclability aspects as a salient features for sustainable applications also fulfilled.

3.4 TiO$_2$ surface-modified nanomaterials

In recent years, heterogeneous catalysis has shown its potential in many fields to become an emerging aspect for designing sustainable protocol processes for obtaining advanced natured molecules. The current catalysis field has been precisely named as an efficient and ecofriendly research sector since it analyze the possible and maximize utilization of nonhazardous metals for as atom efficient manner. In this search titanium dioxide (TiO$_2$), has been widely commercialized as a white material after its production in the 1920s. Since TiO$_2$ having the property of strong bright white color, and strong refractive index inertness, nontoxic and along with low price good it is cheap ingredient for ointments, food coloring, sunscreens, sustainable paints, and widely other commercial products too.[67] Since the breakthrough research of the ultraviolet (UV) light-induced photo electrochemical water splitting on its surfaces in 1972 came in existence, this research has been exclusively used for various energy-based chemical transformation and applications.[67] Such energy-based applications cover a wide area such as hydrogen production via photocatalytic way from water or organic chemicals, decomposition and formulation of organic molecules, pollutants and dye degradation, fuel small unit like methanol or ethanol from CO$_2$, oxygen reduction, and uses in appropriate solar cells synthesis, rechargeable batteries-super capacitor and future based biomedical devices.[68] The photocatalytic properties of titania are governed by various key points, such as its crystal morphology-based atom arranged structure, tunable surface area, and so forth. To extend the application area of TiO$_2$, postsurface modification tethering is a much appreciable procedure. Some of them are described in this chapter.

Takahara and coworkers[69] demonstrated grafting of organic moieties was carried out over the surface of TiO$_2$ NPs (Fig. 11.12; schematic diagram of TiO$_2$ grafting). Surface-modified TiO$_2$ NPs were screened for the direct polymerization of polystyrene (PS), and

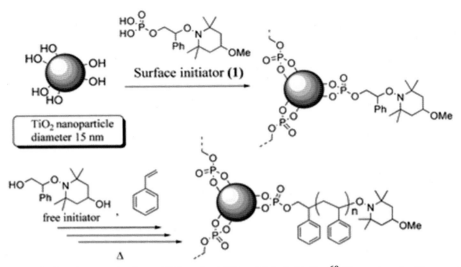

Figure 11.12 Covalent anchoring of PS-grafted-TiO$_2$—PS hybrid film.[69] *(Reproduced with permission from Ref. 69. Copyright 2006 Royal Society of Chemistry.)*

along with it the dispersibility of formulated PS-grafted-TiO$_2$ was investigated under various solvents. The PS-grafted-TiO$_2$—PS hybrid films (prepared via PS-grafted-TiO$_2$ particles and PS matrix powder in CHCl$_3$ solvent casting) exhibited appreciable transmittance under the visible light region too which might be due to the fine dispersion of NPs in the PS matrix. Since PS-grafted-TiO$_2$ particles were evenly dispersed over the surface so could be the strong reason that the difference in absorbance between UV and visible regions was exceptionally large. Such polymer hybrid thin films could be used in medical, 3D printing industry, and paint industry as well.

In another appreciable strategy to anchor the homogeneous complex over photoactive material layer, Meyer and coresearcher[70,71] have formulated an easy and simple protocol that includes immobilization of supramolecular systems composite of Ru sensitizers with Ru-based water oxidation heterogeneous catalysts over the surface of thin TiO$_2$ layer (Fig. 11.13A). Such formulation was eventually used to save the molecules from desorption. Such formulated devices exhibited water splitting into hydrogen and oxygen under visible light-driven with a stabilized photocurrent density of 100 µA/cm^2 (0.2 V vs. NHE; pH 4.6).[70] Additionally, Wu along with his coworkers[72] formulated an exciting sensitizer-catalyst dyad assembly based supermolecular photocatalyst. Such combination is achieved via binding the cobaloxime catalyst over a pyridine-substituted Ru sensitizer which was effectively anchored over the alumina thin film coated NiO. The fabricated device, resulted into photoelectrode which exhibited appreciable photocurrent of 10 mA/cm^2 (0.1 V vs. NHE; pH 7) with noticeable faradaic efficiency of 68% for H$_2$ evolution. Such results were exhibited after 2.5 h run under visible light, which

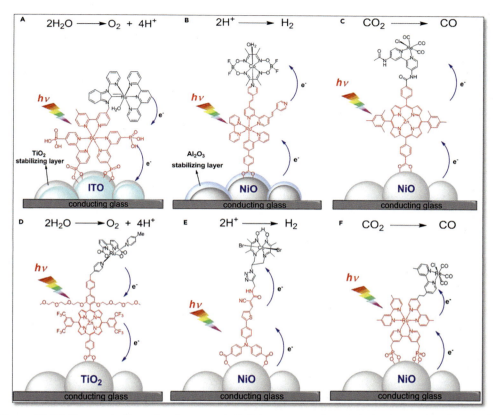

Figure 11.13 Demonstration of various covalent linkage group for catalysts via covalent bonds. (A) Catalyst-Ru-based covalent bonded to TiO_2 conducting glass for water oxidation reaction; (B) catalyst-Ru anchored over NiO conducting glass for H_2 generation reaction application; (C) catalyst-Zn porphyrin anchored covalently to NiO conducting glass for effective CO_2 reduction process; (D) covalent anchoring of catalyst-Zn porphyrin to TiO_2 for water oxidation process; (E) covalent anchoring of Co-based catalyst—Organic to NiO conducting glass in H_2 generation process; (F) covalent anchoring of Ru-based catalysts for CO_2 reduction process. *(Reproduced with permission from Ref. 70. Copyright 2015 Royal Wiley-VCH Verlag GmbH & Co. KGaA, Weinheim.)*

is quite impressive result in hydrogen evolution (Fig. 11.13B). Inoue and his coworkers[73] formulated a device fabrication in which the photocathode in a photoelectrocatalytic (PEC) cell which is utilized for the effective CO_2 reduction. Notably, a Zn–porphyrin which is a sensitizer unit and at the same time, Re catalyst unit connected via an amide bond (Fig. 11.13C).[73] The sensitizer-based catalyst with the supermolecular dyad assembly exhibited hole–injection capacity of 98.2% (confirmed via fluorescence quenching experiments). In such an assembly the photocathode achieved a photocurrent of −20 mA/cm^2 range under 430 nm light illumination with appreciable TON for CO_2 reduction with a Faradaic efficiency of 6.2% [incident photon conversion efficiency

(IPCE) = 0.91% and an absorbed photon conversion efficiency of 2.25%]. Additionally Imahori and his coworkers[74] reported a water-oxidizing dye-sensitized photoelectrochemical cell DS-PEC device. In which, a Zn-porphyrin sensitizer and an [Ru-bda] catalyst attached covalently as a "molecular dyad" (Fig. 11.13D). In addition; Artero and his coresearchers[75] established the protocol for the formulation of cobalt-diimine-dioxime complex; directly attached to a NiO electrode (Fig. 11.13E).[75] Ishitani and his coworkers formulated an Ru sensitizer-Re catalyst supramolecular metal complex after immobilization over the NiO as electrode for feasible CO_2 reduction under visible light (Fig. 11.13F).[76] Herein, we have summarized recent outcome in fabrication protocol for different water-splitting devices formulated by simply taken in to account molecular engineering via ligand anchoring over photoactive materials TiO_2. The earlier-described research can be a key plan for the modern industrial application simultaneously open the gateway of further improved industrial-based energy applications.

4. Conclusion

The encouraging physical and chemical properties of the aforementioned nanomaterials have inspired various future research on developing efficient catalytic activities at industrial level owns large specific surface area, tunable material modification. Today, in every sector globally, nanomaterials are slowly but with strong moves being effectively commercialized and are emerging as protocol technologies. Since the nanomaterials and engineered nanomaterials with specific functional groups are the future generation smart materials with enormous possible applications in the future industrially important reactions such as oxidation, isomerization, polymerization and reduction, hydrogenation, and so forth. In the present chapter, authors elucidate the present scientific needs, with the scientific perceptiveness along with the wide scientific divination for the effective functionalization of nanomaterials and its possible industrial application in various catalytic field. Many research findings are still far behind commercialization and will require substantial improvement regarding costs, simplification in synthesis protocol, greener production of value-added products, and so forth.

References

1. Hussain CM. *Handbook of nanomaterials for industrial applications*. Elsevier; 2018.
2. Predoi D, Motelica-Heino M, Le Coustumer P. Advances in functionalized materials research. *J Nanomater* 2015;**2015**:412690.
3. Richards RM. Introduction to nanoscale materials in chemistry, edition II. In: *Nanoscale materials in chemistry*. John Wiley & Sons, Ltd; 2009. p. 1–14.
4. Osman TM, Rardon DE, Friedman LB, Vega LF. The commercialization of nanomaterials: today and tomorrow. *JOM* 2006;**58**:21–4.
5. Vázquez CI, Iglesias RA. Chapter 38 - engineered nanomaterials in energy production industry. In: Mustansar Hussain C, editor. *Handbook of nanomaterials for industrial applications*. Elsevier; 2018. p. 713–23.

6. Alonso F, Beletskaya IP, Yus M. Non-conventional methodologies for transition-metal catalysed carbon–carbon coupling: a critical overview. Part 1: the Heck reaction. *Tetrahedron* 2005;**61**: 11771–835.
7. Alice Abigail E. Nanotechnology in herbicide resistance. In: *Nanostructured materials. Ramalingam Chidambaram ED1 - Mohindar Singh Seehra) Ch. 11*. IntechOpen; 2017.
8. Hermes P-H, et al. Carbon nanotubes as plant growth regulators: prospects. In: Patra JK, Fraceto LF, Das G, Campos EVR, editors. *Green nanoparticles: synthesis and biomedical applications*. Springer International Publishing; 2020. p. 77–115.
9. Amna, Alharby HF, Hakeem KR, Qureshi MI. Weed control through herbicide-loaded nanoparticles. In: Husen A, Iqbal M, editors. *Nanomaterials and plant potential*. Springer International Publishing; 2019. p. 507–27.
10. Pirollo KF, Chang EH. Does a targeting ligand influence nanoparticle tumor localization or uptake? *Trends Biotechnol* 2008;**26**:552–8.
11. Kim MW, Kwon S-H, Choi JH, Lee A. A promising biocompatible platform: lipid-based and bio-inspired smart drug delivery systems for cancer therapy. *Int J Mol Sci* 2018;**19**:3859.
12. Das S, Bhardwaj A, Pandey LM. Functionalized biogenic nanoparticles for use in emerging biomedical applications: a review. *Curr. Nanomater.* 2021;**6**:119–39.
13. Nghia Tran K, Ha-Lien Tran P, Van Vo T, Truong-Dinh Tran T. Design of fucoidan functionalized - iron oxide nanoparticles for biomedical applications. *Curr Drug Deliv* 2016;**13**:774–83.
14. Subbiah R, Veerapandian M, Yun KS. Nanoparticles: functionalization and multifunctional applications in biomedical sciences. *Curr Med Chem* 2010;**17**:4559–77.
15. Xie X, et al. Challenges and opportunities from basic cancer biology for nanomedicine for targeted drug delivery. *Curr Cancer Drug Targets* 2019;**19**:257–76.
16. Dhar A, Vekariya RL, Bhadja P. n-Alkane isomerization by catalysis—a method of industrial importance: an overview. *Cogent Chem* 2018;**4**:1514686.
17. Torborg C, Beller M. Recent applications of palladium-catalyzed coupling reactions in the pharmaceutical, agrochemical, and fine chemical industries. *Adv Synth Catal* 2009;**351**:3027–43.
18. Shende VS, Saptal VB, Bhanage BM. Recent advances utilized in the recycling of homogeneous catalysis. *Chem Rec* 2019;**19**:2022–43.
19. San Kong P, Aroua MK, Daud WMAW. Catalytic esterification of bioglycerol to value-added products. *Rev Chem Eng* 2015;**31**:437–51.
20. Kiparissides C. Polymerization reactor modeling: a review of recent developments and future directions. *Chem Eng Sci* 1996;**51**:1637–59.
21. Bendjeriou-Sedjerari A, Pelletier JDA, Abou-hamad E, Emsley L, Basset J-M. A well-defined mesoporous amine silica surface via a selective treatment of SBA-15 with ammonia. *Chem Commun* 2012; **48**:3067–9.
22. Rao PC, Mandal S. Potential utilization of metal–organic frameworks in heterogeneous catalysis: a case study of hydrogen-bond donating and single-site catalysis. *Chem Asian J* 2019;**14**:4087–102.
23. Anandan S, Yoon M. Photocatalytic activities of the nano-sized TiO_2-supported Y-zeolites. *J Photochem Photobiol C Photochem Rev* 2003;**4**:5–18.
24. Geckeler KE, Samal S. Syntheses and properties of macromolecular fullerenes, a review. *Polym Int* 1999;**48**:743–57.
25. Crudden CM, Sateesh M, Lewis R. Mercaptopropyl-modified mesoporous silica: a remarkable support for the preparation of a reusable, heterogeneous palladium catalyst for coupling reactions. *J Am Chem Soc* 2005;**127**:10045–50.
26. Zhang L, He R, Gu H-C. Oleic acid coating on the monodisperse magnetite nanoparticles. *Appl Surf Sci* 2006;**253**:2611–7.
27. Yinghuai Z, et al. Supported ultra small palladium on magnetic nanoparticles used as catalysts for suzuki cross-coupling and Heck reactions. *Adv Synth Catal* 2007;**349**:1917–22.
28. Kasprzak A, Bystrzejewski M, Poplawska M. Sulfonated carbon-encapsulated iron nanoparticles as an efficient magnetic nanocatalyst for highly selective synthesis of benzimidazoles. *Dalton Trans* 2018;**47**: 6314–22.

29. Hosokawa M, Nogi K, Naito M, Yokoyama T, editors. *Application 41-surface modification of inorganic nanoparticles by organic functional groups in nanoparticle technology handbook.* Elsevier; 2008. p. 593—6.
30. Li N, Binder WH. Click-chemistry for nanoparticle-modification. *J Mater Chem* 2011;**21**:16717—34.
31. Sala RL, et al. Evaluation of modified silica nanoparticles in carboxylated nitrile rubber nanocomposites. *Colloids Surf. Physicochem. Eng. Asp* 2014;**462**:45—51.
32. Vidal-Iglesias FJ, Gómez-Mingot M, Solla-Gullón J. Surface treatment strategies on catalytic metal nanoparticles. In: Aliofkhazraei M, editor. *Handbook of nanoparticles.* Springer International Publishing; 2016. p. 1101—25.
33. Bahrami Z, Badiei A, Ziarani GM. Carboxylic acid-functionalized SBA-15 nanorods for gemcitabine delivery. *J Nanopart Res* 2015;**17**:125.
34. Zhang G, Zhao P, Hao L, Xu Y. Amine-modified SBA-15(P): a promising adsorbent for CO_2 capture. *J CO_2 Util* 2018;**24**:22.
35. Morales V, Idso MN, Balabasquer M, Chmelka B, García-Muñoz RA. Correlating surface-functionalization of mesoporous silica with adsorption and release of pharmaceutical guest species. *J Phys Chem C* 2016;**120**:16887—98.
36. Sánchez-Vicente Y, et al. A new sustainable route in supercritical CO_2 to functionalize silica SBA-15 with 3-aminopropyltrimethoxysilane as material for carbon capture. *Chem Eng J* 2015;**264**:886—98.
37. Croissant JG, Butler KS, Zink JI, Brinker CJ. Synthetic amorphous silica nanoparticles: toxicity, biomedical and environmental implications. *Nat Rev Mater* 2020;**5**:886—909.
38. Das A, Maji K, Naskar S, Manna U. Facile optimization of hierarchical topography and chemistry on magnetically active graphene oxide nanosheets. *Chem Sci* 2020;**11**:6556—66.
39. Bordoni AV, Lombardo MV, Regazzoni AE, Soler-Illia GJAA, Wolosiuk A. Simple thiol-ene click chemistry modification of SBA-15 silica pores with carboxylic acids. *J Colloid Interf Sci* 2015;**450**:316—24.
40. Sharma P, Rathod J, Singh AP, Kumar P, Sasson Y. Synthesis of heterogeneous Ru(ii)-1,2,3-triazole catalyst supported over SBA-15: application to the hydrogen transfer reaction and unusual highly selective 1,4-disubstituted triazole formation via multicomponent click reaction. *Catal. Sci. Technol* 2018;**8**:3246—59.
41. Singh R, Bapat R, Quen L, Feng H, Polshettiwar V. Atomic layer deposited (ALD) TiO_2 on fibrous nano-silica (KCC-1) for photocatalysis: nanoparticle formation and size quantization effect. *ACS Catalal* 2016;**6**:2770—84.
42. Yuan N, Zhang X, Wang L. The marriage of metal—organic frameworks and silica materials for advanced applications. *Coord Chem Rev* 2020;**421**:213442.
43. Zhao X, Wang Y, Li D-S, Bu X, Feng P. Metal—organic frameworks for separation. *Adv Mater* 2018;**30**:1705189.
44. Farrusseng D, Aguado S, Pinel C. Metal—organic frameworks: opportunities for catalysis. *Angew Chem Int Ed* 2009;**48**:7502—13.
45. He W, et al. Structural evolution and compositional modulation of ZIF-8-derived hybrids comprised of metallic Ni nanoparticles and silica as interlayer. *Inorg Chem* 2019;**58**:7255—66.
46. Mitchell SC, Waring RH. Aminophenols. In: *Ullmann's encyclopedia of industrial chemistry.* American Cancer Society; 2000.
47. Karmakar A, Prabakaran V, Zhao D, Chua KJ. A review of metal-organic frameworks (MOFs) as energy-efficient desiccants for adsorption driven heat-transformation applications. *Appl Energy* 2020;**269**:115070.
48. Yan Q, Gozin M, Zhao F, Cohen A, Pang S. Highly energetic compositions based on functionalized carbon nanomaterials. *Nanoscale* 2016;**8**:4799—851.
49. Fang B, Chang D, Xu Z, Gao C. A review on graphene fibers: expectations, advances, and prospects. *Adv Mater* 2020;**32**:1902664.
50. Pumera M. Electrochemistry of graphene, graphene oxide and other graphenoids: review. *Electrochem Commun* 2013;**36**:14—8.
51. Thanh NTK, Maclean N, Mahiddine S. Mechanisms of nucleation and growth of nanoparticles in solution. *Chem Rev* 2014;**114**:7610—30.

52. Pyun J. Graphene oxide as catalyst: application of carbon materials beyond nanotechnology. *Angew Chem Int Ed* 2011;**50**:46–8.
53. Peyvandi A, Soroushian P, Abdol N, Balachandra AM. Surface-modified graphite nanomaterials for improved reinforcement efficiency in cementitious paste. *Carbon* 2013;**63**:175–86.
54. García-Merino B, Bringas E, Ortiz I. Synthesis and applications of surface-modified magnetic nanoparticles: progress and future prospects. *Rev Chem Eng* 2021. https://doi.org/10.1515/revce-2020-0072.
55. Gawande MB, Branco PS, Varma RS. Nano-magnetite (Fe$_3$O$_4$) as a support for recyclable catalysts in the development of sustainable methodologies. *Chem Soc Rev* 2013;**42**:3371–93.
56. Shylesh S, Schünemann V, Thiel WR. Magnetically separable nanocatalysts: bridges between homogeneous and heterogeneous catalysis. *Angew Chem Int Ed* 2010;**49**:3428–59.
57. Villa S, Riani P, Locardi F, Canepa F. Functionalization of Fe$_3$O$_4$ NPs by silanization: use of amine (APTES) and thiol (MPTMS) silanes and their physical characterization. *Materials* 2016;**9**.
58. Wang H, Zhang W, Shentu B, Gu C, Weng Z. Immobilization of copper(II)-poly(N-vinylimidazole) complex on magnetic nanoparticles and its catalysis of oxidative polymerization of 2,6-dimethylphenol in water. *J Appl Polym Sci* 2012;**125**:3730–6.
59. Rossi LM, et al. A magnetically recoverable scavenger for palladium based on thiol-modified magnetite nanoparticles. *Appl Catal Gen* 2007;**330**:139–44.
60. Baig N, Varma RS. Magnetically retrievable catalysts for organic synthesis. *Chem Commun* 2012;**48**:2582–4.
61. Yoon T-J, Lee W, Oh Y-S, Lee J-K. Magnetic nanoparticles as a catalyst vehicle for simple and easy recycling. *New J Chem* 2003;**27**:227–9.
62. Abu-Dief AM, Abdel-Fatah SM. Development and functionalization of magnetic nanoparticles as powerful and green catalysts for organic synthesis. *Beni-Suef Univ J Appl Sci* 2018;**7**:55–67.
63. Shen Y-M, Duan W-L, Shi M. Chemical fixation of carbon dioxide Co-catalyzed by a combination of Schiff bases or phenols and organic bases. *Eur J Org Chem* 2004;**2004**:3080–9.
64. Shylesh S, et al. Nanoparticle supported, magnetically recoverable oxodiperoxo molybdenum complexes: efficient catalysts for selective epoxidation reactions. *Adv Synth Catal* 2009;**351**:1789–95.
65. Hirakawa T, et al. A magnetically separable heterogeneous deallylation catalyst: [CpRu(η3-C3H5)(2-pyridinecarboxylato)]PF6 complex supported on a ferromagnetic microsize particle Fe$_3$O$_4$@SiO$_2$. *Eur J Org Chem* 2009:789–92.
66. Xu H-J, Wan X, Shen Y-Y, Xu S, Feng Y-S. Magnetic nano-Fe$_3$O$_4$-supported 1-Benzyl-1,4-dihydronicotinamide (BNAH): synthesis and application in the catalytic reduction of α,β-epoxy ketones. *Org Lett* 2012;**14**:1210–3.
67. He J, Kumar A, Khan M, Lo IMC. Critical review of photocatalytic disinfection of bacteria: from noble metals- and carbon nanomaterials-TiO$_2$ composites to challenges of water characteristics and strategic solutions. *Sci Total Environ* 2021;**758**:143953.
68. Wang Y, Saitow K. Mechanochemical synthesis of red-light-active green TiO$_2$ photocatalysts with disorder: defect-rich, with polymorphs, and No metal loading. *Chem Mater* 2020;**32**:9190–200.
69. Matsuno R, Otsuka H, Takahara A. Polystyrene-grafted titanium oxide nanoparticles prepared through surface-initiated nitroxide-mediated radical polymerization and their application to polymer hybrid thin films. *Soft Matter* 2006;**2**:415–21.
70. Li F, Yang H, Li W, Sun L. Device fabrication for water oxidation, hydrogen generation, and CO$_2$ reduction via molecular engineering. *Joule* 2018;**2**:36–60.
71. Song W, et al. Photoinduced stepwise oxidative activation of a chromophore–catalyst assembly on TiO$_2$. *J Phys Chem Lett* 2011;**2**:1808–13.
72. Ji Z, He M, Huang Z, Ozkan U, Wu Y. Photostable p-type dye-sensitized photoelectrochemical cells for water reduction. *J Am Chem Soc* 2013;**135**:11696–9.
73. Kou Y, Nakatani S, Sunagawa G, Tachikawa Y, Masui D, Shimada T, et al. Visible light-induced reduction of carbon dioxide sensitized by a porphyrin-rhenium dyad metal complex on p-type semiconducting NiO as the reduction terminal end of an artificial photosynthetic system. *J Catal* 2014;**310**:57–66.

74. Yamamoto M, Nishizawa Y, Chabera P, Li F, Pascher T, Sundstrom V, et al. Visible light-driven water oxidation with a subporphyrin sensitizer and a water oxidation catalyst. *Chem Commun* 2016;**52**: 13702—5.
75. Kaeffer N, et al. Covalent design for dye-sensitized H_2-evolving photocathodes based on a cobalt diimine—dioxime catalyst. *J Am Chem Soc* 2016;**138**:12308—11.
76. Sahara G, et al. Photoelectrochemical CO_2 reduction using a Ru(ii)—Re(i) multinuclear metal complex on a p-type semiconducting NiO electrode. *Chem Commun* 2015;**51**:10722—5.

CHAPTER 12

Assessment of health, safety, and economics of surface-modified nanomaterials for catalytic applications: a review

Sushil R. Kanel[1], Mallikarjuna N. Nadagouda[1,2], Amita Nakarmi[3], Arindam Malakar[4], Chittaranjan Ray[4] and Lok R. Pokhrel[5]

[1]Department of Chemistry, Wright State University, Dayton, OH, United States; [2]Department of Mechanical and Materials Engineering, Wright State University, Dayton, OH, United States; [3]Department of Chemistry, University of Arkansas at Little Rock, Little Rock, AR, United States; [4]Nebraska Water Center, Part of the Robert B. Daugherty Water for Food Global Institute, University of Nebraska, Lincoln, NE, United States; [5]Department of Public Health, The Brody School of Medicine, East Carolina University, Greenville, NC, United States

1. Introduction

Nanoparticles or nanomaterials (NMs) are discrete nanometer (10^{-9} m size in at least one dimension) scale assemblies of atoms.[1] These NMs show properties that are not found in their bulk form of the same material. A recent review reported that NMs > 30 nm do not, in general, show properties that would require regulatory scrutiny beyond that required for their bulk materials.[2] NMs are used as catalytic material in chemical engineering, electrical engineering, and medicine.[3] Different types of NMs are used as catalytic materials, for example, carbon (carbon nanotubes, fullerene, graphene, carbon black, graphite, inorganic nanotubes), metal and metal oxides (iron, cerium oxide, cobalt, aluminum), and others (quantum dots, clays).[4] The surface-modified and pristine NMs may elicit toxicity to humans under certain conditions may elicit toxicity to humans under certain conditions. Lee et al. reported the application, health, and safety of NMs in the construction industry.[5] Asmatulu et al. reported NMs application and their safety in the automobile industry, and they reported that NMs are may elicitly toxic to humans under certain conditions to human cells and organs.[6] There are no regulations on NMs except bulk (>30 nm) form though they can enter humans through lungs, intestinal tract, skins.[7] Harmful NMs should be regulated for environmental, health, and safety.[2]

Catalysts work by lowering the activation energy, leading to easier completion of the reaction and saving energy—an essential commodity. Catalysts can thereby effectively make the reactions more efficient, and products can be formed at lower temperatures and pressure, directly decreasing the cost in large-scale productions.[8] There are a wide variety of catalysts. Among them, the nanocatalysts derived from NMs seem to be excellent substitutes to other conventional catalysts because they can provide larger surface area, sturdiness, and stability.[9,10] The larger surface area of NM-based catalysts provides larger active surfaces than bulk catalysts, enhancing the multi-fold efficiency of catalytic processes. The faster reaction rates provided by larger active surfaces can yield a faster reaction rate and promote reaction completion with higher efficiency. However, regardless of the benefits of the NMs catalyst, there are difficulties separating the catalysts from the final product, and when the product is utilized in the environment or by humans, it may be detrimental to human health and safety (H&S).[11]

Herein, we reviewed the open literature on the health, safety, and economics of surface-modified nanomaterials (SNMs) for application as catalysts. This review paper's main objectives are to (1) summarize surface-modified nanocatalysts synthesis and their applications published in peer-reviewed journals, (2) report H&S of the SNMs, and (3) summarize the cost of SNMs synthesis and compare them with commercial catalysts SNMs, SMNMs.

1.1 NMs categories and nomenclature

Based on the forms, NMs can be categorized as (1) bulk (one phase of multiphase); (2) surface (NMs on the surface or as a film attached on another surface); and (3) particles, e.g., (a) surface-bound NMs, (b) NMs suspended in liquids, (c) NM suspended in solids, and (d) airborne NMs.[12] Lin et al. reported NMs such as graphene oxide, graphene-hemin nanocomposites, carbon nanotubes, carbon nanodots, mesoporous silica-encapsulated gold nanoparticles, gold nanoclusters, and nanoceria are used as enzymes to accelerate catalytic reactions up to 10^{19} times.[13]

1.2 Synthesis of SNMs with the application as catalyst

SNMs are synthesized using either bottom-up or top-down technology. A review has already been reported on nanocatalysts, e.g., TiO_2, CdS, WO_3, SnS, and ZnO synthesis, and their photocatalytic efficiencies to remove organic pollutants from water.[14]

Nanomaterials are being applied to catalyze many reactions more efficiently. Surface modification is often required that can be achieved incorporating onto their surfaces at least a second element by various means[15,16] to maximize these potentials. For example, bromide ions can be used to effectively reduce the number of low coordination sites on PdNPs.[16–18] Because adsorbates tend to adsorb more strongly on low-coordination atoms than on terrace sites, edges can be removed by controlling the adsorption and desorption of strong adsorbates. Furthermore, the result is bromide-treated PdNPs (Br-PdNPs) with a moderate enhancement in both activity and stability.[17]

Another molecule that could successfully surface modify nanoparticles is carbon monoxide (CO).[19] Surface modification of nanoparticles with CO leads to fewer irregularities than those that were not pretreated with a saturated CO solution.[19] CO annealing has also been used to surface modify bimetallic nanoparticles by subjecting similar Pt_3CONPs to two different CO treatments.[20] First was a gas phase treatment, during which the sample was placed into a rotary evaporator and repeatedly evacuated and filled with CO to eliminate residual oxygen. The distiller was then filled with an ambient pressure of CO and heated to 200°C for 3 h, a process that produced a surface aggregation profile without having corrosion on the carbon support. The second process entailed electrochemical CO annealing in which the catalyst was placed onto a rotating disk electrode and subjected to a potential cycling in a CO-saturated alkaline solution for 60 min. Due to higher adsorption enthalpy of CO on Pt than on Cobalt (Co), the Pt segregates to the surface displacing the Co to the core.[16,20]

Another surface modification strategy aims at the formation of a Pt overlayer because of the removal of a nonnoble metal from a Pt alloy using electrochemical dissolution/dealloying protocols. Samples formed are usually thick, but corrugated layers known as Pt skeleton form on the surfaces. This surface contains an essential number of low-coordinated atoms with potential for use in catalysis.[16]

NMs have shown great promise in regenerative medicine due to their chemical and physical properties. Encapsulation, surface modification, and purification of carbon nanotubes (CNTs) open up more prospects for their biomedicine application, particularly in drug and gene delivery, where their selectivity, low toxicity, and low distribution in the cells may make them a good candidate. Also, surface modification of CNTs increases their usefulness in artificial implants and tissue engineering.[21]

Another method that could lead to the synthesis of PtNPs is thermal treatments. Thermal treatments could induce crystallization of amorphous nanoparticles, making them display physical properties unlike bulk.[16] Small size PtNPs could be functionalized onto multiwalled carbon nanotubes (MWCNTs) to improve the reactivity of PtNPs toward oxygen reduction and methanol and carbon monoxide oxidations.[16,22]

1.3 Different types of SNMs

Different types of NMs and SNMs are used in chemical engineering, water, wastewater, soil treatment/amendments, and medicines. Selected NMs used in the field of chemical and environmental engineering are reported below and in Fig. 12.1.

1.3.1 Carbonaceous SNMs

The carbon nanomaterials combine the exceptional sp^2 hybridized carbon bonds with exceptional chemical, mechanical, and electrical properties. Carbon NMs are used in many fields, including environmental remediation, filtering pollutants, delivering medications, and supercapacitors. The roles of carbonaceous carbon nanomaterials would become more and more significant in a future life.[27]

Yaganeh et al. demonstrated concentrations of airborne particles during the production of fullerenes and carbon nanotubes in a commercial nanotechnology facility in 2008. The authors measured $PM_{2.5}$, submicrometer size distribution, and photoionization potential, an indicator of the particles' carbonaceous content, at three locations inside a

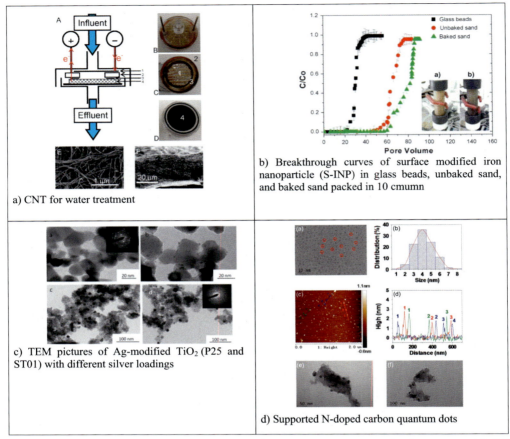

Figure 12.1 Different types of NMs and SNMs used in water, wastewater, and soil treatment: (A) CNT for water treatment[23]; (B) Breakthrough curves of surface-modified iron nanoparticle (S-INP) in glass beads, unbaked sand, and baked sand packed in 10 cm column[24]; (C) TEM pictures of Ag-modified TiO$_2$ (P25 and ST01) with different silver loadings 0.5% and 2%: (a)–(d) samples Ag–P25 (0.5), Ag–P25 (2), Ag-ST01(0.5), and Ag-ST01(2), respectively[25]; (D) Supported N-doped carbon quantum dots (a) TEM image and (b) the corresponding size distribution of Supported N-doped carbon quantum dots (NCQDs), (c) AFM image and (d) the corresponding height distribution of NCQDs; (e), (f) TEM images of p-NCQD-700.[26]

nanotechnology facility; inside the fume hood, outside the fume hood, and in the background. They found that an average concentration of PM$_{2.5}$ and particle number were lower inside the facility than outdoors. They also concluded that nanomaterials were associated with increased air pollution levels caused by nanomaterials' processing or production.[28]

Turkevich et al. conducted concentration and ignition scans on various carbonaceous nanoobjects, including fullerene, SWCNT, carbon black, MWCNT, graphene, CNF, and graphite. They investigated the minimum explosive concentration (MEC), minimum ignition energy (MIE), and minimum ignition temperature (MIT) for the chemical compounds. Nanocarbons are showing similar combustion characteristics to coal and graphite. Fullerene nanofiber is confirmed to be borderline St^{-1} fullerene nanofiber at KSt 200 bar-m/s. They also determined that there was an MIE for aluminum. Although explosions of nanocarbons and their equivalents are comparable (in severity), they are much less likely to ignite than are their coals (i.e., the nanocarbons have lower MIEs than do the graphite).[29]

In consideration of the selected studies on nanotechnology's effects on the environment and human health, the researchers assess the state of the science. The researchers focused on the application of emerging nanotechnology to textiles and face coatings. These are the most available, commercially available, and excellent nanomaterials such as nano-TiO_2, nano-ZnO, layered silica, and carbon nanotubes. They have provided criteria that can be useful in analytically interpreting the state of the art on nanomaterials' effects. These are the worst problems that affect the environment. There was a concern that the chemicals in the substances could be harmful to people's health. Some nanomaterials may cause much more hazards to humans than other species, such as skin, gastrointestinal or respiratory tract.[30]

There are more effects on human health because of occupational exposure to nanomaterials. There is no Occupational Safety and Health Administration (OSHA) standard regarding nanomaterials exposure. There are also significant gaps in nanomaterials hazard knowledge. Preliminary data suggest that multiwalled carbon nanotubes (MWCNT) lead to lung inflammation, while carbon nanofibers (CNF) do not lead to lung inflammation. Carbonaceous nanomaterials (CNMs) have been associated with various lung disorders in laboratory animals.[31] SNMs containing carbons are categories in these NMs, such as carbon nanotubes, fullerene, graphene, carbon black, graphite, and inorganic nanotubes. A study reported the assessment of the current Environmental Health and Safety Regulation regarding handling carbon nanotubes.[32] Carbonaceous NMs are divided into different materials as follows:

1.4 Carbon nanotubes

Carbon NMs contain carbons, for example such as in carbon nanotubes, graphene, carbon black, graphite, fullerene, and inorganic nanotubes.[32]

Carbon nanotubes (CNTs) and carbon allotropes play an important role in developing new chemical and physical products. There are single-walled carbon nanotubes (SWCNTs: 1–5 nm diameter and 1 nm–1 mm tube length), double-walled carbon nanotubes (DWCNTs), and MWCNTs (MWCNTs: 5–50 nm diameter). Some

CNTs have different lengths, shapes, surfaces, purity, and propensity to form agglomerates and aggregates. The Co, Fe, Ni, and Mo used in nanotube catalyst production may help remove metal contaminants. CNTs may be functionalized or coated with proteins, polymers, or metals to improve dispersion in solvents or obtain specific functions. These factors influence the chemical and properties of CNTs. The ultrafine particles are smaller than larger particles. They can behave like asbestos and other needle-shaped fibers. They are expected to be biologically persistent.[33]

Maynard et al.[34] examined the potential toxicity of SWCNT—a particular allotropic form. The physical nature of the aerosol formed by SWCNT during mechanical agitation was investigated on a laboratory basis. A field study was conducted to examine airborne and dermal exposures to SWCNT during unrefined material handling. Whether they are refined or unrefined, this will happen. Estimates of airborne nanotube material concentrations generated during handling indicate that in all cases, concentrations were below 53 g/m^3. SWCNT handling glove deposits were estimated at between 0.2 and 6 mg per hand.[34]

Tian et al.[35] assessed the toxicity of five nanomaterials on human fibroblasts. They analyzed the physicochemical characteristics of these nanomaterials in order to determine their toxic effects. Cell survival and attachment assays were evaluated using different amounts of carbon nanotubes, active carbon, carbon black, multiwall carbon nanotubes, and carbon graphite. The refined nanoparticle caused the strongest toxic effect compared to its unrefined version. They tested many variables such as physical dimensions, surface areas, dosages, aspect ratios, and surface chemistry. The variables that predict these carbon nanomaterials' potential toxicity, in which SWCNTs caused the strongest cellular apoptosis and necrosis. Both the unrefined and refined SWCNTs were found to be toxic. There are physical changes in dispersed nanomaterials due to their surface chemistry changes, and thereupon cell detachment and apoptosis/necrosis.[35]

Ryman-Rasmussen et al.[36] revealed that MWCNTs reach the subpleural of mice after a single inhaled exposure of 30 mg/m^3 for 6 h. Nanotubes were embedded within subpleural vasculature and mural cells. Mononuclear cell aggregates increased in number and size in the lung after 1 day, and these aggregates contained semitransparent nanotube-containing macrophages. Subpleural fibrosis triggered by inhaling nanotechnology increased after two weeks and six weeks after inhalation. These harmful effects could be unique to this type of CNT, which may or may not continue when CNT is repeated in low doses. Due to this uncertainty, precautions should be taken to minimize inhalation during handling and using MWCNTs.[36]

Clift et al.[37] investigated how five different patented CNTs affect the viability of human monocyte-derived macrophages (MDM) and their ability to cause an oxidative stress reaction an inflammatory response in the different immune cells over two days. None of the CNTs exhibited significant cytotoxicity after 24 h. Only those long, MWNCTs are the source of harmful reactive oxygen species after 24 h incubation at 0.005–0.02 mg/mL. Therefore, the industrial MWCNTs studies found that the individual CNTs demonstrated hazardous effects in vitro that may be linked to their physical properties.

The same research group (Clift et al.) conducted experiments with SWCNTs and MWCNTs on lung epithelial cells within 24 h. They confirmed no cytotoxic effects to any of the cells were tested except a significant increase in TNF-α after SWCNT and MWCNT exposure. They found a significant increase in TNF-α and IL-8 concentration for 16HBE14° epithelial cells and TCC-C at 0.02 mg/mL after carbon black exposure. However, TCC-C leads to a higher release of IL-8 compared to epithelial cells. The reduced glutathione (GSH) content showed a significant difference ($p < .05$) between SWCNT monoculture and TCC-C (0.005–0.02 mg/mL). It was concluded that in comparison with monoculture systems, multicellular systems could be used as predictive in vitro screening tools for determining the potentially deleterious effects associated with CNTs.[38]

1.5 Fullerene (nC$_{60}$)

Fullerenes, or C$_{60}$, were first isolated by Kroto et al.[38a] The C$_{60}$ fullerene consists of 60 carbon atoms with approximately 0.7 nm and a molecular weight of 720 g/mol. Thirty carbon double bonds are present in the molecule, which makes it vulnerable to free radicals. Fullerenes have interesting properties which make them very useful in engineering and medicine. Some isomers which are improved have been said to have increased biological activity. Fullerenes' current applications include targeted drug delivery systems, lubricants, energy applications, catalysis, and polymer modifications. Application in the consumption industry includes antifriction, cosmetic, and sporting goods. Apart from fullerenes' production, their use in the markets is also limited but expected to grow significantly over the next decade.[39]

Rouse et al. subjected the human epidermal keratinocytes (HEK) to fullerene-based amino acid (Baa) solutions at concentrations of 0.4–0.00004 mg/mL in a humidified 5% CO$_2$ atmosphere in 37°C in order to understand the biological activity of functional C$_{60}$. MTT cell viability immediately after the following transfection decreased significantly at 0.4 and 0.04 mg/mL. In parallel, human cytokines (IL-6, IL-8, TNF-a, IL-1b, and IL-10) concentrations ranged from 0.5 to 0.004 mg/mL. Media was collected at 1, 4, 8, 12, 24, and 48 h. IL-8 concentrations at 0.04 mg/mL showed significantly higher ($p < .05$) than all other doses at 8, 12, 24, and 48 h. IL-6 activities and IL-1b activities were observed at 24 and 48 h for the pathogen concentrations of 0.4 and 0.04 mg/mL. There was no significant TNF-α or IL-10 activity at any time points for any of the concentrations. These results indicate that concentrations below 0.04 mg/mL induce less cytokine production and retain cell viability. Bisphenol A (BPA) damages cells and causes inflammation in health organs.[40]

Yamawaki and Iwai treated an endothelial cell (ECs) with 1−100 g/mL of the super microparticles [$C_{60}(OH)_{24}$; mean diameter, 7.1 nm]. $C_{60}(OH)_{24}$ induced cytotoxic vacuole formation and decreased cell density in a dose-dependent manner to explore the direct effects of nanomaterials on endothelial toxicity. The LDH assay found that a maximum dose of $C_{60}(OH)_{24}$ induced cytotoxicity. A proliferation assay showed $C_{60}(OH)_{24}$ to be a potent EC inhibitor. $C_{60}(OH)_{24}$ did not induce apoptosis but caused the accumulation of polyubiquitinated proteins and facilitated autophagy. Autophagosomes were confirmed using Western blot analysis with a specific marker, light chain 3 antibody, and electron microscopy. Low-dose $C_{60}(OH)_{24}$ (10 g/mL for eight days) inhibited cell attachment and delayed EC growth. Yamawaki and Iwai did for the first time examined the toxicity of water-soluble fullerenes to ECs. Although fullerenes changed morphology in a dose-dependent manner, only maximal doses caused cell death and inhibited cell growth. EC death was caused by activating ubiquitin-autophagy cell death pathways.[41]

Fiorito et al. have examined C- SWNTs and C-fullerenes highly purifiable to in vitro induce an inflammatory reaction of the murine and human macrophage cells. They evaluated C-SWNTs and C-fullerenes' ability to induce the release of NO by murine cells, stimulate the phagocytic behavior of human macrophage cells, and be cytotoxic against such cells in order to determine the capacity of those C-derivatives as biological inducers of inflammatory reactions. The authors showed that SWNTs-C-nanotubes do not stimulate NO releases via murine macrophage cells in culture when highly purified, C-fullerenes, and their absorptions from humans macrophage cells are very poor and toxic to human macrophage cells.[42]

1.6 Graphene-based nanomaterials

Graphene-based NMs are used in a different field that guarantees the study of H&S of these materials. A review discussed H&S of graphene materials to human health through in vitro and in vivo model systems.[43]

Graphene is the youngest member of the family of carbon nanomaterials that are found in single atom and planar sheet. Graphene is a member of the covalent family of compounds that are made up of carbon. Both CNT and graphene materials possess outstanding physical properties with high anisotropy and a tunable layer that has contributed to their wide array of uses. Graphene and carbon nanotubes will be used in biomedical research in the years to come. Graphene and carbon nanotubes seem to be developed and maturing faster than other carbon-based nanomaterials. Because of graphene's widespread use in many industries, its influence on air and other environmental variables need to be studied.[44]

Ahmed and Rodrigues expressed concern about the acute toxicity effect of graphene oxide (GO) on the microbial functions related to biological wastewater treatment. The results of the tests showed that GO was toxic to organisms at high and moderate concentrations. The biological removal of nutrients, such as organics, nitrogen, and phosphorus, was significantly reduced with GO in the waste treatment process. On the other hand, the presence of GO can have a substantial influence on the water quality of the effluent. By microscopic examinations, researchers confirmed the influx of algal blooms in and around activated sludge flocs. The results demonstrated that the interaction of the GO with the wastewater resulted in a significant amount of reactive oxygen species, which is one of the mechanisms responsible for the toxic effect of GO.[45]

Drasler et al.[46] modified an aerosolization system combined with 3D human lung model to study the potential side effects of graphene-related materials (GRM). Two different types of GRM were injected into each individual. Six endpoints or procedures were evaluated, such as cell viability, morphology, barrier integrity, induction of (pro-) inflammation, and oxidative stress reactions. The results did not show a harmful effect to the 3D lung model at the two different doses (300 and 1000 ng/cm^2) during acute exposure.[46]

The effect of GO on Suwannee River humic and fulvic acids (SRHA and SRFA), alginate and aluminum oxide were investigated using a quartz crystal microbalance with dissipation monitoring (QCM-D). Deposition behavior shows that GO is highest on alginate, followed by SRFA, SRHA, and aluminum oxide. This shows glycol binding on positively charged poly-L-lysine coated surfaces. GO with has hydroxyl, epoxy, and carboxyl functional groups have increased interactions with NOM. Ionic strength (IS) and ionic valence (IV) affected interactions with environmental surfaces. The NOM coating makes the surface conductive. Even when releasing, different surfaces vary differently. Low IS water surface releases ionic silver into all surfaces. IS is used to release NOM from surfaces. Alginate-covered surfaces trap GO particles in the sticky layer of the alginate (Table 12.1).[47]

1.6.1 Metal and metal oxides

In the past decades, metal and metal-oxide NMs have generated immense attention. The unique physical and chemical characteristics attributed to metal and metal-oxide NMs due to their nano size are widely used in various catalysis industry sectors. The most common metal-oxide NMs include Al_2O_3, CuO, Fe_3O_4, MgO, MnO_2, NiO, TiO_2, ZnO, ZrO_2, and SiO_2[52] (Table 12.2).

Currently, metal and metal-oxide–based NMs are being used to catalyze various kinds of organic reactions. The high selectivity and reusable property make them cost-effective and economical, and these properties help metal-based NMs fulfill the requirements for novel catalysts.[88] One of the primary concerns of using metal or metal oxide NMs is their inherent tendency to agglomerate.

The high surface area-to-volume ratio generates high surface energy, resulting in easier agglomeration. The surface modification of NMs can help in the reduction of high surface energy and make them more efficient. Modification of NMs surface has

Table 12.1 Chemical properties of carboneous NMs catalysts.

NMs	Types	Size	Catalytic application	Toxicity	References
Carbon nanotubes	Single-wall CNT (SWCNT) Pristine-SWCNT Multiwall CNT (MWCNT)	1.4 nm 10–15 nm		▸ Pulmonary toxicity and biomarkers of toxicity in CNT-treated skin cell cultures ▸ Fibrosis, and biochemical/toxicological changes in the lungs ▸ Mitochondrial function and proteasome formation ▸ Reactive oxygen species (ROS) generation and reduction of cell viability ▸ Primary human alveolar macrophages ▸ Lysosomal membrane destabilization ▸ Apoptotic, inflammatory, and fibrogenic effects in lung ▸ Apoptosis via mitochondrial pathway and scavenger receptor	48–57
Fullerene	Fullerene		▸ Antitumor growth properties ▸ Electronic and quantum properties	▸ Antioxidant and radical scavenging properties ▸ Reproductive toxicity and carcinogenic effect ▸ Respiratory system (inhalation and intratracheal administration), gastrointestinal tract (peroral and intraperitoneal administration), after dermal and parenteral (intraperitoneal, intravenous) administration.	39,48,50,52,58,59

Graphene-based NM	Graphene Graphene oxide	8–25 nm	Filtration	▸ Lung inflammation and injury in lavage fluid and tissue gene expression ▸ Various living systems such as microbes, mammalian cells, and animal models ▸ Autophagosome accumulation and lysosome impairment in	60–62
Carbon black	Graphene-like		Biomedical and bioelectronic field	Perturbations in the different biological parameters	63
Graphite	Graphite oxide		Fire safety	Smoke toxicity	64
M-Xenes	M-Xenes		Biosensor applications	Interaction with natural biomacromolecule	65

Table 12.2 Chemical properties of metal and metal-oxide NMs used as catalysts.

Nanomaterials (NMs)	Size (nm)	Catalytic application	Toxicity	References
Aluminum oxide	8–160	Catalytic activity in synthesis of 2-aminothiazole	Decreased cell viability, mitochondrial functions; increases oxidative stress; alter protein expression; DNA damage	66–69
Cerium oxide or nanoceria	25 ± 1.5	Fuel oxidation catalysis	Cytotoxicity; oxidative stress; genotoxicity	70
Copper oxide	<50	Selective oxidation of hydrocarbons; efficient ectorcatalyst for oxygen reduction reactions	Decreased cell viability; increased lipid peroxidation	71
Cobalt oxide	20–90	Utilized in chloralkali electrocatalysis systems	Impacts human immune cells; affect pulmonary systems; induce reactive oxygen species (ROS) formation	72,73
Gold	3–20	Catalytic reduction of nitroarenes and carbon monoxide oxidation	Causes oxidative stress; cardiac hypertrophy	74,75
Iron oxides	14–150	Phenolic and aniline catalytic oxidation	Decreased cell viability	76,77
Palladium	2–60	Catalysis of carbon cross-coupling reactions	Cytotoxicity; proinflammatory; ROS production	78
Silver	13–15	Catalytic oxidation of tryptophan	Damage to cell membrane, disrupt ATP, DNA replication, cell viability release oxidative ROS	75,79–81
Silica based NMs	15–46	Saturated and unsaturated hydrocarbon oxidation	Increase in oxidative stress, ROS, lactate dehydrogenase, malondialdehyde; mitochondrial damage	82

Table 12.2 Chemical properties of metal and metal-oxide NMs used as catalysts.—cont'd

Nanomaterials (NMs)	Size (nm)	Catalytic application	Toxicity	References
Titanium dioxide	20–160	Photocatalytic activity utilized for water purification	DNA fragmentation, adduct formation, and damage; genotoxicity; cytotoxicity; increase in oxidative stress	75,83–85
Zinc oxide	<20–419	Photocatalytic degradation of organic and inorganic pollutants for waste water treatment	DNA and mitochondrial damage; decreased cell viability; release of lactate dehydrogenase; increased oxidative stress; inflammatory biomarkers; apoptosis	86
Zirconium dioxide	31.9 ± 1.9	Catalyst for synthesis of dimethyl carbonate, CO hydrogenation	Affect osteogenesis; cytotoxicity	87

been utilized to minimize their tendency to form aggregates.[89] Surface modification can also reduce the toxicity of metal/metal oxide NMs.[75]

Ideally, surface modification of NMs should improve surface properties without altering their primary catalytic activity. Physical and chemical methods can be utilized for surface modification of metal-based NMs (Fig. 12.2). Generally, surfactants or macromolecules are employed for physical modification by adsorption of these molecules on the surface of the NMs. The electrostatic interactions between the polar groups of the surfactants and NMs can make adsorption feasible.[91] In the case of chemical modification of NMs surface, the surface modifier are bonded to NMs covalently. For example, covalent grafting can couple polymers, carboxylic acids or organophosphorus molecules.[89]

The metal coating is also utilized for surface modification of metal-oxide NMs, enhancing the nanocatalyst property.[92,93] The metal-based nanocatalyst coating onto the surface of other NMs can support the nanocatalyst and deliver unique characteristics, which can aid in their separation after the reaction occurs.[94] For example, Fe_3O_4, a magnetic NM, has been utilized to support palladium NMs for easy separation of Pd NMs after hydrogenation catalytic reactions.[95] Researchers have utilized this magnetic

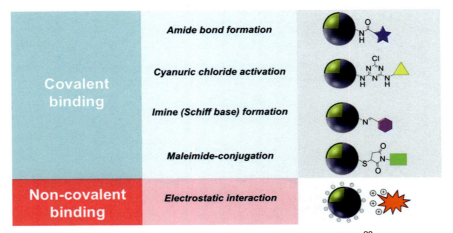

Figure 12.2 Summary of surface modification methods.[90]

property of Fe_3O_4 for producing easily recoverable Pd NM catalysts, which were immobilized on the surface of silica-coated surface-functionalized Fe_3O_4 (Fig. 12.3).[94]

Other properties can also be tweaked or enhanced by surface modification of NMs. The iron-based coating has also been studied to study Suzuki coupling reactions where palladium was deposited on the surface of magnetite shell with carbon core (Pd–Fe_3O_4@C) was investigated for bromobenzene and phenylboronic acid reactions. The Fe_3O_4@C presented better catalytic activity than other catalysts for similar reaction systems.[96]

A study reported Ag coated nanoscale nickel ferrite ($NiFe_2O_4$) in catalytic reactions and studies its antibacterial and antifungal properties to bacteria (*Bacillus subtilis* [gram-positive] and *Pseudomonas syringae* [gram-negative]) and fungi (*Alternaria solani* and

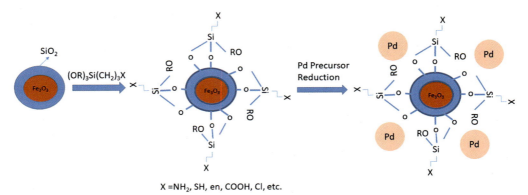

Figure 12.3 Magnetically recoverable Pd NPs immobilized on the surface of silica-coated iron oxide NPs functionalized with organosilanes.[94]

Fusarium oxysporum). They reported that bacteria and fungi growth was inhibited entirely 24 h after treatment with Ag—NiFe$_2$O$_4$. Authors also synthesized Mo—NiFe$_2$O$_4$ and efficiently degraded cis-cyclooctene and a wide variety of alkenes.[97] Recently, NMs such as copper nanoparticles are used in cleaning face masks to treat viral aerosols in the fight against the COVID-19 pandemic.[98]

Metal and oxide NMs have been surface-modified and stabilized through organic coupling chemistry or electrostatic interactions with various organic moieties, metal complexes, and biomolecules. The chemical modifiers, which can be covalently attached through direct functionalization reaction or use of bifunctional cross-linkers, have reduced the inherent toxicity of NMs. However, there are existing knowledge gaps, and research is needed to improve the NMs surface functionalization to enhance the catalytic property and impart nontoxicity to human health. Efficiency of metal-based surface support of SNMs can also be improved. There are varieties of metal surface modification over metal-oxide (metal@metal oxide) NM systems, but surface modification has been limited to SiO$_2$ and TiO$_2$ metal-oxide systems. It is worth noting that transition metal/metal-oxide NMs and lanthanide metal-oxide NMs have potential in various catalytic processes such as water-splitting catalysis and hydrogenation.[99]

It is important to realize that nanocatalysts are still hard to separate from their final product due to the size limitation. The release of these metal or metal-oxide NMs with the final product, which is utilized in various sectors, can have a detrimental impact on the natural ecosystem and be toxic to human health.[85] Therefore, it is essential to understand nanocatalysts' inherent toxicity with respect to safeguarding human health and the environment when the generated product interacts and is released to the environment. Further studies are needed to validate the separation processes of the metal or metal-oxide NMs.

1.7 Other nanomaterials
1.7.1 Quantum dots

Quantum dots (QDs) are NMs that show the quantum effect as they are generally less than 10 nm in diameter.[100] They are made with transition metals, most commonly cadmium and zinc. The core consists of either CdSe or ZnSe and has a Zn or Cd shell. They have a bright fluorescence and are increasingly being used as a substitute for fluorescent dyes in bioimaging. They are usually synthesized in nonpolar solvents and made water-soluble through encapsulation by amphiphilic polymers.[101]

QDs also contain heavy metals, e.g., chromium, lead, zinc, cadmium, and selenium, among others.[102] The leaching of heavy metals from QDs (core-shell or nonshell structure quantum dots) has been implicated in the toxicity to mammals, bacteria, and viruses.[5,103] In eukaryotic cells, QDs led to oxidative stress, nucleic acid damage, and cytotoxicity.[104] Chemically unstable nanoparticles can be oxidized, reduced, and dissolved in biological media, leading to the release of toxic ions. Nanoparticles that show a higher solubility in cellular growth media (such as ZnO nanoparticles) show stronger toxicity to mammalian cells than nanoparticles with low solubility (such as TiO$_2$). Solubility was found to strongly influence the cytotoxic response.[105]

A study reported palladium/magnetite nanocatalysts toxicity on human skin (HaCaT) and human colon (CaCo-2) cell lines and a cell line from rainbow trout gills (RTgill-W1) and found that there are minor effects.[106] As these NMs are used in water and wastewater treatment, the trace amount remaining in the system will not affect much.

The toxicity of QDs can be caused by cadmium and selenium in the QD core, which may be toxic to cells and live organisms. Another reason for toxicity includes ROS generation, which can damage enzymes and membranes.[107] The toxicity of QDs has been studied in microorganisms, other lower-order organisms, plants, and mammalian cells.

To study the effects of QDs on bacteria, Mahendra et al.[101] used weathered versus intact QD coatings. When *E. coli*, *Bacillus subtilis*, and *Pseudomonas aeruginosa* were exposed to the two types of QDs, weathered QDs killed nearly 100% of bacteria at concentrations above 20 nM while coated QDs were less toxic. For Gram-positive bacteria, there was a lower growth rate and growth yield. For Gram-negative bacteria, there was no statistically significant decrease in growth yield.[101] This means QDs are more bacteriostatic than bactericidal.

For plants, specifically in *Arabidopsis thaliana*, QDs were not uptaken by plants within seven days of exposure.[108] This is due to defense mechanisms already in place, such as the cell wall, which can exclude particles it does not want and regulate different plant proteins. Because of this, there was high adsorption on the plant root surfaces. Although there were no physical changes to the plant, the GSH/GSSG ratio's decreased levels were concerning.[108]

Studies have shown that in mammalian cells, QDs containing CdSe core may lead to the accumulation of toxic metals in the cells, with the surface coating playing a role in the toxicity. Specifically, in frog and rat cells, long-term QD toxicity was reported at concentrations greater than 1 g/L.[101] Another study looked at the differences in the effect of CdSe QDs on rat hepatocytes in vitro versus in vivo (i.e., mice).[109] The in vitro results showed that the CdSe QDs were not cytotoxic when delivered in a sterile setting. When the QDs were exposed to air for 30 min, oxidation of the surface occurred, which led to cytotoxic effects. Further testing with different types of surface coatings showed that cytotoxicity was virtually eliminated. Mice data showed that oxidation can still occur after injecting with QDs, even with multiple surface coatings, leading to cytotoxicity.[109]

Vanadium (V) is a trace element that is created during the combustion of fossil fuels. As technologies are advancing, vanadium oxide nanoparticles (VO$_2$NPs) are being developed for various purposes, including catalysts and medical devices. Humans could be exposed to VONPs as they are being released into the environment.[110] In vivo and in vitro studies of VO$_2$ and vanadium pentoxide (V$_2$O$_5$) NPs were studied using animal models. Through nose-only inhalation, rats were exposed to VO$_2$NPs via bronchoalveolar lavage fluid (BAL) and showed elevated levels of stress indicators including lactase dehydrogenase (LDH), alkaline phosphatase (ALKP), and gamma-glutamyl transpeptidase (GGT) in the exposed subjects.[111] Histopathological analysis of alveolar tissue of exposed animals revealed fibrotic lesions, suggesting collagen degradation. At the cellular level, VO$_2$NPs triggered cytotoxicity of intrahepatic biliary epithelial cells.[112]

Intracellular reactive oxygen species (ROS) are reported to be generated following exposure to V$_2$O$_5$NPs, leading to apoptosis in cancer cells by upregulating p53 proteins, and downregulating proto-oncogene surviving that regulates cell division and apoptosis. Mice with melanoma administered V$_2$O$_5$NPs showed increased survival compared to the nontreated control group.[113]

Antibacterial properties of metal-oxide NPs are dependent on their crystalline structure. Highly crystalline VO$_2$NP is more toxic than industrial and normal VO$_2$. When comparing the antibacterial properties, the crystalline VO$_2$NP was more toxic and was attributed to the physical interactions between the bacteria and the NPs.[113a] When tested against common bacteria that cause nosocomial infections, it was revealed that crystallites that are more disorganized and have defects were more toxic.[114]

The short-term (1 day) and long-term (20 days) pulmonary toxicities of VO$_2$NPs were studied using A549 lung cells. Different size particles were compared: NPs with 20–30 mm (denoted as C-VO$_2$) and microparticles with a size of approximately 1 μm (denoted as M-VO$_2$). In short-term exposure, cell viability decreased at 2.5 μg/mL C-VO$_2$ concentration. At this concentration, C-VO$_2$ caused more cell loss compared to M-VO$_2$. Cell viability in long-term exposure decreased to less than 50% after exposure to C-VO$_2$ at a concentration of 0.3 μg/mL. M-VO$_2$ under the same condition did not show toxicity. Therefore, NPs of VO$_2$ are more toxic than microparticles. 0.05 μg/mL was the higher nontoxic dose tested against the A549 cells.[112]

V$_2$O$_5$NPs of 300 nm diameter were synthesized to study their antibacterial properties. *E. coli* was cultured on agar containing 0.3, 0.5, and 1.0 mg/mL of V$_2$O$_5$NPs and a control plate. Less growth was observed at 0.3 and 0.5 mg/mL of V$_2$O$_5$NPs compared to the control. At 1.0 mg/mL V$_2$O$_5$NPs, no bacterial growth occurred.[115]

Nanoclays are silicate particles that have nano-sized pores. In agriculture, nano clay is being considered for application as nano fertilizer, regulating the release of nutrients, decreasing the quantity of fertilizer, and preventing leaching. Nanoclay has the potential to treat surface water, wastewater, and groundwater that are contaminated. Nanomaterials are of interest because of their size and efficiency in absorbing pollutants. Nanoremediation uses reactive materials in detoxification.[116] Remediation methods include absorptive, reactive, in situ and ex-situ. Removal of contaminants by sequestration involves absorptive technologies, while differential reactivity may impact the degradation of contaminants. In situ technologies treat contaminants at the site, while ex situ treats the contamination off-site.[116]

The most commonly used nano clay minerals for environmental purposes are montmorillonite, kaolinite, halloysite, and palygorskite. Nanobubble-trapped nanoclays have been used instead of nanomaterials in remediation efforts to decrease anoxia in nutrient-rich polluted water.[117] These hypoxic conditions are harmful to aquatic organisms residing near these dead zones, where the sediment is generating N and P. Zeolite, the oxygen nanobubble (ONB) clay, has been developed to supply oxygen to the sediment.[117]

Clay nanotubes are developed as a low-cost pesticide carrier.[118] Clay nanotubes enhance pesticide effectivity, reduce the concentration of pesticides, and decrease pesticide leaching. Nanoclay pesticides control insects by physisorption to the pests' cuticular lipids. Other types of nanoclay gutbusters are insecticides that are formulated with a controlled release of substance when exposed to alkaline environments, such as the stomach of an insect.[118]

2. Human health and safety consequences of SMNs

Chemically unstable NPs can be oxidized, reduced, and dissolved in biological media, potentially leading to toxic ions release. NPs with higher solubility in cellular growth media, such as ZnONPs, show stronger toxicity to mammalian cells than the NPs with a low solubility such as TiO$_2$NPs, suggesting that solubility could strongly influence the cytotoxic response.[105] Fig. 12.4 shows the exposure pathway, circulation, redistribution, and final excretion of nanomaterials inside the human body.

A study reported Palladium/magnetite nanocatalysts had low toxicity on human skin (HaCaT) and human colon (CaCo-2) cells and RTgill-W1 cells.[106] As these NMs are used in water and wastewater treatment, the trace amount of metals in the water may not be of health and safety concern.

Several SNMs reportedly produce ROS. Studies have shown that ROS generation is involved in the toxicity of nanoparticles, including CeO_2, TiO_2, nC_{60}, Fe_3O_4, and FeO NPs.[2]

Nano zero-valent iron (nZVI)-based products are used as electron donors to transform contaminants in soil and groundwater through reduction or oxidation. They are of great interest in environmental remediation and are being studied in at least 80 studies.[119] There are various methods to synthesize nZVI, with the chemical reduction being the most popular. Other methods of synthesis include precision milling, carbothermal reduction, ultrasonic wave assistance, electrolysis, and green synthesis, which is the most environmentally friendly method.[120]

The toxicity of nZVI is widely studied on microorganisms and lower organisms such as crustaceans, fish larvae, arthropods, annelids, and plants. There is limited information on its effect on mammalian cells. The toxicity of nZVI has been associated with its adsorption onto the cell membranes. Such adsorption can interfere with cellular

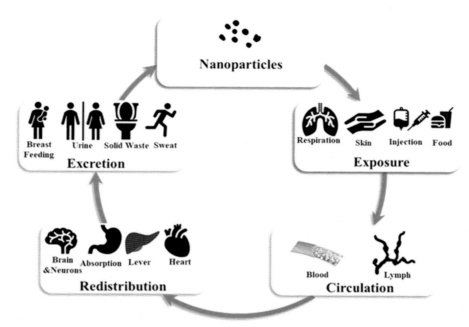

Figure 12.4 Exposure pathway, circulation, redistribution, and final excretion of nanomaterials in the human body.[85]

signalings, cause structural changes to the membranes, or prevent mobility and nutrient intake, potentially leading to cell death.[121] Also, the toxicity of nZVI has been associated with the formation of ROS due to oxidation. ROS has the potential to cause lipid peroxidation and DNA damage.[121]

The majority of studies investigating the toxicity of nZVI toward microorganisms tested *Escherichia coli (E. coli)*, which showed more sensitivity to Fe^{2+} and nZVI than other types of bacteria, and overall had a bactericidal effect toward *E. coli*.[122,123] Increased ROS generation caused damage to both proteins and lipids in the bacteria and showed signs of oxidative stress. However, in Gram-positive bacteria, the application of nZVI stimulated growth.[123] Toxicity has also been reported in various aquatic and soil organisms such as *Heterocypris incongreuens* and *Oryzias latipes*, that was sensitive to nZVI coated with carboxymethyl cellulose (CMC).[120] Phytoplanktons are negatively affected by nZVI as well. A commercial form of nZVI, called Nanofer 25S, was tested on three types of phytoplanktons: *Isochrysis galbana*, *Dunaliella tertiolecta*, and *Thalassiosira pseudonana*. All three species showed decreased growth when exposed to the commercial nZVI at concentrations above 6 mg/L. Similar results were also observed in zooplankton, specifically *Daphnia magna*, where Nanofer particles at concentrations above 1 mg/L caused significant die-offs within the first 24–48 h.[119]

The toxicity of nZVI toward plants is ambiguous. The effects of nZVI are based on the type of plant, the concentration of nZVI, and its properties. On one hand, with *Arabidopsis thaliana*, there were no toxic effects to its growth at nZVI concentrations up to 100 mg/L. Similar results were seen in *Medicago sativa* and *Typha latifolia*.[123] On the other hand, plants such as *Lolium perenne*, *Hordeum vulgare*, and *Linum usitatissimum* showed negative effects on germination and growth due to nZVI blocking water and nutrient uptake in the roots because of the accumulation of nZVI particles on the root surface. There is evidence of the decay of older leaves and lack of growth of younger leaves.[120] This ambiguity reveals that nZVI can be used in the environment without a significant negative impact on plants when used at certain concentrations.

The toxicity of nZVI toward mammalian cells is significant. When tested on human bronchial epithelium cells, the presence of nZVI killed them. Similar results were seen in mice nerve cells. They suffered oxidative stress when exposed to nZVI. However, there is a potential solution to decrease the toxicity. By modifying the surface of the nanoparticles with compounds such as polyasparaginate, the toxicity to mammalian cells can be reduced due to limited direct contact to cells.[120,122]

3. Economic aspects of NMs used as catalysts

3.1 Nano zerovalent iron (nZVI)

Among different NMs, nano zero-valent iron is one of the most used NM in groundwater remediation.[24,124–126] The cost of this NM was very expensive as it was initially started using in 90s, but the cost decreased as time went on. Similarly, other NM catalyst costs may be decreased in the future.

3.2 CFM@PDA/Pd composite nanocatalyst

Xi et al. utilized polydopamine (PDA) secreted by marine mussels to coat cotton microfiber, which was used as a reducing agent for Pd nanoparticle growth. Packing the CFM@PDA/Pd composite into a cylinder, they were able to catalyze the reduction of 4-nitrophenol more efficiently than the conventional batch reaction systems. In addition to the CFM@PDA/Pd composite system being recyclable up to nine times without a loss of efficacy, the system also had a higher turnover frequency (TOF) than the conventional batch reaction system TOF of 0.643 min^{-1}, compared to the conventional system with TOF of 1.587 min^{-1}. Due to its simple and economical preparation and its increased efficiency over the conventional fixed-bed system, this system is an economical alternative to the conventional method.[127]

3.3 FeOx/C nanocatalyst

Morales et al. investigated a Fenton-type nanocatalyst (FeOx/C) and hydrogen peroxide for potential inactivation of *Ascaris* eggs in water. Advanced oxidation processes (AOP)

Table 12.3 Comparison of conventional methods versus Fenton-type nanocatalyst (FeOx/C) + hydrogen peroxide for inactivation of *Ascaris* eggs in water.

Items	Nanocatalyst production	Disinfection process
Production	342 ton/year	23 m^3/s
Raw materials (USD/day)	$148,089.00	$182,727.00
Equipment (USD)	$92,683.00	$4,615,000.00
Wages (USD/day)	$488,049.00	$258,878.00
Operational and management cost (USD/year)	$10,569.00	$273,351.00
Cost per unit	$186,492.00/ton	$0.128/m^3

Adapted from Ref. 128.

produce ROS, which can degrade the protective layers of the eggs, are the main type of method currently used for the inactivation of *Ascaris* eggs in water.[128] In Fenton-like reactions, the catalyst forms a surface complex with hydrogen peroxide. ROS are then produced as a result of reversible electron transfer from the ligand to the metal. The catalytic activity of these processes can also be improved using nanoscale iron, as the reduction in particle size leads to an increase in specific contact areas. Using the Fenton-type nanocatalyst and hydrogen peroxide, a high inactivation rate of ascaris eggs was observed (Table 12.3). This method was found to be more cost-effective than the traditional AOP disinfection processes.[128]

3.4 UV/Ni–TiO$_2$ nanocatalyst

Pirkarami et al. carried out a cost-analysis of photocatalysis of dyes: Reactive Red 19 (RR 19), Acid Orange 7 (AO 7), and Acid Red 18 (AR 18), from water using UV/Ni–TiO$_2$ nanocatalyst. In addition to finding effective removal rates, the cost was determined as Operating cost (USD/m^3) = aCenergy + bCphotocatalyst + cCelectrolyte; whereby, Cenergy is the energy required by UV irradiation and electrode operation (kWh/m^3), Cphotocatalyst is the amount of photocatalyst consumed (kg/m^3), and Celectrolyte is the amount of electrolyte consumed (kg/m^3). The letters a, b, and c represent unit prices for the Iranian market in July 2012, as follows: (a) electrical energy price: 0.042 USD/kWh, (b) nano-Ni–TiO$_2$ price: 220 USD/kg, and (c) electrolyte NaCl price: 5.70 USD/kg. The cost was 9.42USD[129] to remove 1 m^3 of dye in the experiment.

Transesterification: Nanocatalysts can be used to catalyze the transesterification process using microalgae as a feedstock to produce biodiesel. Nanocatalysts have several unique benefits when used in this process, such as high surface area, high catalytic activity, surface rigidity, and resistance to saponification.[130] These features make nanocatalysts more appealing than the traditional heterocatalysts due to the downsides of such catalysts, such as mass transfer resistance, time consumption, inefficiency, and quicker deactivation. The use of nanocatalysts has also shown an increase in biodiesel yield from microalgae compared with conventional methods such as homogenous, heterogeneous, and enzymatic catalysts. This increase in the yield of biodiesel could make the transesterification process more economical,[130].

3.5 FePt–Ag nanocatalysts

Liu and colleagues synthesized FePt–Ag nanocatalysts as an alternative to the traditional methods to reduce the dyes methyl orange (MO) and rhodamine B (RhB).[131] Effluents of dyes such as MO and RhB from various industries are a significant source of water pollution. This is a significant issue, as the dyes can block sunlight from entering the water, which can decrease oxygen levels and inhibit aquatic life growth.[131] Also, many dyes are carcinogenic. Current traditional methods for removing dyes from water include absorption, electrochemical methods, photochemical methods, and ultrafiltration. Due to the complex structure of these dyes, many of these methods are not popularized. A newer catalytic reduction that uses noble metal nanocatalysts has recently been investigated, but the difficulty in recycling the nanocatalysts postuse was identified as one of its shortcomings.[131] Recent research has demonstrated the potential for bimetallic FePt nanocrystals as catalysts for the degradation of dyes. Liu et al. also developed a FePt–Ag composite nanocatalyst using polyethyleneimine dithiocarbamate (PEI-DTC) as a polymer for the seed deposition process. They investigated the impact of the amount of Ag seeds used on the degradation efficacy of MO and RhB. They found that these nanocatalysts were highly efficient for the degradation of dyes as they could be reused six times.[131]

3.6 Other nanocatalysts

Nanocatalysts are useful in the production of biodiesel. The types of nanocatalysts used include metal oxides, hydrocalcite-based catalysts, zeolite, and carbon-based materials. The benefits of using nanocatalysts stem from their small size, higher surface area, and surface functionalizations, all of which could favor increased reactivity.[132] The use of nanocatalysts in biodiesel production could be more cost-effective. The most common method for biodiesel production, which occurs due to carbon fixation, is homogenous and heterogeneous catalysis by the transesterification method. Due to their small size and the ability to modify their physical and chemical properties, nanocatalysts make them a potentially cost-saving means of producing biodiesel.[132] The efficacy of biodiesel production is influenced by the relative weight, density, and amount of the catalyst used. The use of nanocatalysts could optimize these parameters and more efficient biodiesel production.[132]

4. Conclusions and future perspectives

This paper offers a critical discussion on the health, safety, and economic analyses of SNMs used in catalysis. Various scientific studies show continual advancement in NM catalysts development and their applications; however, the interactions of these NMs with cells and tissues are still unknown. There is a need to understand better SNM's health effects on humans, plants, and animals in real-world conditions.

Acknowledgments

Authors like to thank the initial review support by C. Suarez, S. Sabu, J. Williams and C. Ubah.

References

[1] Yuan G. Natural and modified nanomaterials as sorbents of environmental contaminants. *J Environ Sci Health A* 2004;**39**(10):2661–70.

[2] Auffan M, Rose J, Bottero J-Y, Lowry GV, Jolivet J-P, Wiesner MR. Towards a definition of inorganic nanoparticles from an environmental, health and safety perspective. *Nat Nanotechnol* 2009; **4**(10):634–41.

[3] Centi G, Perathoner S. Creating and mastering nano-objects to design advanced catalytic materials. *Coord Chem Rev* 2011;**255**(13):1480–98.

[4] Sharma N, Ojha H, Bharadwaj A, Pathak DP, Sharma RK. Preparation and catalytic applications of nanomaterials: a review. *RSC Adv* 2015;**5**(66):53381–403.

[5] Lee J, Mahendra S, Alvarez PJJ. Nanomaterials in the construction industry: a review of their applications and environmental health and safety considerations. *ACS Nano* 2010;**4**(7):3580–90.

[6] Asmatulu R, Nguyen P, Asmatulu E. In: Asmatulu R, editor. *Nanotechnology safety*. Amsterdam: Elsevier; 2013. p. 57–72.

[7] Albrecht MA, Evans CW, Raston CL. Green chemistry and the health implications of nanoparticles. *Green Chem* 2006;**8**(5):417–32.

[8] Narayan N, Meiyazhagan A, Vajtai R. Metal nanoparticles as green catalysts. *Materials* 2019;**12**(21): 3602.

[9] Hunt ST, Milina M, Alba-Rubio AC, Hendon CH, Dumesic JA, Román-Leshkov Y. Self-assembly of noble metal monolayers on transition metal carbide nanoparticle catalysts. *Science* 2016;**352**(6288):974–8.

[10] Yang Q, Xu Q, Jiang H-L. Metal–organic frameworks meet metal nanoparticles: synergistic effect for enhanced catalysis. *Chem Soc Rev* 2017;**46**(15):4774–808.

[11] Majedi SM, Lee HK. Recent advances in the separation and quantification of metallic nanoparticles and ions in the environment. *Trac Trends Anal Chem* 2016;**75**:183–96.

[12] Foss Hansen S, Larsen BH, Olsen SI, Baun A. Categorization framework to aid hazard identification of nanomaterials. *Nanotoxicology* 2007;**1**(3):243–50.

[13] Lin Y, Ren J, Qu X. Catalytically active nanomaterials: a promising candidate for artificial enzymes. *Acc Chem Res* 2014;**47**(4):1097–105.

[14] Sudha D, Sivakumar P. Review on the photocatalytic activity of various composite catalysts. *Chem Eng Process: Process Intensif* 2015;**97**:112–33.

[15] Iijima M, Kamiya H. Surface modification for improving the stability of nanoparticles in liquid media. *KONA Powder Part J* 2009;**27**:119–29.

[16] Vidal-Iglesias FJ, Gómez-Mingot M, Solla-Gullón J. In: Aliofkhazraei M, editor. *Handbook of nanoparticles*. Springer; 2016.

[17] Cai Y, Ma C, Zhu Y, Wang JX, Adzic RR. Low-coordination sites in oxygen-reduction electrocatalysis: their roles and methods for removal. *Langmuir* 2011;**27**(13):8540–7.

[18] Zhao H, Ding G, Xu L, Cai M. A phosphine-free heterogeneous Suzuki–Miyaura reaction of aryl bromides catalyzed by MCM-41-supported tridentate nitrogen palladium complex under air. *Appl Organomet Chem* 2011;**25**(12):871–5.

[19] Arenz M, Mayrhofer KJJ, Stamenkovic V, Blizanac BB, Tomoyuki T, Ross PN, et al. The effect of the particle size on the kinetics of CO electrooxidation on high surface area Pt catalysts. *J Am Chem Soc* 2005;**127**(18):6819–29.

[20] Meier JC, Galeano C, Katsounaros I, Witte J, Bongard HJ, Topalov AA, et al. Design criteria for stable Pt/C fuel cell catalysts. *Beilstein J Nanotechnol* 2014;**5**:44–67.

[21] Eatemadi A, Daraee H, Karimkhanloo H, Kouhi M, Zarghami N, Akbarzadeh A, et al. Carbon nanotubes: properties, synthesis, purification, and medical applications. *Nanoscale Res Lett* 2014;**9**.

[22] Lee SW, Chen SO, Sheng WC, Yabuuchi N, Kim YT, Mitani T, et al. Roles of surface steps on Pt nanoparticles in electro-oxidation of carbon monoxide and methanol. *J Am Chem Soc* 2009;**131**(43):15669–77.

[23] Vecitis CD, Schnoor MH, Rahaman MS, Schiffman JD, Elimelech M. Electrochemical multiwalled carbon nanotube filter for viral and bacterial removal and inactivation. *Environ Sci Technol* 2011;**45**(8):3672–9.

[24] Kanel SR, Nepal D, Manning B, Choi H. Transport of surface-modified iron nanoparticle in porous media and application to arsenic(III) remediation. *J Nanoparticle Res* 2007;**9**(5):725–35.

[25] Grabowska E, Zaleska A, Sorgues S, Kunst M, Etcheberry A, Colbeau-Justin C, et al. Modification of titanium(IV) dioxide with small silver nanoparticles: application in photocatalysis. *J Phys Chem C* 2013;**117**(4):1955–62.

[26] Hou J, Li H, Tang Y, Sun J, Fu H, Qu X, et al. Supported N-doped carbon quantum dots as the highly effective peroxydisulfate catalysts for bisphenol F degradation. *Appl Catal B Environ* 2018;**238**:225–35.

[27] Shan SJ, Zhao Y, Tang H, Cui FY. A mini-review of carbonaceous nanomaterials for removal of contaminants from wastewater. *IOP Conf Ser Earth Environ Sci* 2017;**68**:012003.

[28] Yeganeh B, Kull CM, Hull MS, Marr LC. Characterization of airborne particles during production of carbonaceous nanomaterials. *Environ Sci Technol* 2008;**42**(12):4600–6.

[29] Turkevich LA, Dastidar AG, Hachmeister Z, Lim M. Potential explosion hazard of carbonaceous nanoparticles: explosion parameters of selected materials. *J Hazard Mater* 2015;**295**:97–103.

[30] Som C, Wick P, Krug H, Nowack B. Environmental and health effects of nanomaterials in nanotextiles and façade coatings. *Environ Int* 2011;**37**(6):1131–42.

[31] Yanamala N, Desai IC, Miller W, Kodali VK, Syamlal G, Roberts JR, et al. Grouping of carbonaceous nanomaterials based on association of patterns of inflammatory markers in BAL fluid with adverse outcomes in lungs. *Nanotoxicology* 2019;**13**(8):1102–16.

[32] Golin CB, Bougher TL, Mallow A, Cola BA. Toward a comprehensive framework for nanomaterials: an interdisciplinary assessment of the current Environmental Health and Safety Regulation regarding the handling of carbon nanotubes. *J Chem Health Saf* 2013;**20**(4):9–24.

[33] Aschberger K, Johnston HJ, Stone V, Aitken RJ, Hankin SM, Peters SAK, et al. Review of carbon nanotubes toxicity and exposure—appraisal of human health risk assessment based on open literature. *Crit Rev Toxicol* 2010a;**40**(9):759–90.

[34] Maynard AD, Baron PA, Foley M, Shvedova AA, Kisin ER, Castranova V. Exposure to carbon nanotube material: aerosol release during the handling of unrefined single-walled carbon nanotube material. *J Toxicol Environ Health A* 2004;**67**(1):87–107.

[35] Tian F, Cui D, Schwarz H, Estrada GG, Kobayashi H. Cytotoxicity of single-wall carbon nanotubes on human fibroblasts. *Toxicol In Vitro* 2006;**20**(7):1202–12.

[36] Ryman-Rasmussen JP, Cesta MF, Brody AR, Shipley-Phillips JK, Everitt JI, Tewksbury EW, et al. Inhaled carbon nanotubes reach the subpleural tissue in mice. *Nat Nanotechnol* 2009;**4**(11):747–51.

[37] Clift MJD, Frey S, Endes C, Hirsch V, Kuhn DA, Johnston BD, et al. Assessing the impact of the physical properties of industrially produced carbon nanotubes on their interaction with human primary macrophages in vitro. *BioNanoMaterials* 2013b;**14**(3–4):239–48.

[38] Clift MJD, Endes C, Vanhecke D, Wick P, Gehr P, Schins RPF, et al. A comparative study of different in vitro lung cell culture systems to assess the most beneficial tool for screening the potential adverse effects of carbon nanotubes. *Toxicol Sci* 2013a;**137**(1):55–64.

[38a] Kroto HW, Heath JR, O'Brien SC, Curl RF, Smalley RE. C_{60}: Buckminsterfullerene. *Nature* 1985; **318**:162–3.

[39] Aschberger K, Johnston HJ, Stone V, Aitken RJ, Tran CL, Hankin SM, et al. Review of fullerene toxicity and exposure — appraisal of a human health risk assessment, based on open literature. *Regul Toxicol Pharmacol* 2010b;**58**(3):455–73.

[40] Rouse JG, Yang J, Barron AR, Monteiro-Riviere NA. Fullerene-based amino acid nanoparticle interactions with human epidermal keratinocytes. *Toxicol In Vitro* 2006;**20**(8):1313–20.

[41] Yamawaki H, Iwai N. Cytotoxicity of water-soluble fullerene in vascular endothelial cells. *Am J Physiol Cell Physiol* 2006;**290**(6):C1495–502.

[42] Fiorito S, Serafino A, Andreola F, Bernier P. Effects of fullerenes and single-wall carbon nanotubes on murine and human macrophages. *Carbon* 2006;**44**(6):1100–5.

[43] Fadeel B, Bussy C, Merino S, Vázquez E, Flahaut E, Mouchet F, et al. Safety assessment of graphene-based materials: focus on human health and the environment. *ACS Nano* 2018;**12**(11):10582–620.

[44] Bussy C, Ali-Boucetta H, Kostarelos K. Safety considerations for graphene: lessons learnt from carbon nanotubes. *Acc Chem Res* 2013;**46**(3):692–701.

[45] Ahmed F, Rodrigues DF. Investigation of acute effects of graphene oxide on wastewater microbial community: a case study. *J Hazard Mater* 2013;**256–257**:33–9.

[46] Drasler B, Kucki M, Delhaes F, Buerki-Thurnherr T, Vanhecke D, Korejwo D, et al. Single exposure to aerosolized graphene oxide and graphene nanoplatelets did not initiate an acute biological response in a 3D human lung model. *Carbon* 2018;**137**:125–35.

[47] Chowdhury I, Duch MC, Mansukhani ND, Hersam MC, Bouchard D. Interactions of graphene oxide nanomaterials with natural organic matter and metal oxide surfaces. *Environ Sci Technol* 2014; **48**(16):9382–90.

[48] Dong P-X, Wan B, Wang Z-X, Guo L-H, Yang Y, Zhao L. Exposure of single-walled carbon nanotubes impairs the functions of primarily cultured murine peritoneal macrophages. *Nanotoxicology* 2013;**7**(5):1028–42.

[49] Fujita K, Fukuda M, Endoh S, Maru J, Kato H, Nakamura A, et al. Size effects of single-walled carbon nanotubes on in vivo and in vitro pulmonary toxicity. *Inhal Toxicol* 2015;**27**(4):207–23.

[50] Jia G, Wang H, Yan L, Wang X, Pei R, Yan T, et al. Cytotoxicity of carbon nanomaterials:single-WallNanotube, multi-WallNanotube, and fullerene. *Environ Sci Technol* 2005;**39**(5):1378–83.

[51] Kang S, Kim J-E, Kim D, Woo CG, Pikhitsa PV, Cho M-H, et al. Comparison of cellular toxicity between multi-walled carbon nanotubes and onion-like shell-shaped carbon nanoparticles. *J Nanoparticle Res* 2015;**17**(9):378.

[52] Mallakpour S, Madani M. A review of current coupling agents for modification of metal oxide nanoparticles. *Prog Org Coating* 2015;**86**:194–207.

[53] Park E-J, Zahari NEM, Kang M-S, Lee Sj, Lee K, Lee B-S, et al. Toxic response of HIPCO single-walled carbon nanotubes in mice and RAW264.7 macrophage cells. *Toxicol Lett* 2014;**229**(1): 167–77.

[54] Sweeney S, Hu S, Ruenraroengsak P, Chen S, Gow A, Schwander S, et al. Carboxylation of multi-walled carbon nanotubes reduces their toxicity in primary human alveolar macrophages. *Environ Sci Nano* 2016;**3**(6):1340–50.

[55] Tahara Y, Nakamura M, Yang M, Zhang M, Iijima S, Yudasaka M. Lysosomal membrane destabilization induced by high accumulation of single-walled carbon nanohorns in murine macrophage RAW 264.7. *Biomaterials* 2012;**33**(9):2762–9.

[56] van Berlo D, Wilhelmi V, Boots AW, Hullmann M, Kuhlbusch TAJ, Bast A, et al. Apoptotic, inflammatory, and fibrogenic effects of two different types of multi-walled carbon nanotubes in mouse lung. *Arch Toxicol* 2014;**88**(9):1725–37.

[57] Wang X, Guo J, Chen T, Nie H, Wang H, Zang J, et al. Multi-walled carbon nanotubes induce apoptosis via mitochondrial pathway and scavenger receptor. *Toxicol In Vitro* 2012;**26**(6):799–806.

[58] Hendrickson OD, Zherdev AV, Gmoshinskii IV, Dzantiev BB. Fullerenes: in vivo studies of biodistribution, toxicity, and biological action. *Nanotechnol Russ* 2014;**9**(11):601—17.

[59] Sergio M, Behzadi H, Otto A, van der Spoel D. Fullerenes toxicity and electronic properties. *Environ Chem Lett* 2013;**11**(2):105—18.

[60] Lalwani G, D'Agati M, Khan AM, Sitharaman B. Toxicology of graphene-based nanomaterials. *Adv Drug Deliv Rev* 2016;**105**:109—44.

[61] Roberts JR, Mercer RR, Stefaniak AB, Seehra MS, Geddam UK, Chaudhuri IS, et al. Evaluation of pulmonary and systemic toxicity following lung exposure to graphite nanoplates: a member of the graphene-based nanomaterial family. *Part Fibre Toxicol* 2016;**13**(1):34.

[62] Wan B, Wang Z-X, Lv Q-Y, Dong P-X, Zhao L-X, Yang Y, et al. Single-walled carbon nanotubes and graphene oxides induce autophagosome accumulation and lysosome impairment in primarily cultured murine peritoneal macrophages. *Toxicol Lett* 2013;**221**(2):118—27.

[63] d'Amora M, Alfe M, Gargiulo V, Giordani S. Graphene-like layers from carbon black: in vivo toxicity assessment. *Nanomaterials* 2020;**10**(8):1472.

[64] Zhou Y, Chu F, Qiu S, Guo W, Zhang S, Xu Z, et al. Construction of graphite oxide modified black phosphorus through covalent linkage: an efficient strategy for smoke toxicity and fire hazard suppression of epoxy resin. *J Hazard Mater* 2020;**399**:123015.

[65] Rozmyslowska-Wojciechowska A, Szuplewska A, Wojciechowski T, Pozniak S, Mitrzak J, Chudy M, et al. A simple, low-cost and green method for controlling the cytotoxicity of MXenes. *Mater Sci Eng C* 2020;**111**.

[66] Alshatwi AA, Vaiyapuri Subbarayan P, Ramesh E, Al-Hazzani AA, Alsaif MA, Alwarthan AA. Al$_2$O$_3$ nanoparticles induce mitochondria-mediated cell death and upregulate the expression of signaling genes in human mesenchymal stem cells. *J Biochem Mol Toxicol* 2012;**26**(11):469—76.

[67] Balasubramanyam A, Sailaja N, Mahboob M, Rahman MF, Hussain SM, Grover P. In vivo genotoxicity assessment of aluminium oxide nanomaterials in rat peripheral blood cells using the comet assay and micronucleus test. *Mutagenesis* 2009;**24**(3):245—51.

[68] Chen L, Yokel RA, Hennig B, Toborek M. Manufactured aluminum oxide nanoparticles decrease expression of tight junction proteins in brain vasculature. *J Neuroimmune Pharmacol* 2008;**3**(4):286—95.

[69] Radziun E, Dudkiewicz Wilczyńska J, Książek I, Nowak K, Anuszewska EL, Kunicki A, et al. Assessment of the cytotoxicity of aluminium oxide nanoparticles on selected mammalian cells. *Toxicol In Vitro* 2011;**25**(8):1694—700.

[70] Kumari M, Singh SP, Chinde S, Rahman MF, Mahboob M, Grover P. Toxicity study of cerium oxide nanoparticles in human neuroblastoma cells. *Int J Toxicol* 2014;**33**(2):86—97.

[71] Assadian E, Zarei MH, Gilani AG, Farshin M, Degampanah H, Pourahmad J. Toxicity of copper oxide (CuO) nanoparticles on human blood lymphocytes. *Biol Trace Elem Res* 2018;**184**(2):350—7.

[72] Abudayyak M, Gurkaynak TA, Ozhan G. In vitro evaluation of cobalt oxide nanoparticle-induced toxicity. *Toxicol Ind Health* 2017;**33**(8):646—54.

[73] Chattopadhyay S, Dash SK, Tripathy S, Das B, Mandal D, Pramanik P, et al. Toxicity of cobalt oxide nanoparticles to normal cells; an in vitro and in vivo study. *Chem Biol Interact* 2015;**226**:58—71.

[74] Sahu SC, Hayes AW. Toxicity of nanomaterials found in human environment: a literature review. *Toxicol Res Appl* 2017;**1**. 2397847317726352.

[75] Sajid M, Ilyas M, Basheer C, Tariq M, Daud M, Baig N, et al. Impact of nanoparticles on human and environment: review of toxicity factors, exposures, control strategies, and future prospects. *Environ Sci Pollut Control Ser* 2015;**22**(6):4122—43.

[76] Liu G, Gao J, Ai H, Chen X. Applications and potential toxicity of magnetic iron oxide nanoparticles. *Small* 2013;**9**(9-10):1533—45.

[77] Pawelczyk E, Arbab AS, Chaudhry A, Balakumaran A, Robey PG, Frank JA. In vitro model of bromodeoxyuridine or iron oxide nanoparticle uptake by activated macrophages from labeled stem cells: implications for cellular therapy. *Stem Cells* 2008;**26**(5):1366—75.

[78] Leso V, Iavicoli I. Palladium nanoparticles: toxicological effects and potential implications for occupational risk assessment. *Int J Mol Sci* 2018;**19**(2):503.

[79] Foldbjerg R, Dang DA, Autrup H. Cytotoxicity and genotoxicity of silver nanoparticles in the human lung cancer cell line, A549. *Arch Toxicol* 2011;**85**(7):743—50.

[80] Haase A, Tentschert J, Jungnickel H, Graf P, Mantion A, Draude F, et al. Toxicity of silver nanoparticles in human macrophages: uptake, intracellular distribution and cellular responses. *J Phys Conf* 2011;**304**:012030.

[81] Hussain SM, Hess KL, Gearhart JM, Geiss KT, Schlager JJ. In vitro toxicity of nanoparticles in BRL 3A rat liver cells. *Toxicol In Vitro* 2005;**19**(7):975–83.

[82] Sun L, Li Y, Liu X, Jin M, Zhang L, Du Z, et al. Cytotoxicity and mitochondrial damage caused by silica nanoparticles. *Toxicol In Vitro* 2011;**25**(8):1619–29.

[83] Bhattacharya K, Davoren M, Boertz J, Schins RPF, Hoffmann E, Dopp E. Titanium dioxide nanoparticles induce oxidative stress and DNA-adduct formation but not DNA-breakage in human lung cells. *Part Fibre Toxicol* 2009;**6**(1):17.

[84] Liu R, Yin L, Pu Y, Liang G, Zhang J, Su Y, et al. Pulmonary toxicity induced by three forms of titanium dioxide nanoparticles via intra-tracheal instillation in rats. *Prog Nat Sci* 2009;**19**(5):573–9.

[85] Malakar A, Kanel SR, Ray C, Snow DD, Nadagouda MN. Nanomaterials in the environment, human exposure pathway, and health effects: a review. *Sci Total Environ* 2021;**759**:143470.

[86] Sharma V, Singh P, Pandey AK, Dhawan A. Induction of oxidative stress, DNA damage and apoptosis in mouse liver after sub-acute oral exposure to zinc oxide nanoparticles. *Mutat Res Genet Toxicol Environ Mutagen* 2012;**745**(1):84–91.

[87] Ye M, Shi B. Zirconia nanoparticles-induced toxic effects in osteoblast-like 3T3-E1 cells. *Nanoscale Res Lett* 2018;**13**(1):353.

[88] Ganachari SV, Hublikar L, Yaradoddi JS, Math SS. In: Martínez LMT, Kharissova OV, Kharisov BI, editors. *Handbook of Ecomaterials*. Cham: Springer International Publishing; 2019. p. 2357–68.

[89] Ahangaran F, Navarchian AH. Recent advances in chemical surface modification of metal oxide nanoparticles with silane coupling agents: a review. *Adv Colloid Interf Sci* 2020;**286**:102298.

[90] Talebzadeh S, Queffelec C, Knight DA. Surface modification of plasmonic noble metal-metal oxide core-shell nanoparticles. *Nanoscale Adv* 2019;**1**(12):4578–91.

[91] Sakeye M, Smått J-H. Comparison of different amino-functionalization procedures on a selection of metal oxide microparticles: degree of modification and hydrolytic stability. *Langmuir* 2012;**28**(49): 16941–50.

[92] Cartwright A, Jackson K, Morgan C, Anderson A, Britt DW. A review of metal and metal-oxide nanoparticle coating technologies to inhibit agglomeration and increase bioactivity for agricultural applications. *Agronomy* 2020;**10**(7):1018.

[93] Lenne Q, Leroux YR, Lagrost C. Surface modification for promoting durable, efficient, and selective electrocatalysts. *ChemElectroChem* 2020;**7**(11):2345–63.

[94] Rossi LM, Costa NJS, Silva FP, Gonçalves RV. Magnetic nanocatalysts: supported metal nanoparticles for catalytic applications. *Nanotechnol Rev* 2013;**2**(5):597–614.

[95] Rossi LM, Nangoi IM, Costa NJS. Ligand-assisted preparation of palladium supported nanoparticles: a step toward size control. *Inorg Chem* 2009;**48**(11):4640–2.

[96] Li R, Zhang P, Huang Y, Zhang P, Zhong H, Chen Q. Pd–Fe$_3$O$_4$@C hybrid nanoparticles: preparation, characterization, and their high catalytic activity toward Suzuki coupling reactions. *J Mater Chem* 2012;**22**(42):22750–5.

[97] Golkhatmi FM, Bahramian B, Mamarabadi M. Application of surface modified nano ferrite nickel in catalytic reaction (epoxidation of alkenes) and investigation on its antibacterial and antifungal activities. *Mater Sci Eng C* 2017;**78**:1–11.

[98] Kumar S, Karmacharya M, Joshi SR, Gulenko O, Park J, Kim G-H, et al. Photoactive antiviral face mask with self-sterilization and reusability. *Nano Lett* 2021;**21**(1):337–43.

[99] Zuo Q-Q, Feng Y-L, Chen S, Qiu Z, Xie L-Q, Xiao Z-Y, et al. Dimeric core–shell Ag$_2$@TiO$_2$ nanoparticles for off-resonance Raman study of the TiO$_2$–N719 interface. *J Phys Chem C* 2015; **119**(32):18396–403.

[100] Wei YL, Ebendorff-Heidepriem H, Zhao JB. Recent advances in hybrid optical materials: integrating nanoparticles within a glass matrix. *Adv Opt Mater* 2019;**7**(21):34.

[101] Mahendra S, Zhu H, Colvin VL, Alvarez PJ. Quantum dot weathering results in microbial toxicity. *Environ Sci Technol* 2008a;**42**(24):9424–30.

[102] Yu WW, Chang E, Falkner JC, Zhang J, Al-Somali AM, Sayes CM, et al. Forming biocompatible and nonaggregated nanocrystals in water using amphiphilic polymers. *J Am Chem Soc* 2007;**129**(10):2871−9.

[103] Wang Z, Tang M. The cytotoxicity of core-shell or non-shell structure quantum dots and reflection on environmental friendly: a review. *Environ Res* 2021;**194**:110593.

[104] Lin P, Chen J-W, Chang LW, Wu J-P, Redding L, Chang H, et al. Computational and ultrastructural toxicology of a nanoparticle, quantum dot 705, in mice. *Environ Sci Technol* 2008;**42**(16):6264−70.

[105] Brunner TJ, Wick P, Manser P, Spohn P, Grass RN, Limbach LK, et al. In vitro cytotoxicity of oxide nanoparticles:comparison to asbestos, silica, and the effect of particle solubility. *Environ Sci Technol* 2006;**40**(14):4374−81.

[106] Hildebrand H, Kühnel D, Potthoff A, Mackenzie K, Springer A, Schirmer K. Evaluating the cytotoxicity of palladium/magnetite nanocatalysts intended for wastewater treatment. *Environ Pollut* 2010;**158**(1):65−73.

[107] Nikazar S, Sivasankarapillai VS, Rahdar A, Gasmi S, Anumol PS, Shanavas MS. Revisiting the cytotoxicity of quantum dots: an in-depth overview. *Biophys Rev* 2020;**12**(3):703−18.

[108] Navarro DA, Bisson MA, Aga DS. Investigating uptake of water-dispersible CdSe/ZnS quantum dot nanoparticles by *Arabidopsis thaliana* plants. *J Hazard Mater* 2012;**211−212**:427−35.

[109] Derfus AM, Chan WCW, Bhatia SN. Probing the cytotoxicity of semiconductor quantum dots. *Nano Lett* 2004;**4**(1):11−8.

[110] Park E-J, Lee G-H, Yoon C, Jeong U, Kim Y, Chang J, et al. Tissue distribution following 28 day repeated oral administration of aluminum-based nanoparticles with different properties and the in vitro toxicity. *J Appl Toxicol* 2017;**37**(12):1408−19.

[111] Kulkarni A, Kumar GS, Kaur J, Tikoo K. A comparative study of the toxicological aspects of vanadium pentoxide and vanadium oxide nanoparticles. *Inhal Toxicol* 2014;**26**(13):772−88.

[112] Xi W-S, Song Z-M, Chen Z, Chen N, Yan G-H, Gao Y, et al. Short-term and long-term toxicological effects of vanadium dioxide nanoparticles on A549 cells. *Environ Sci Nano* 2019;**6**(2):565−79.

[113] Das S, Roy A, Barui AK, Alabbasi MMA, Kuncha M, Sistla R, et al. Anti-angiogenic vanadium pentoxide nanoparticles for the treatment of melanoma and their in vivo toxicity study. *Nanoscale* 2020;**12**(14):7604−21.

[113a] Wu D, Su QQ, Li Y, Zhang C, Qin X, Liu YY, Xi WS, Gao YF, Cao AN, Liu XG, Wang HF. Wang Toxicity assessment and mechanistic investigation of engineered monoclinic VO2 nanoparticles. *Nanoscale* 2018;**10**(20):9736−46.

[114] Perelshtein I, Lipovsky A, Perkas N, Gedanken A, Moschini E, Mantecca P. The influence of the crystalline nature of nano-metal oxides on their antibacterial and toxicity properties. *Nano Res* 2015;**8**(2):695−707.

[115] Raj S, Kumar S, Chatterjee K. Facile synthesis of vanadia nanoparticles and assessment of antibacterial activity and cytotoxicity. *Mater Technol* 2016;**31**(10):562−73.

[116] Prabhakar V, Tahira B. Nanotechnology, future tools for remediation. *Int J Emerg Technol Adv Eng* 2013;**3**(7):54−9.

[117] Biswas B, Warr LN, Hilder EF, Goswami N, Rahman MM, Churchman JG, et al. Biocompatible functionalisation of nanoclays for improved environmental remediation. *Chem Soc Rev* 2019;**48**(14):3740−70.

[118] Constantinescu F, Boiu Sicuia OA. In: Thajuddin N, Mathew S, editors. *Phytonanotechnology*. Elsevier; 2020. p. 245−87.

[119] Keller AA, Garner K, Miller RJ, Lenihan HS. Toxicity of nano-zero valent iron to freshwater and marine organisms. *PLoS One* 2012;**7**(8).

[120] Stefaniuk M, Oleszczuk P, Ok YS. Review on nano zero-valent iron (nZVI): from synthesis to environmental applications. *Chem Eng J* 2016;**287**:618−32.

[121] Chen X, Ji D, Wang X, Zang L. Review on nano zero-valent iron (nZVI): from modification to environmental applications. *IOP Conf Ser Earth Environ Sci* 2017;**51**:012004.

[122] Keane E. *Fate, transport and toxicity of nanoscale zero-valent iron (nZVI) used during superfund remediation*. Duke University; 2010.

[123] Pasinszki T, Krebsz M. Synthesis and application of zero-valent iron nanoparticles in water treatment, environmental remediation, catalysis, and their biological effects. *Nanomaterials* 2020;**10**(5):917.

[124] Kanel SR, Choi H, Kim J-Y, Vigneswaran S, Shim WG. Removal of arsenic(III) from groundwater using low-cost industrial by-products-blast furnace slag. *Water Qual Res J* 2006;**41**(2):130–9.

[125] Kanel SR, Manning B, Charlet L, Choi H. Removal of arsenic(III) from groundwater by nanoscale zero-valent iron. *Environ Sci Technol* 2005;**39**(5):1291–8.

[126] Raji M, Mirbagheri SA, Ye F, Dutta J. Nano zero-valent iron on activated carbon cloth support as Fenton-like catalyst for efficient color and COD removal from melanoidin wastewater. *Chemosphere* 2021;**263**:127945.

[127] Xi J, Xiao J, Xiao F, Jin Y, Dong Y, Jing F, et al. Mussel-inspired functionalization of cotton for nano-catalyst support and its application in a fixed-bed system with high performance. *Sci Rep* 2016;**6**(1):21904.

[128] Morales AA, Schouwenaars R, Pfeiffer H, Ramírez-Zamora RM. Inactivation of Ascaris eggs in water using hydrogen peroxide and a Fenton type nanocatalyst (FeOx/C) synthesized by a novel hybrid production process. *J Water Health* 2013;**11**(3):419–29.

[129] Pirkarami A, Olya ME, Raeis Farshid S. UV/Ni–TiO$_2$ nanocatalyst for electrochemical removal of dyes considering operating costs. *Water Resour Ind* 2014;**5**:9–20.

[130] Akubude VC, Nwaigwe KN, Dintwa E. Production of biodiesel from microalgae via nanocatalyzed transesterification process: a review. *Mater Sci Energy Technol* 2019;**2**(2):216–25.

[131] Liu Y, Zhang Y, Kou Q, Chen Y, Sun Y, Han D, et al. Highly efficient, low-cost, and magnetically recoverable FePt–Ag nanocatalysts: towards green reduction of organic dyes. *Nanomaterials* 2018;**8**(5):329.

[132] Bano S, Ganie AS, Sultana S, Sabir S, Khan MZ. Fabrication and optimization of nanocatalyst for biodiesel production: an overview. *Front Energy Res* 2020;**8**(350).

CHAPTER 13

Future of SMNs catalysts for industry applications

Ajaysing S. Nimbalkar[1,2], Dipali P. Upare[1], Nitin P. Lad[3] and Pravin P. Upare[2]

[1]Green Chemistry Division, University of Science of Technology, Daejeon, South Korea; [2]Green Carbon Catalysis Research Group, Korea Research Institute of Chemical Technology, Daejeon, South Korea; [3]Organic Chemistry Research Centre, Department of Chemistry, A.M. Science College, Nashik, Maharashtra, India

1. Introduction

Nanomaterials are used in numerous prospective research areas, such as catalysis, pharmaceutical, fine chemicals, agricultural, biomedical, textiles, cosmetics, optics, food industries, battery industry, semiconductor devices, aerospace, and construction for plentiful applications.[1–3] In catalyst and chemical industry, nanoparticle-based materials shown great potential in industrial catalytic utilizations for the manufacturing, energy conversion/storage industries, research, and development.[2–5] Surface-modified nanomaterials (SMNs) has emerged as a new class of nanomaterials with promising improvement in the performance compared to outdated micro and macromaterials. Hence, it is expected to advance the field of catalytic- and chemistry-related applications thanks to the distinguished and desired nanolevel interaction between the surface functionalities and diverse structures, which has brought attention to SMNs.[2,6,7] The most important property of nanomaterials is their surface morphology and surface functionality. There is insufficient literature on SMNs as a process to improve desired surface properties.[2,6,7] Considering the requisite safety, economic viability, ecofriendliness of manufacturing, utilization, and degradation and efficiency, the potential application of SMNs has yet to be elaborated on a broad spectra. Surface modification of nanomaterials can be employed prior to target application in order to obtain well-defined morphologies, chemical structures, and compositions. In a number of catalytic processes the surface morphology of SMNs directs the effectivity of active surface area and pore structure, which is marginally larger than the macroscopic geometrical area. In addition to injection molding, extrusion, casting, and rolling, various advanced techniques has been applied from lab-scale to industrial-scale production of materials, which possibly enable limited abilities toward increased surface area well above the geometric one.[6,8,9]

Nanomaterials can be incorporated into polymeric nanocomposites. However, the incorporation of surface functionality in nanomaterials can be done by applying various hydroxyl, carboxylic groups, thiols, amines, amides, sulfonic, and other organic functionalizations. For the surface modification nanomaterials, several techniques such as (1) copolymerization of functional organosilanes, macromonomers, and metal alkoxides;

(2) functionalization of organic compounds using in situ sol–gel-derived silica or metallic oxide synthesis; and (3) incorporating inorganic nanoparticles within polymer matrix can significantly affect the properties of the matrix; and (4) organic functionalization of nanotubes, nanosheets, or other compounds with lamellar structures.[10] After modification, SMNs might exhibit improved physiochemical properties such as surface chemical properties, acidity, basicity, diffusibility, thermal and mechanical properties, electrical, and optical properties. Heterogeneous nanomaterial exhibits individual features in size, shape, and surface sites leading to variable particle-specific reactivity. The outstanding properties of SMNs depend on the type of nanoparticles that are incorporated, their size, and shape. Fig. 13.1 represents the surface modification of nanomaterials and their properties for a variety of catalytic applications.

Heterogeneous catalysis is the type of catalysis where the catalyst substance exists in a different phase to the reaction mixture, which provide easier separation and recovery procedure.[11] Typically, heterogeneous catalysis involves reactant diffusion phenomenon on their surface, followed by a surface-based reaction between the adsorbed species, and desorption of products. The key of this alternative pathway to the noncatalytic conversions is the lower activation energy than that required for the unanalyzed reaction. The

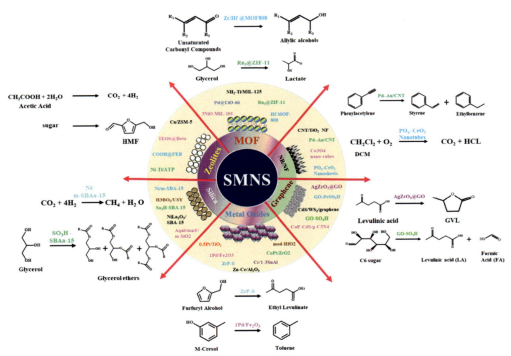

Figure 13.1 Surface modification of nanomaterials and their properties for variety of catalytic applications.

global market of catalyst is projected to reach USD 48.0 billion by 2027, expanding with CAGR of 4.4%, according to the latest report by Grand View Research Inc.[12] Approximately 30% of gross domestic product (GDP) and catalytic processes manufacture more than 90% of chemicals utilize catalysts for various processes for about 90% of chemicals and substances.[13] Adding to this, the present guidelines supporting the catalyst development and utilization is expected to enhance market growth over the next eight years. According to one report,[12] major players including Albemarle Corporation and Royal Dutch Shell PLC mainly regulate the catalysis market; these agencies offer companies with profits starting for pretreatment of raw materials to the final product manufacturing, which enhances the economic values of scale level to finest level. Developing markets of Asia Pacific nations including India and China are projected for growth in manufacturing of chemicals and polymers and petrochemical catalysts. Whereas the catalyst for refining industries exists an attractive segment in fossil fuel–rich countries such as Oman, Kuwait, and the Kingdom of Saudi Arabia.

Inorganic solid materials, especially nanomaterials (nanocatalysts), have received great attention for their utilization in the production of fuels, fine chemicals, and fertilizers. However, catalysts and their particular sites are often specific for particular reactions; for example, metal-supported materials are highly recommended for hydrogenation, hydrogenolysis, and oxidation; acidic and basic catalysts are favorable for hydrolysis, alcoholysis, dehydration, etherification, esterification, and so forth.[3,9,11] According to reports,[2,6,7] variation in desired product distribution and desired product could be obtained through varying functionalities (active sites) and surface properties of the catalyst material. However, reaction parameter optimization is an important relevant story. Value-added, chemical-based industries are the result of designing required physiochemical properties with materials. Current industrial development nanomaterials and their surface modification extend possible techniques for difficult conversions. Researchers are exploring how catalysts work and how to improve their effectiveness with extensive fundamental and applied research. This chapter focuses on the existing industrial applications of SMNs and further emphasizes their importance in rapidly emerging applications in chemical industries through research and development.

2. Nanoparticles catalysts

Despite the advantages of homogeneous catalysts and their significant utilizations in a variety of applications, the aforementioned systems were found to be less commercialized owing to the difficult separation of the catalyst from the final reaction product. Several strategies have been employed to resolve the separation issues in homogeneous catalysis, and the use of heterogeneous catalyst systems were found to be the most logical solutions.[14] Usually, most of the industry have considered the heterogeneous catalyst with the high surface area on which active sites are strongly dispersed and attached. The beauty of heterogeneity in a reaction medium allows for easy separation procedures from reaction or product mixture, which could significantly affect the cost of product isolation and purification.

Catalysts with nanoparticles are considered as sustainable alternatives to conventional heterogeneous catalysts due to their higher surface area and more reachable active components, which could offer the space for reacting the surface active sites and reaction component, by which nanocatalysts act like the homogenous catalysis and increase the reaction rate as well as efficiency. Polshettiwar et al., have discussed about nanocatalysts and their needs in catalysis in their advanced review.[11] Along with the excellent separation compatibility, controllable activity, and selectivity through their morphological and chemical modification is the most important feature of nanocatalysts, which can make it more efficient and offer great potential as well as scope in industrial-scale conversion process. However, advanced nanotechnology techniques makes systematic productions of nanomaterials with controlled shapes, size, morphology, composites, surface properties, and surface functionality. Fig. 13.2 represents the schematic representation for necessity of heterogeneous nanocatalysts over homogenous catalysts.

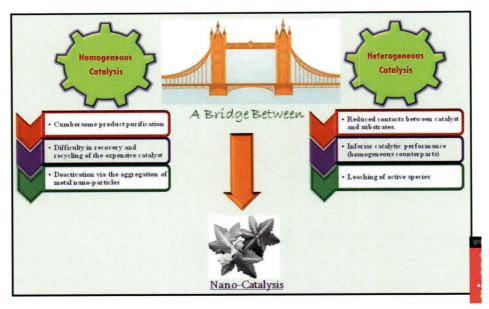

Figure 13.2 Nanocatalysts the bridge between heterogeneous and homogenous catalysis. *(Adopted from Ref. 11 with copyright permission.)*

3. Catalytic applications of nanomaterial

In nanomaterial catalysis, most of the researchers are focusing on design and synthesis of various nanomaterials based on silica. Metal oxide, metal based, MOF, polymer composites, and their utilization in emerging trends of application in sustainable nanocatalysts. As discussed earlier, nanoparticle-based materials can substitute conventional heterogeneous and homogenous catalysts and serve as active and stable heterogeneous catalyst materials in industrially important chemical conversions. Current ongoing interest in nanoparticles catalysts is for the development of efficient and more sustainable protocol for demanding reactions like C—H activations, hydrogenolysis, hydrogenation, C—C coupling, oxidation, hydrolysis, dehydration, CO_2 capture as well as CO_2 conversion into value-added chemicals, environmental conversions, biomass conversion processes, refinery processes, photocatalysis, electrocatalysis, water splitting, hydrogen productions, and so forth.[2,7] In brief, development in nanocatalysis requires modified concept of heterogeneous and homogenous catalysis, material and surface science, organometallic catalysis, and bioorganic catalysis, by which more efficient and sustainable green nanocatalysts will come out to address the most important and challenging issues in human society. By considering the economic importance of catalysis-based industries, the economic values depend on the development of four important catalysis sectors (polymer industry, coal, oil, gas refine industry, chemical manufacturing, and environmental application).[3] Hence, extensive research is necessary on development of advanced nanoparticle catalysts and their activity improvement through their surface and structural modifications.

3.1 Surface modification of nanoparticles and techniques

SMNs are materials of great scientific interest with a variety of catalytic applications, already discussed earlier and nicely summarized by Li et al.[7] The widespread research on SMNs has offered the ability to synthesize nanomaterials with specifically controlled morphology, structures, and compositions.[2] In literature,[7] several reports are found, and these are related to the interfaces of metal oxides necessary for the catalytic activity in gas-phase and liquid-phase conditions; these interfaces can be modified in a systematic way to selective-catalyzed subsequent coupled reactions. Consequently, the boundaries between ingredients of SMNs deeply affect their properties, and this indirectly tunes their reactivity as nanocatalysts. In brief, the surface of the nanomaterial and interfaces are important factors for efficient nanocatalyst development. However, detailed characterization of the catalyst surface will remain an important task for further development of efficient nanocatalysts.

Plenty of methods such as chemical, biological, physical, and hybrid techniques are recommended for the synthesis of SMNs.[15,16] The common material synthesis methods (i.e., precipitation, impregnation, hydrothermal, electrochemical, sol-gel, solvothermal synthesis, sonochemical, plasma, arc discharge, atomic layer deposition methods, microwave synthesis, physical treatment, in situ synthesis, chemical vapor deposition method, biochemical methods, and so forth), which are in regular practice can be utilize for the formation of advanced SMNs, these common techniques could benefit the synthesis efficiency after necessary additional improvements. SMNs is an advanced technology, which refers to the use of chemical, physical, biological, and other related methods to deal with nanoparticle surface. The modification techniques can be employed according to the need of use,[17] for example, change or addition of functional group moieties, surface modification of nanoparticles, physical properties and morphology, surface energy, surface wettability, electric properties, reaction characteristics and protocols, electrical and photochemical properties, modification in sorption properties, and so forth. These modifications are usually applied on solid materials, yet it is possible to find some instances of the modification to the surface of specific liquids. The surface modification is the act of modifying the surface of a material by bringing physical, chemical, or biological characteristics that are different from the ones originally found on the surface of a material.

No doubt, almost all the industrial catalysts were developed by pragmatic methods, which have taken more than 100 years of dedicated effort. The combined efforts in the related research areas of material development, surface science, computational tools, and characterizations are navigating the development of catalysis development of advanced catalysis. Of course, developing the advanced and efficient catalysts will require a depth of understanding in mechanistic behavior of catalyst surface and structural characteristic. Nevertheless, most of the industrial catalytic processes are operated under pressure conditions and over the well-dispersed nanoparticle-based catalysts. However, shape and size of nanoparticles are the important features, which are governing the catalytic properties of nanomaterials (Fig. 13.3).

The size dependence of catalysis can be attributed to the surface chemistry and structure of a catalyst nanoparticle that could be essentially determines catalytic performance. Surface adsorptions on nanoparticle surface is a very fundamental step through which solid–liquid and solid–gas interface preformed through which reaction happened on

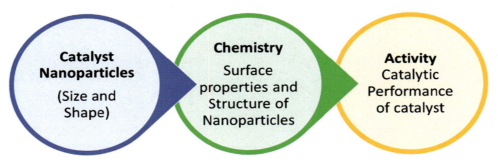

Figure 13.3 Correlation of nanoparticles, surface chemistry, and activity.

the catalyst surface. In particular, reactant molecules, reaction intermediates, dissociated species should be interact with atoms of nanoparticles and active sites, which could be ensemble of single or multiple atoms on catalyst surface of nanoparticle and it could offer a desired reaction pathway through different binding configuration and strength. Thus, the size and shapes of SMNs can influence the catalytic performance, which could be essentially carried out by altering chemical and structural factors of catalyst surfaces. This book chapter emphasizes the effect of size and shape on the performance of nanocatalysts for various industrially important reactions including deoxygenation, oxidation, Fischer-Tropsch synthesis, isomerization, water-gas shift reaction, hydrogenation, and so forth. Furthermore, some important studies on SMNS of different catalysts materials for their catalytic applications in important chemicals conversions processes are well summarized, and it is presented in Table 13.1.

4. Shape and size dependent catalysts and reactions

The Fischer—Tropsch (FT) synthesis is a well know reaction for conversion of gas into long-chain hydrocarbons.[59-63] The FT synthesis is highly sensitive to the structure of the catalysts particles. Nevertheless, selective FT synthesis is highly challenging task. However, number of reports have been claimed with diverse research approaches to an efficient nanocatalysts based on cobalt metal.[63-67] For instance, Jong and coworkers obtained FT synthesis highest activity for cobalt catalyst with 6 nm, whereas the activity was decreased with the smaller and higher cobalt particles size of 6.0 nm.[68] Fig. 13.4 represents size dependent performance of Cobalt and Fe carbide particles on FT conversions, which could suggest the hypothesis for highest FT activity on catalysts of particular size.

In addition, they investigated effect of particle size on the selectivity to the C5+ product. However, Borg and coworkers obtained highest selectivity for C5+ products with the particles size of 7—8 nm of cobalt.[69] However, several reports are existed for the other metals based catalysts such as Fe-based,[59,63,70,71] Ru-based,[72-74] for FT synthesis and their nanoparticles effects on the activity. As compared to the Co- and Fe-based catalysts in FT synthesis, due to their high activities at lower temperatures Ru-based catalysts offers promising potential for FT synthesis. Carballo and coworkers investigated the TOF of Ru nanoparticles catalyzed CO consumption with 10 nm and greater size, they observed that Ru particle with smaller size (<10.0 nm) can blocked due to strong adsorption of CO molecules.[72]

4.1 Shape dependency of activity in water-gas shift

Considering the promising application in purification of H_2 gas for low temperature fuel cell water-gas shift reaction (WGS) was extensively studied in last few decades.[75,76] Metallic nanoparticles supported reducible oxide supports was mainly consider in this industrially potential conversion processes. In reports, Au nanoparticle supported catalysts

Table 13.1 Examples for surface modifications of nanocatalysts for catalytic applications.

Sr. No.	Catalyst	Surface modification	Catalytic process	References
1	CoPt/ZrO$_2$	Plasma treatment on ZrO$_2$ support	Fischer–Tropsch catalysis	18
2	1%Pd/Fe$_2$O$_3$	Pd nanoparticles supported reduced Fe$_2$O$_3$	Hydrodeoxygenation of biomass	19
3	Pd/HfO$_2$	Pd nanoparticles supported on surface-modified metal oxides	Catalytic oxidation of Lean methane	20
4	La$_{0.8}$Sr$_{0.2}$CoO$_3$	Surface modification of La$_{0.8}$Sr$_{0.2}$CoO$_3$ USING diluted oxalic acid	Propane oxidation	21
5	2Pd/Sn–Al$_2$O$_3$	Surface Lewis acid sites of Sn-doped Pd/Al$_2$O$_3$	Chemoselective reductive N-acetylation of nitrobenzene	22
6	Zn–Ca/Al$_2$O$_3$	Calcium promoter	Transesterification of palm oil	23
7	Cr/1-3SnAl	Foamed Sn-modified alumina	Propane dehydrogenation	24
8	a-Cr$_2$WO$_6$/WO$_3$	Cr$_2$WO$_6$-modified WO$_3$ nanowires	Oxidative desulfurization	25
9	50% Sn$_2$STA/Ta$_2$O$_5$	Sn-modified silicotungstic acid supported on Ta$_2$O$_5$	Conversion of levulinic acid to ethyl levulinate	26
10	ZrP–S	Sulfonation of zirconium phosphate	Furfural alcohol to ethyl levulinate	27
11	0.5Pt/TiO$_2$	Pt loading	Photocatalytic application	28
12	Aquivion/m-SiO$_2$	Aquivion PFSA modified mesoporous silica	Etherification of 5-HMF to biofuel	29
13	USY	H$_3$BO$_3$ modified USY	Catalytic cracking	30
14	Rh/TPP	surface embedded on amine-functional mesoporous molecular sieves	Hydrogenation of biomass-derived levulinic acid to biofuel additive γ-GVL	31

Table 13.1 Examples for surface modifications of nanocatalysts for catalytic applications.—cont'd

Sr. No.	Catalyst	Surface modification	Catalytic process	References
15	Ni/m-SBA-15	Ni dispersion over mesoporous SBA-15	Desulfurization of sulfur compounds and gasoline	32
16	SO_3H-SBA-15	Sulfonic-modified m-SiO_2	Glycerol esterification	33
17	TiMecapSBA15-O_2	Surface, modification with functionalized protic molecules	Epoxidation of 1-octene	34
18	Ni–La_2O_3/SBA-15	La-modified Ni/SBA-15	CO_2 methanation	35
19	Silane Treated BETA	Surface-modified Beta zeolite	Production of lighter fuels by steam-assisted catalytic cracking from heavy oil	36
20	Surface-modified ZSM-5	Controlled copper deposition on surface-modified ZSM-5	Partial oxidation of methane	37
21	Ferrite nanoparticles	Surface modification of ferrite nanoparticles with dicarboxylic acids	Synthesis of 5-HMF	38
22	Ni–Ti/ATP	Ti-modified Ni/Attapulgite	Steam reforming of acetic acid	39
23	Ir/Px-TIO-7	Phosphorus modification of TiO_2	Synthesis of benzimidazoles	40
24	Pd–Au/CNT	Surface modification with nanosized Pd–Au bimetallic phases on carbon nanotubes	Phenylacetylene hydrogenation	41
25	GO-PrSO$_3$H	Propyl-SO_3H functionalization of graphene oxide	Acid-catalyzed esterification and acetalization reactions	42
26	GO-SO_3H	Sulfonic acid functionalization of graphene oxide	Selective decomposition of hexose sugar	43
27	AgZrO$_2$@GO	Ag ZrO_2 nanocomposite with graphene oxide	Levulinic acid to GVL	44

Continued

Table 13.1 Examples for surface modifications of nanocatalysts for catalytic applications.—cont'd

Sr. No.	Catalyst	Surface modification	Catalytic process	References
28	TCPIL/CuFe$_2$O$_4$/BNONS	Triphosphonated ionic liquid-CuFe$_2$O$_4$-modification of boron nitride	Catalytic reduction of nitroanilines and dyes	45
29	Ni/NC-x	Nanoclay surface supported Ni	Carbon dioxide reforming of methane	46
30	CNP-0.1	Phosphor-doped graphitic carbon nitride-supported Pd as a highly efficient catalyst for styrene hydrogenation	Styrene hydrogenation	47
31	CNT/TiO$_2$ NF	TiO$_2$ surface modification of CNF	Photocatalytic reduction Cr	48
32	HC-PA/ZnCl/KOH	Surface modification of activated carbon	CO$_2$ adsorption	49
33	CoP-modified CdS/g-C$_3$N$_4$	Surface Co-catalyst modification to synergistically enhance the photocatalytic hydrogen	Photocatalytic decomposition of water	50
34	CdS/WS$_2$/graphene composites	Hierarchical layered S$_2$/Graphene-modified CdS nanorods	Photocatalytic H$_2$ evolution	51
35	COA-2.5	Oxamide-modification of g-C$_3$N$_4$ nanostructures	Visible light photocatalysis	52
36	Co$_3$O$_4$ nanocubes	Surface modification of Co$_3$O$_4$ nanocubes with TEOS	Fischer–Tropsch catalysis	53
37	PO$_x$-CeO$_2$-nanosheets	Phosphate modification of CeO$_2$ nanosheets	Oxidation of dichloromethane	54
38	Ru$_3$@ZIF-11	Cluster-encapsulated ZIF-11	Simultanious conversion of Glycerol and CO$_2$	55

Table 13.1 Examples for surface modifications of nanocatalysts for catalytic applications.—cont'd

Sr. No.	Catalyst	Surface modification	Catalytic process	References
40	3NiO-MIL-101	Mesoporous chromium terephthalate MIL-101 was modified with NiO via atomic layer deposition (ALD) process using bis(cyclopentadienyl) nickel (Ni(Cp)2)	CO oxidation	56
41	NH_2—Ti/MIL-125	Porous Amino-functionalized titanium terephthalate MIL-125	Syngas purification	57
42	6wt%Pd@UiO-66	Synergistic effect of Pd NPs over UiO-66	Carbon dioxide methanation	58

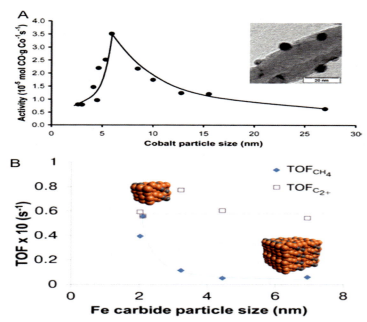

Figure 13.4 Fischer—Tropsch (FT) synthesis; an activity versus particle size of nanocatalysts. (A) Activity of different size of Co-nanoparticles. Inset: TEM image of Co/carbon nanofibers. Reproduced with permission from ref. (B) TOFs of a series of iron carbide particles with different sizes. *(Reproduced with permission from Ref. 7. Copyright 2012 American Chemical Society.)*

were found to be highly efficient in low temperature WGS conversions.[77–79] Si et al.,[80] have investigated the effect of different CeO_2 shape as a support for Au, where rod like CeO_2 showed best WGS activity as compare to the activity obtained over cube and polyhedron shape. They demonstrated that both of the CeO_2 planes (110) and (100) are important and these planes are more dominant in rod like CeO_2 than other shapes, because it needs lower activation barrier for the formation of oxygen vacancies on (110) plane of CeO_2 nanoparticles of rod-type shapes.

4.2 Oxidation

Catalytic oxidation of CO is important reaction for almost every industrial processes, for CO removal from automobile exhaust gas, and fuel cell related issues.[81–84] The pioneering studies of Haruta and coworkers demonstrated the Au size dependent Co oxidation, where Au particle size of 5 nm showed highest activity.[85] However, Goodman and coworkers observed that TOF of CO oxidation varies with the Au particles ranging from 1 to 6 nm.[2,86,87] This caused due to the quantum size effects of nano-sized Au catalyst. Beyond Au-based catalysts, Pt, Pd, Rh, and Ru nanoparticles are also effective in CO oxidation and their size and shape can affects the catalytic performance.[2] However, metal support interactions and metal support charge transfer modification can affect the catalytic activity. Studies of Weiher et al.[88] stated that the active phase of Au nanoparticle supported catalysts is sole dependent on Au nanoparticles. The theoretical study of CO oxidation on Au nanocluster supported rutile TiO_2 depicted the reactivity of Au nanoparticle due to low coordinated surface sites of Au atoms, where essential adsorption energy of reactant is higher than Au atoms of high coordination numbers (Fig. 13.5).[89]

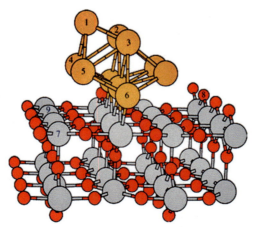

Figure 13.5 A model of Au cluster (Au_{10}) supported on TiO_2(110). Yellow: Au; Gray: Ti; Red: O. *(Reproduced with permission from Ref. 89. Copyright 2005, Wiley-VCH.)*

Interestingly, surface-modified Au nanoparticles with chemically bonded thiolate ligand exhibits higher activity compare to the activity obtained over fresh Au nanoparticles. However, improvement of Co oxidation seen with styrene modified Au nanocluster supported on CeO_2, TiO_2 and Fe_2O_3 at low temperature.[90,91] For instance, Au25(SR)18/CeO_2, the interface of Au nanoclusters and CeO_2 was suspected as an active sites in CO oxidation;[90,91] where CO molecules found to be adsorbed near to interface on Au atoms; while molecular oxygen is activated on CeO_2 near to interface, and the coupling of CO and oxygen forms CO_2 molecules. Similarly, for other oxidation reactions, significant improvement in catalytic activity was observed after surface modification of nanoparticles. Effect of Pt nanoparticle size was extensively studied for CH_4 oxidation, and smaller Pt nanoparticles supported catalysts (supports ZrO_2, Al_2O_3, ZrO_2-CeO_2) enhances the oxidation activity.[83,92,93] The size and shape dependent activity of Pt was well rationalized with the growth of fraction of undercoordinated Pt atoms along with the decline of size of Pt nanoparticles. However, several efforts have been done in development of efficient nanocatalysts for complete CH_4 oxidation.

In addition, size and surface-dependent activity of Pt nanoparticles was extensively investigated for oxidation of n-alkanes (C1—C6).[94,95] The coordinated Pt nanoparticles in Pt supported catalyst are important in this conversion, because noncoordinated Pt particles can easily oxidized during oxidation process and affect the catalytic activity. Cao and coworkers had significant effector and wrote the excellent review, which have covered most of the industrially important oxidation reaction and their activity correlation with shape and size effect of nanoparticles. These oxidation reactions are oxidation of glucose, cyclohexane, benzylic alcohol, oxidation, n-hexane, methanol to ethanol, propylene epoxidation, propane oxidation, and so forth.

4.3 Hydrogenations

Catalytic hydrogenation of organic compound is a very promising possibility for future biorefineries and petroleum refineries; it is a vital applied technique for defunctionalization in a refinery processes in the manufacturing of value-added chemicals, fine chemicals, pharmaceuticals, food additives, fuel, and so forth.[96] In literature,[97,98] several hydrogenation conversions have been shown to be surface and structure sensitive. However, transition, precious, noble metal-supported nanocatalysts are extensively used for selective hydrogenation.[7] Few best studies of Somorjai et al., discussed the size and surface effect of Pt nanoparticles supported SBA-15 catalysts in hydrogenation ethylene, pyrrole, benzene to toluene and 1,3-butadine.[4,99–105] In their studies, surface modification nanoparticles was done by different techniques such as alcohol reductions and dendrimer template methods. However, all of these conversions are Pt size and surface dependents. In specific, these activities are high for smaller particle of Pt less than 4 nm and which is due to the presence of coordinately unsaturated surface atoms.

In case of ring opening of pyrrolidine, surface properties of Pt on SBA-15 was nicely explained, the ring opening activity was found to be greatly enhanced with the increase of Pt particles (<2 nm).[101,105] They estimated surface interaction of reactant with smaller Pt nanoparticles, which favor for complete conversion of pyrrole into n-butylamine. In addition, hydrogenation of 1,3-butadine reaction pathway can be altered by changing the Pt particles size as well as Pt surface properties.[102]

Interestingly, several strategies for various important conversions are available, which involve size- and surface-dependent hydrogenation activity of different metal-supported catalysts. These conversions involved hydrogenation of saturated and unsaturated aldehydes, carboxylic acid, aromatics, nitro aromatics, and olefins. In addition to these important hydrogenations, similar types of strategies are applied for hydrogenolysis (HMF, Glycerol, furfural, etc.), cross-coupling reactions, hydrogen peroxide decompositions, isomerization, oxygen reduction, photocatalytic reactions, and several refinery conversion processes, and so forth.[2,7]

Recently Hwang et al. have developed $Cu-SiO_2$ nanocomposite catalyst,[96,106,107] which are very active for hydrogenation of industrially important carboxylic acid and their esters into corresponding lactones and alcohols, related schematic representation is shown in Fig. 13.6. In general, Cu-based catalysts are rarely considered as a robust catalyst in selective carboxylic acid hydrogenations due to copper sintering and leaching issues. Hence this study presented a new protocol for copper-silica nanocomposite materials with very high loading of copper (>80%) via precipitation deposition method. This simple and approach provide robust nanocomposite copper-silica catalyst, where copper particle size of 20 nm were highly dispersed in a matrix of 7 nm silica nanoparticles. Interestingly, their nanocomposite catalyst showed their excellent vapor phase hydrogenation activity in industrially important biomass-based platform chemicals into corresponding value-added products. The suggested biobased platform chemicals are levulinic acid, lactic acid, HMF, furfural alcohol, butyric acid, adipic acid, succinic acid, maleic acid, and so forth. Importantly, same $Cu-SiO_2$ catalyst have showed efficient hydrogenation in challenging conversion of biobased sugars (fructose and glucose) into corresponding alcohols with repetitive activity under liquid phase conditions. Similarly, HMF hydrogenation into dimethyl furan and dihydroxymethylfuran was successfully performed over these unique catalysts systems.[108]

Same group have developed nanoalloy Ru—Sn/ZnO catalyst, where surface of Ru nanoparticle was modified with Sn and produced active Ru_3Sn_7 alloy phases on ZnO support. By considering the same objective, which was discussed above for pure carboxylic acid hydrogenation, robust Ru_3Sn_7 alloy supported ZnO catalyst was developed.[109] This surface modified bimetallic Ru—Sn/ZnO catalysts displays superior performance over unmodified Ru/ZnO in vapor-phase hydrogenation fermentation derived butyric acid into n-butanol with high yield (>98%) for more than 3500 h of reaction time without any deactivation even in presence of corrosive acetic acid impurity with butyric acid feed.[109] Time on stream profile together with transmission electron microscopy (TEM) images of reduced 1Ru—2Sn/ZnO are shown in Fig. 13.7.

Future of SMNs catalysts for industry applications   333

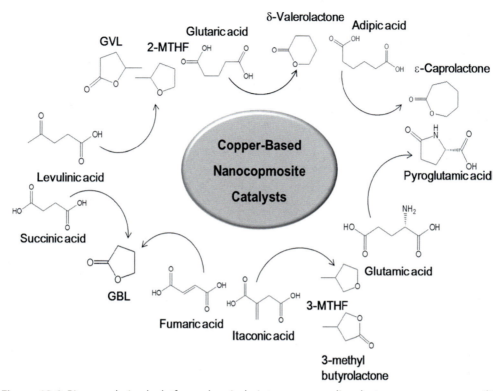

Figure 13.6 Biomass-derived platform chemicals into corresponding lactones over copper—silica nanocomposite catalysts. *(Adopted from Ref. 55. Copyright permission from Wiley-VCH.)*

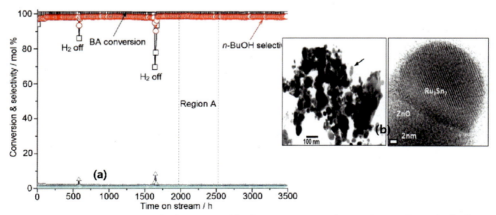

Figure 13.7 (A) Time on stream for butyric acid hydrogenation to n-butanol over 1Ru—2Sn/ZnO catalyst. Region A: activity using fermentation derived bio-butyric acid; (B) TEM images of 1Ru—2Sn/ZnO catalysts, where presence of the Ru_3Sn_7 nanoalloy phase on ZnO is clearly shown. *(Adopted from Ref. 109. Copyright permission from Wiley-VCH.)*

TEM image illustrate nanoparticles of the Ru—Sn alloy-rich phases are clearly seen on TEM images (Ru$_3$Sn$_7$). These well-dispersed surface-modified Ru$_3$Sn$_7$ nanoalloy particles are distinctively larger than metallic ruthenium nanoparticles on Ru/ZnO. In additions to butyric acid hydrogenation, these surface-modified Ru$_3$Sn$_7$ alloy catalyst was found to be very active in selective hydrogenation of 2, 5-hydroxumethylfurfural into dimethyl furan and related conversions.[110] However, DMF also exhibits almost similar properties of fuel as biobutanol.

5. Metal-isolated single atoms

Metal-isolated single atoms are an emerging class of nanocatalysts with maximized atom utilization morphologies along with defined active centers. Specifically, particle size of active surface decreases sharply which enhance the active surface on catalysts until active sites shrinks into a single atoms.[7] These single atoms are highly unstable, which require further stabilization steps. The concept of isolated single-atom site is not only related to the metallic site, it also includes metal-support interface. Noteworthy, catalysis behavior of isolated single-atom site can be different from parent metallic nanoparticles, and types of support material as well as functionalities can influence it. According to related findings, ISAS benefits almost 100% of atom utilizations in particular reaction, which can be great prospects particularly for expensive noble and precious metals.[7] Along with its great advantages, new catalytic features as high resistance to unfavorable side reactions is generated. Interestingly, same catalytic behaviors was observed during formic acid conversion into fuel cell on ISAS Pt sites, whereas CO formation was observed on parent Pt metals.[111] Oxidative-addition indorsed Pd-leaching mechanism was stated for the Suzuki-Miyaura reaction by using a Pd-nanostructure design by Li and coworkers.[112] Li et al. also developed cobalt ISAS on nitrogen-doped carbon support material by systematic MOF pyrolysis, resultant catalyst material provide high metallic loading with high temperature stability.[113]

Zeolite, metal oxide, and hydroxides are the most important class for industrially important heterogeneous catalysts in many important industrial processes.[114–118] New energy-efficient strategies need to be develop based on the high atom exploitation of ISAS catalyst to upsurge the energy competence and reduce the production viability of whole process. However, these synthesis strategies relies on the detailed structural and properties information. While ISAS synthesis, two important strategies need to be consider such as coordination of ISAS and aggregation of ISAS into nanoparticles with the support material.[114,119,120] In case of ISAS fabricated catalysts, the interaction of metal to support should be stronger than metal-metal bond. For ISAS catalysts synthesis and design, some techniques are recommended, and these discuss below.

5.1 Atomic layer deposition method (ALD)

The atomic layer deposition method (ALD) method is almost analogous with the chemical vapor deposition method used to deposit gaseous atomic precursors on the surface of solid substrate. Using this technique, thin film of a variety of active materials can be produced. ALD offer consecutive, self-limiting reactions with excellent conformity on high characteristic ratio structures, fine level control on thickness, and tunable composition. These outstanding advantages of ALD can offer many industrial and research applications. The main advantage of the ALD technique is that the same deposited layer can be deposited repeated times to achieve desired active sites as well as properties of material. This technique was efficiently utilized for metal ISAS synthesis.[121–123] Lu et al. had made an effort to develop an advanced ALD process for fabricating active catalyst system in which FeOx species are atomically dispersed on Pt nanoparticles.[121] The schematic representation is shown in Fig. 13.8. They choose Pt/SiO$_2$ as a parent material for surface modification through ALD, in which ferrocene (FeCp2) and oxygen was alternatively exposed, which results selective deposition of FeOx on the surface of Pt nanoparticles instead of inert SiO$_2$ support. However, varying of FeOx deposition cycles can be can results the different concentration of FeOx species on Pt nanoparticles. It could directly affect the CO oxidation activity and it was enhanced than the activity for Pt/FeOx. No doubt, ALD is very efficient technique for ISAS synthesis and which can be utilized beyond oxide/hydroxide materials. In another study of Lu et al. Pd ISAS deposited on graphene oxide catalyst was prepared by ALD treatment using Pd(hfac)$_2$ precursors,

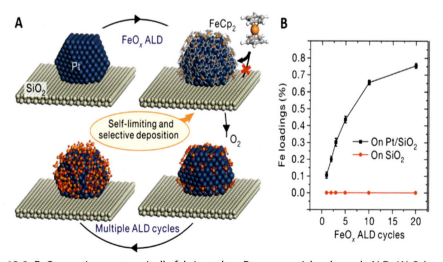

Figure 13.8 FeOx species are atomically fabricated on Pt nanoparticles through ALD. (A) Scheme: selective ALD of FeOx on a Pt/SiO$_2$ catalyst. (B) ALD cycles of Fe loadings on the Pt/SiO$_2$ and SiO$_2$ samples. *(Reproduced with permission from Ref 121. Copyright 2019 Nature Publishing Group.)*

and this surface-modified catalyst exhibits high and stable activity in selective hydrogenation of 1,3-butadiene.[124] ALD technique has been efficiently applied by researchers for surface modification of nanocatalysts in several applications.[7]

The other representative techniques are also investigated for the synthesis of ISAS catalyst, which includes nanoparticles to ISAS transformation, heterogenization of homogenous organic catalysis, NH_3 annealing, polymer pyrolysis, spatial isolation, and so forth.[7] Nitrogen-doped carbon supported ISAS material synthesized by carbon annealing with NH_3. Paralyzing of organic precursor containing nitrogen can also use for the N-doped carbon material. Using NH_3 annealing procedure, Huang et al. have synthesized CO, Fe and Ni ISAS on N-doped graphene.[125] Furthermore, this study demonstrated the molecular-level study and morphologies of the Ni1, Fe1, Co1/NG catalysts and implantation of CON_4C_4, FeN_4C_4 and NiC_4N_4 in corresponding graphene lattices. Also, Tour et al., reported a postdecoration strategy to prepare Co_2/CN catalysts by NH_3 annealing procedure.[126] In case of nanoparticles to ISAS transformation, stripping of metal precursor using different techniques such as reduction, pyrolysis, and so forth. Along with immobilization of the metal ISAS on support it represents the vastly applied procedure for the manufacture of ISAS catalysts.

Datye and coworkers have developed a method to transforms Pt NP to Pt ISAS using volatile platinum oxide intermediates.[127,128] On the other hand, these types of strategies are challenging because of rare availability of volatile oxides of noble metals. Li et al., have developed Pd, Pt, Au ISAS catalysts by calcination of metal encapsulated ZIF-8 metal organic framework. The calcination was performed at 900–1000°C, where metal nanoparticles found to be vanished and transformed into ISAS. The resultant materials appeared as stable catalyst materials, which is highly stable at very high temperature, more than 900°C under reactive atmosphere. As compared to the other ISAS synthesis methods nanoparticle to ISAS transformation method have great benefits, such as effective synthesis way, this technique can be applied for the most of the metal components, production of thermally stable catalyst. Wu et al. had significant effort and established a sequence of thermal atomization approaches to convert Cu,[129] Ni,[130] and Pt nanoparticles[131] into their analogous ISAS species.

Metal-supported carbon catalyst are quite famous in homogenous catalysis due to the unreactive nature. In industries, Pd and Pt nanoparticle supported carbon catalysts are widely applied in the manufacturing of fine chemicals and other important organic transformations. As discussed above, ISAS metal-supported catalyst is expected to sustainably enhance catalytic efficiency.[132–134] For instance, Suzuki-Miyura is an important C—C coupling reaction, in which Pd based catalysts has been significantly investigated. Recently, Perez-Ramirez et al. have developed catalyst Pd ISAS grafted exfoliated graphitic carbon nitride (ECN) that provide exceptional catalytic activity (TOF = 549 h^{-1}) and stability for Suzuki—Miyaura reaction, which is superior than that of homogeneous catalysts (34 h^{-1}).[135] This modified material topographies electron rich properties and accessible N-site functionalities which are very conductive for the stabilization of metal species. In addition, Pd1/ECN was found to be robust catalysts and shown stable activity provided stable activity for the Suzuki–Miyaura coupling of bromobenzene with phenylboronic acid pinacol ester exist as much superior as compare to homogeneous Pd(PPh$_3$)$_4$, graft molecular complex (palladium acetate anchored on silica functionalized by 3-mercaptopropylethyl sulfide, PdAc-MPES/SiO$_2$) and Pd/C. In another study of Lee et al., ISAS Platinum supported thiolated multiwalled nanotubes (Pt/SMWNT) was superior and stable as compare to homogenous Pt precursors.[136] ISAS Pd supported on hallo ZIF (H-ZIF) was developed by Wu and coworkers, which exhibited excellent performance in conversion of aryl acetylenes derivatives to synthesize conjugated dienes.[134] Pd ISAS/H-ZIF catalyst system presents promising motif for the heterogenization of homogeneous catalysis, with the reasonable benefits of mentioned catalysts.

5.2 Isolated cluster site catalysts (ICSC)

ICSC represents another important class of surface-modified nanocatalysts. Metal nanocluster generated by aggregations of a few metal atoms, which could exhibits distinct catalytic properties than parent metal atom.[137–139] Number of responsible factors such as, high ratio of surface to aggregated atoms, symmetrical as well as electronic shell closings, geometrical confinement, superatomic character, and so forth implies the corresponding outputs. Activity of cluster-based catalysts are tunable by rearranging the atomic compositions as well as size, these modifications can be directly proportional to the activity of cluster catalysts materials.[137–139] Whereas, these clusters are easily deposited and dispersed on variety of active support materials. Due to the great catalytic performances of well-defined clusters, various techniques are efficiently explored. So far, industrial-scale applicability of cluster catalysts are very limited due to the difficulty in the scale-up of catalysts synthesis at industrial-scale. Hence, significant investigation on the development cluster catalysis being necessary to welcome it in the line of industrially potential nanocatalysts systems.

Clusters can be classified into di- and multiatomic. Structure of cluster is determining through high resolution HAADF-STEM, DFT calculations, and XAFS spectra stimulations. He et al. synthesized first Fe$_2$ cluster embedded into graphene vacancies.[140] Yan et al., have developed Pt$_2$ cluster deposited graphene, which is very efficient for hydrolytic dehydrogenation of ammonia borane.[141] In this study, they estimated lower adsorption energies of ammonia borane and hydrogen on dimeric Pt cluster as compare to ISAS Pt and Pt nanoparticles through DFT studies. In another study of Mon et al.,[142] well-define Pt$_2$ nanocluster confined with the functional channels of MOF were developed; this unique material exhibits excellent electronic and catalytic properties, which are appealing for various industrially important reactions. Triatomic, tetratomic, and hexatomic clusters are stable form of clusters, which are utilized in variety of catalytic reactions; for instance, Os_3, Ir_4, Ir_6, Pd_6, and other clusters are systematically developed and investigated.[7] In addition, bimetallic cluster are important class of cluster catalysis, which imparts substantial properties in catalytic processes as second metal can control the catalytic properties of a monometallic catalyst through geometric, electronic, lattice strain, or bifunctional parameters.[143,144]

5.3 Plasma

Plasma catalysis is an emerging technology, which can create more opportunities in existing technologies for a variety of applications, such as synthesis of SMN, ammonia productions, hydrogen productions, air purification.[145,146] The benefits of plasma modified catalysis over conventional catalysis are enabled by combining the high activity of catalysts material with the high reactivity of the plasma.[145–147] Plasma treatment on nanomaterials can generate more reactive species in the plasma, which can also react at the catalyst surface; this synergetic combination can effectively react and efficiently produces desired products with high yield. However, some studies are based on the synergistic effects plasma and combination.[146] So far, synergy effect in plasma considered as the improvement in some process descriptor, such as selectivity, yield, or energy efficiency, as observed in the plasma catalytic processes. For instance, Zhang et al. reported plasma/Cu–Ni/c-Al_2O_3 synergy in dry reforming of methane (DRM).[148] They obtained 69% conversion of CH_4, 75% for CO_2 conversion over plasma combination with catalysts, those conversions are significant as compare to the activity obtained over plasma, and catalysts itself. It was suggested that the activity improvement originated from the surface adsorption of reactive plasma species, followed by recombination of the adsorbed species. In brief, the plasma may heat up the catalyst surface, thereby enhancing the desorption of surface species. However, adverse effects of these types of synergy are also reported by researchers.[146]

In brief, plasma catalysis also suffers from some challenges compared to thermal catalysis and plasma only processing, for example; the additional energy cost for initiating the plasma and the efficient integration of the catalyst in the discharge chamber, respectively. In hydrocarbon reforming, the obtained energy efficiencies were found to remain significantly below the thresholds required for commercial application. Plasma catalysis is currently nowhere near to competing with steam reforming of methane for hydrogen production or for synthesis of synthetic fuels.[145–147]

6. Conclusion and outlook

Heterogeneous catalysis with surface-modified materials will put forwards new prospects for the improvement of advanced catalytic processes in industrially feasible chemical conversions. No doubt, significant progress have been seen in precise controlling the surface and catalyst morphology from the nanoscale to the atomic scale. For instance, doping of oxide moieties opens the promising routes to modify the morphology and electronic properties of supported metals, which could induce dissociation and reaction processes of molecules adsorb on the oxide surface. However, modified nanoparticles and the atomic flexibility of metallic cluster can provide resultant impact on adsorption and catalytic properties. Systematically varying the surface and structural constraints of nanomaterials make it more efficient to structure-sensitive catalytic reactions. Variations in particle sizes of catalyst materials can be tuned to the properties of surface sites. However, the share of metal-support interactions can benefits the reactivity and stability to the metallic nanomaterials in several ways.

SMNS are one of the important classes of potential industrially heterogeneous catalysts due to their differences in shapes and size, variable in surface sites, and particle-specific catalytic activity, which could preferably preferred and vital to the development of efficient nanocatalysts. Significant equipment and techniques have been technologically advanced to monitor catalysis on single nanoparticles and surface or interface of nanomaterials, where reaction takes place. In recent decades, the development of SMNs in catalyst materials have fundamentally changed the path in which heterogeneous catalysts are designed. The nanocatalysts with controlled compositions, morphology, and structural properties have been widely used for heterogeneous catalysis in several industrially important reactions. In addition, shape and size dependency on catalytic performances on SMN catalysts are related with different coordination behavior of catalyst atoms of facets specific for different shapes and sizes of nanomaterials, with subsequent activity and selectivity. Thus, altogether, the field of SMN catalysis is an attractive one, as demonstrated by the nanocatalysts and its attractive perspectives in several chemical transformations. At its current stage, the SMN catalysis growing dramatically, and it is expected that the key challenges will be addressed in near future, and will include more opportunities at the industrial-scale, catalysis-based conversion processes. It is evident from the current chapter that the degree of improvement in the activity and stability vary for the different nanomaterials upon utilizing well-defined SMN nanocatalysts. It is necessary to investigate nanocatalysts that are more advanced and improve the synthesis conditions with the positive intention of achieving an efficient catalytic activity. However, more diverse and better synthesis strategies are available; thus, the accelerating interest in the development of SMN material catalytic applications, the expanding number of SMN nanocatalysts is likely to extent advanced catalytic applications.

References

[1] Behr A. *Angewandte homogene katalyse*. Wiley-VCH; 2008.
[2] Cao S, Tao FF, Tang Y, Li Y, Yu J. Size-and shape-dependent catalytic performances of oxidation and reduction reactions on nanocatalysts. *Chem Soc Rev* 2016;**45**:4747−65.

[3] Heveling J. Heterogeneous catalytic chemistry by example of industrial applications. *J Chem Educ* 2012;**89**:1530—6.
[4] Somorjai G, Rioux R. High technology catalysts towards 100% selectivity: fabrication, characterization and reaction studies. *Catal Today* 2005;**100**:201—15.
[5] Zhou Z-Y, Tian N, Li J-T, Broadwell I, Sun S-G. Nanomaterials of high surface energy with exceptional properties in catalysis and energy storage. *Chem Soc Rev* 2011;**40**:4167—85.
[6] Kango S, Kalia S, Celli A, Njuguna J, Habibi Y, Kumar R. Surface modification of inorganic nanoparticles for development of organic—inorganic nanocomposites—a review. *Prog Polym Sci* 2013;**38**:1232—61.
[7] Li Z, Ji S, Liu Y, Cao X, Tian S, Chen Y, et al. Well-defined materials for heterogeneous catalysis: from nanoparticles to isolated single-atom sites. *Chem Rev* 2019;**120**:623—82.
[8] Mozetič M. *Surface modification to improve properties of materials*. Multidisciplinary Digital Publishing Institute; 2019.
[9] Vengatesan MR, Mittal V. Surface modification of nanomaterials for application in polymer nanocomposites: an overview. *Surface modification of nanoparticle and natural fiber fillers* 2015:1.
[10] Sanchez C, Julián B, Belleville P, Popall M. Applications of hybrid organic—inorganic nanocomposites. *J Mater Chem* 2005;**15**:3559—92.
[11] Polshettiwar V, Varma RS. Green chemistry by nano-catalysis. *Green Chem* 2010;**12**:743—54.
[12] *Catalyst market size, share & trends analysis report by raw material (chemical compounds, Zeolites, metals), by product (heterogeneous, homogeneous), by application, by region, and segment forecasts, 2020 - 2027, research and market, Grand View research, global*. 2020. p. 115.
[13] Hutchings GJ, Davidson MG, Catlow RC, Hardacre C, Turner NJ, Collier P. *Modern developments in catalysis*. World Scientific; 2016.
[14] Lefebvre F, Basset J-M. Recent applications in catalysis of surface organometallic chemistry. *J Mol Catal Chem* 1999;**146**:3—12.
[15] Pandey P, Dahiya M. A brief review on inorganic nanoparticles. *J Crit Rev* 2016;**3**:18—26.
[16] Patra JK, Baek K-H. Green nanobiotechnology: factors affecting synthesis and characterization techniques. *J Nanomater* 2014:2014.
[17] Li N, Binder WH. Click-chemistry for nanoparticle-modification. *J Mater Chem* 2011;**21**:16717—34.
[18] Han Y, Xiao G, Chen M, Chen S, Zhao F, Zhang Y, et al. Effect of support modification and precursor decomposition method on the properties of CoPt/ZrO2 Fischer—Tropsch catalysts. Catalysis Today; 2020.
[19] Hong Y, Zhang H, Sun J, Ayman KM, Hensley AJ, Gu M, et al. Synergistic catalysis between Pd and Fe in gas phase hydrodeoxygenation of m-cresol. *ACS Catal* 2014;**4**:3335—45.
[20] Li C, Li W, Chen K, Ogunbiyi AT, Zhou Z, Xue F, et al. Palladium nanoparticles supported on surface-modified metal oxides for catalytic oxidation of lean methane. *ACS Appl Nano Mater* 2020.
[21] Yang J, Shi L, Li L, Fang Y, Pan C, Zhu Y, et al. Surface modification of macroporous La0. 8Sr0. 2CoO3 perovskite oxides integrated monolithic catalysts for improved propane oxidation. *Catal Today* 2020.
[22] Wang H, Li X, Cui Z, Yang L, Sun S. Tobacco stem-derived N-enriched active carbon: efficient metal free catalyst for reduction of nitroarene. *React Kinet Mech Catal* 2020;**130**:331—46.
[23] Qu T, Niu S, Zhang X, Han K, Lu C. Preparation of calcium modified Zn-Ce/Al2O3 heterogeneous catalyst for biodiesel production through transesterification of palm oil with methanol optimized by response surface methodology. *Fuel* 2021;**284**:118986.
[24] Shao H, Wang X, Gu X, Wang D, Jiang T, Guo X. Improved catalytic performance of CrOx catalysts supported on foamed Sn-modified alumina for propane dehydrogenation. *Microporous Mesoporous Mater* 2021;**311**:110684.
[25] Wang H, Tang M, Shi F, Ding R, Wang L, Wu J, et al. Amorphous Cr2WO6-modified WO3 nanowires with a large specific surface area and rich lewis acid sites: a highly efficient catalyst for oxidative desulfurization. *ACS Appl Mater Interf* 2020;**12**:38140—52.
[26] Ganji P, Roy S. Conversion of levulinic acid to ethyl levulinate using tin modified silicotungstic acid supported on Ta2O5. *Catal Commun* 2020;**134**:105864.
[27] Zhai P, Lv G, Cai Z, Zhu Y, Li H, Zhang X, et al. Efficient production of ethyl levulinate from furfuryl alcohol catalyzed by modified zirconium phosphate. *Chem Select* 2019;**4**:3940—7.

[28] Hidalgo M, Maicu M, Navío J, Colón G. Photocatalytic properties of surface modified platinised TiO$_2$: effects of particle size and structural composition. *Catal Today* 2007;**129**:43—9.

[29] Dou Y, Zhang M, Zhou S, Oldani C, Fang W, Cao Q. Etherification of 5-hydroxymethylfurfural to biofuel additive catalyzed by Aquivion® PFSA modified mesoporous silica. *Eur J Inorg Chem* 2018;**2018**:3706—16.

[30] Feng R, Yan X, Hu X, Qiao K, Yan Z, Rood MJ. High performance of H3BO3 modified USY and equilibrium catalyst with tailored acid sites in catalytic cracking. *Micropor Mesopor Mater* 2017;**243**:319—30.

[31] Anjali K, Venkatesha NJ, Christopher J, Sakthivel A. Rhodium porphyrin molecule-based catalysts for the hydrogenation of biomass derived levulinic acid to biofuel additive γ-valerolactone. *New J Chem* 2020;**44**:11064—75.

[32] Aslam S, Subhan F, Yan Z, Liu Z, Ullah R, Etim U, et al. Unusual nickel dispersion in confined spaces of mesoporous silica by one-pot strategy for deep desulfurization of sulfur compounds and FCC gasoline. *Chem Eng J* 2017;**321**:48—57.

[33] Popova M, Lazarova H, Szegedi A, Mihályi MR, Rangus M, Likozar B, et al. Renewable glycerol esterification over sulfonic-modified mesoporous silicas. *J Serb Chem Soc* 2018;**83**:39—50.

[34] Cordeiro PJ, Tilley TD. Enhancement of the catalytic activity of titanium-based terminal olefin epoxidation catalysts via surface modification with functionalized protic molecules. *ACS Catal* 2011;**1**:455—67.

[35] Wang X, Zhu L, Zhuo Y, Zhu Y, Wang S. Enhancement of CO$_2$ methanation over La-modified Ni/SBA-15 catalysts prepared by different doping methods. *ACS Sustain Chem Eng* 2019;**7**:14647—60.

[36] Khalil U, Muraza O, Kondoh H, Watanabe G, Nakasaka Y, Al-Amer A, et al. Robust surface-modified beta zeolite for selective production of lighter fuels by steam-assisted catalytic cracking from heavy oil. *Fuel* 2016;**168**:61—7.

[37] Sheppard T, Daly H, Goguet A, Thompson JM. Improved efficiency for partial oxidation of methane by controlled copper deposition on surface-modified ZSM-5. *ChemCatChem* 2016;**8**:562—70.

[38] Shaikh M, Sahu M, Khilari S, Kumar AK, Maji P, Ranganath KV. Surface modification of polyhedral nanocrystalline MgO with imidazolium carboxylates for dehydration reactions: a new approach. *RSC Adv* 2016;**6**:82591—5.

[39] Chen M, Hu J, Wang Y, Wang C, Tang Z, Li C, et al. Hydrogen production from acetic acid steam reforming over Ti-modified Ni/Attapulgite catalysts. *Int J Hydrog Energy* 2021;**46**:3651—68.

[40] Yu H, Wada K, Fukutake T, Feng Q, Uemura S, Isoda K, et al. Effect of phosphorus-modification of titania supports on the iridium-catalyzed synthesis of benzimidazoles. *Catalysis Today*; 2020.

[41] Wang S, Xin Z, Huang X, Yu W, Niu S, Shao L. Nanosized Pd—Au bimetallic phases on carbon nanotubes for selective phenylacetylene hydrogenation. *Phys Chem Chem Phys* 2017;**19**:6164—8.

[42] Masteri-Farahani M, Hosseini M-S, Forouzeshfar N. Propyl-SO3H functionalized graphene oxide as multipurpose solid acid catalyst for biodiesel synthesis and acid-catalyzed esterification and acetalization reactions. *Renew Energy* 2020;**151**:1092—101.

[43] Upare PP, Yoon J-W, Kim MY, Kang H-Y, Hwang DW, Hwang YK, et al. Chemical conversion of biomass-derived hexose sugars to levulinic acid over sulfonic acid-functionalized graphene oxide catalysts. *Green Chem* 2013;**15**:2935—43.

[44] Bai X, Ren T, Mao J, Li S, Yin J, Zhou J. A Ag—ZrO$_2$—graphene oxide nanocomposite as a metal-leaching-resistant catalyst for the aqueous-phase hydrogenation of levulinic acid into gamma-valerolactone. *New J Chem* 2020;**44**:16526—36.

[45] Arumugam V, Sriram P, Yen T-J, Redhi GG, Gengan RM. Nano-material as an excellent catalyst for reducing a series of nitroanilines and dyes: triphosphonated ionic liquid-CuFe$_2$O$_4$-modified boron nitride. *Appl Catal B Environ* 2018;**222**:99—114.

[46] Chaisamphao J, Kiatphuengporn S, Faungnawakij K, Donphai W, Chareonpanich M. Effect of modified nanoclay surface supported nickel catalyst on carbon dioxide reforming of methane. *Top Catal* 2021:1—15.

[47] Lin B, Zhang Y, Zhu Y, Zou Y, Hu Y, Du X, et al. Phosphor-doped graphitic carbon nitride-supported Pd as a highly efficient catalyst for styrene hydrogenation. *Catal Commun* 2020;**144**:106094.

[48] Mohamed A, Osman T, Toprak MS, Muhammed M, Yilmaz E, Uheida A. Visible light photocatalytic reduction of Cr (VI) by surface modified CNT/titanium dioxide composites nanofibers. *J Mol Catal Chem* 2016;**424**:45–53.

[49] Sarwar A, Ali M, Khoja AH, Nawar A, Waqas A, Liaquat R, et al. Synthesis and characterization of biomass-derived surface-modified activated carbon for enhanced CO_2 adsorption. *J CO_2 Util* 2021;**46**:101476.

[50] Wang P, Wu T, Wang C, Hou J, Qian J, Ao Y. Combining heterojunction engineering with surface cocatalyst modification to synergistically enhance the photocatalytic hydrogen evolution performance of cadmium sulfide nanorods. *ACS Sustain Chem Eng* 2017;**5**:7670–7.

[51] Xiang Q, Cheng F, Lang D. Hierarchical layered WS2/graphene-modified CdS nanorods for efficient photocatalytic hydrogen evolution. *ChemSusChem* 2016;**9**:996–1002.

[52] Tang H, Wang R, Zhao C, Chen Z, Yang X, Bukhvalov D, et al. Oxamide-modified g-C_3N_4 nanostructures: tailoring surface topography for high-performance visible light photocatalysis. *Chem Eng J* 2019;**374**:1064–75.

[53] Macheli L, Roy A, Carleschi E, Doyle BP, van Steen E. Surface modification of Co_3O_4 nanocubes with TEOS for an improved performance in the Fischer-Tropsch synthesis. *Catal Today* 2020;**343**:176–82.

[54] Dai Q, Zhang Z, Yan J, Wu J, Johnson G, Sun W, et al. Phosphate-functionalized CeO_2 nanosheets for efficient catalytic oxidation of dichloromethane. *Environ Sci Technol* 2018;**52**:13430–7.

[55] Oh K-R, Valekar AH, Cha G-Y, Kim Y, Lee S-K, Sivan SE, et al. In situ synthesis of trimeric ruthenium cluster-encapsulated ZIF-11 and its carbon derivatives for simultaneous conversion of glycerol and CO_2. *Chem Mater* 2020;**32**:10084–95.

[56] Jeong M-G, Kim DH, Lee S-K, Lee JH, Han SW, Park EJ, et al. Decoration of the internal structure of mesoporous chromium terephthalate MIL-101 with NiO using atomic layer deposition. *Micropor Mesopor Mater* 2016;**221**:101–7.

[57] Regufe MJ, Tamajon J, Ribeiro AM, Ferreira A, Lee U-H, Hwang YK, et al. Syngas purification by porous amino-functionalized titanium terephthalate MIL-125. *Energy Fuels* 2015;**29**:4654–64.

[58] Wang X, Liu Y, Zhu L, Li Y, Wang K, Qiu K, et al. Biomass derived N-doped biochar as efficient catalyst supports for CO2 methanation. *J CO_2 Util* 2019;**34**:733–41.

[59] Galvis HMT, Bitter JH, Khare CB, Ruitenbeek M, Dugulan AI, de Jong KP. Supported iron nanoparticles as catalysts for sustainable production of lower olefins. *Science* 2012;**335**:835–8.

[60] Jiao F, Li J, Pan X, Xiao J, Li H, Ma H, et al. Selective conversion of syngas to light olefins. *Science* 2016;**351**:1065–8.

[61] Li J, He Y, Tan L, Zhang P, Peng X, Oruganti A, et al. Integrated tuneable synthesis of liquid fuels via Fischer–Tropsch technology. *Nat Catal* 2018;**1**:787–93.

[62] Navarro V, Van Spronsen MA, Frenken JW. In situ observation of self-assembled hydrocarbon Fischer–Tropsch products on a cobalt catalyst. *Nat Chem* 2016;**8**:929–34.

[63] Torres Galvis HM, Bitter JH, Davidian T, Ruitenbeek M, Dugulan AI, de Jong KP. Iron particle size effects for direct production of lower olefins from synthesis gas. *J Am Chem Soc* 2012;**134**:16207–15.

[64] Den Breejen J, Radstake P, Bezemer G, Bitter J, Frøseth V, Holmen A, et al. On the origin of the cobalt particle size effects in Fischer– Tropsch catalysis. *J Am Chem Soc* 2009;**131**:7197–203.

[65] Harmel J, Peres L, Estrader M, Berliet A, Maury S, Fécant A, et al. hcp-Co nanowires grown on metallic foams as catalysts for fischer–tropsch synthesis. *Angew Chem* 2018;**130**:10739–43.

[66] Melaet G, Lindeman AE, Somorjai GA. Cobalt particle size effects in the Fischer–Tropsch synthesis and in the hydrogenation of CO_2 studied with nanoparticle model catalysts on silica. *Top Catal* 2014;**57**:500–7.

[67] Rane S, Borg Ø, Rytter E, Holmen A. Relation between hydrocarbon selectivity and cobalt particle size for alumina supported cobalt Fischer–Tropsch catalysts. *Appl Catal Gen* 2012;**437**:10–7.

[68] Bezemer GL, Bitter JH, Kuipers HP, Oosterbeek H, Holewijn JE, Xu X, et al. Cobalt particle size effects in the Fischer– Tropsch reaction studied with carbon nanofiber supported catalysts. *J Am Chem Soc* 2006;**128**:3956–64.
[69] Borg Ø, Dietzel PD, Spjelkavik AI, Tveten EZ, Walmsley JC, Diplas S, et al. Fischer–Tropsch synthesis: cobalt particle size and support effects on intrinsic activity and product distribution. *J Catal* 2008;**259**:161–4.
[70] Cheng Y, Lin J, Xu K, Wang H, Yao X, Pei Y, et al. Fischer–Tropsch synthesis to lower olefins over potassium-promoted reduced graphene oxide supported iron catalysts. *ACS Catal* 2016;**6**:389–99.
[71] Wang P, Chen W, Chiang F-K, Dugulan AI, Song Y, Pestman R, et al. Synthesis of stable and low-CO_2 selective ε-iron carbide Fischer-Tropsch catalysts. *Sci Adv* 2018;**4**:eaau2947.
[72] Carballo JMG, Yang J, Holmen A, García-Rodríguez S, Rojas S, Ojeda M, et al. Catalytic effects of ruthenium particle size on the Fischer–Tropsch Synthesis. *J Catal* 2011;**284**:102–8.
[73] Kellner CS, Bell AT. Effects of dispersion on the activity and selectivity of alumina-supported ruthenium catalysts for carbon monoxide hydrogenation. *J Catal* 1982;**75**:251–61.
[74] Smith KJ, Everson RC. Fischer-Tropsch reaction studies with supported ruthenium catalysts: II. Effects of oxidative pretreatment at elevated temperatures. *J Catal* 1986;**99**:349–57.
[75] Andreeva D, dkiev V, Tabakova T, Ilieva L, Falaras P, Bourlinos A, et al. *Catal Today* 2002;**72**:51.
[76] Idakiev V, Tabakova T, Naydenov A, Yuan Z-Y, Su B-L. Gold catalysts supported on mesoporous zirconia for low-temperature water–gas shift reaction. *Appl Catal B Environ* 2006;**63**:178–86.
[77] Acosta B, Smolentseva E, Beloshapkin S, Rangel R, Estrada M, Fuentes S, et al. Gold supported on ceria nanoparticles and nanotubes. *Appl Catal Gen* 2012;**449**:96–104.
[78] Idakiev V, Yuan Z-Y, Tabakova T, Su B-L. Titanium oxide nanotubes as supports of nano-sized gold catalysts for low temperature water-gas shift reaction. *Appl Catal Gen* 2005;**281**:149–55.
[79] Shekhar M, Wang J, Lee W-S, Williams WD, Kim SM, Stach EA, et al. Size and support effects for the water–gas shift catalysis over gold nanoparticles supported on model Al_2O_3 and TiO_2. *J Am Chem Soc* 2012;**134**:4700–8.
[80] Si R, Flytzani-Stephanopoulos M. Shape and crystal-plane effects of nanoscale ceria on the activity of $Au-CeO_2$ catalysts for the water–gas shift reaction. *Angew Chem* 2008;**120**:2926–9.
[81] Burch R, Urbano F. Investigation of the active state of supported palladium catalysts in the combustion of methane. *Appl Catal Gen* 1995;**124**:121–38.
[82] Haruta M. Gold rush. *Nature* 2005;**437**:1098–9.
[83] Muto K-i, Katada N, Niwa M. Complete oxidation of methane on supported palladium catalyst: support effect. *Appl Catal Gen* 1996;**134**:203–15.
[84] Zafiris G, Gorte R. CO oxidation on Pt/α-Al_2O_3 (0001): evidence for structure sensitivity. *J Catal* 1993;**140**:418–23.
[85] Haruta M, Tsubota S, Kobayashi T, Kageyama H, Genet MJ, Delmon B. Low-temperature oxidation of CO over gold supported on TiO_2, α-Fe_2O_3, and Co_3O_4. *J Catal* 1993;**144**:175–92.
[86] Chen M, Goodman D. The structure of catalytically active gold on titania. *Science* 2004;**306**:252–5.
[87] Chen M, Goodman DW. Catalytically active gold on ordered titania supports. *Chem Soc Rev* 2008;**37**:1860–70.
[88] Weiher N, Bus E, Delannoy L, Louis C, Ramaker DE, Miller J, et al. Structure and oxidation state of gold on different supports under various CO oxidation conditions. *J Catal* 2006;**240**:100–7.
[89] Remediakis IN, Lopez N, Nørskov JK. CO oxidation on rutile-supported Au nanoparticles. *Angew Chem Int Ed* 2005;**44**:1824–6.
[90] Nie X, Qian H, Ge Q, Xu H, Jin R. CO oxidation catalyzed by oxide-supported $Au_{25}(SR)_{18}$ nanoclusters and identification of perimeter sites as active centers. *ACS Nano* 2012;**6**:6014–22.
[91] Wu Z, Jiang D-e, Mann AK, Mullins DR, Qiao Z-A, Allard LF, et al. Thiolate ligands as a double-edged sword for CO oxidation on CeO_2 supported $Au_{25}(SCH_2CH_2Ph)_{18}$ nanoclusters. *J Am Chem Soc* 2014;**136**:6111–22.
[92] Liotta L, Di Carlo G, Pantaleo G, Venezia A, Deganello G. Co_3O_4/CeO_2 composite oxides for methane emissions abatement: relationship between Co_3O_4–CeO_2 interaction and catalytic activity. *Appl Catal B Environ* 2006;**66**:217–27.

[93] Wei J, Iglesia E. Mechanism and site requirements for activation and chemical conversion of methane on supported Pt clusters and turnover rate comparisons among noble metals. *J Phys Chem B* 2004;**108**: 4094–103.
[94] Garetto TF, Rincón E, Apesteguía CR. The origin of the enhanced activity of Pt/zeolites for combustion of C2–C4 alkanes. *Appl Catal B Environ* 2007;**73**:65–72.
[95] Gololobov A, Bekk I, Bragina G, Zaikovskii V, Ayupov A, Telegina N, et al. Platinum nanoparticle size effect on specific catalytic activity in n-alkane deep oxidation: dependence on the chain length of the paraffin. *Kinet Catal* 2009;**50**:830–6.
[96] Upare PP, Hwang YK, Lee JM, Hwang DW, Chang JS. Chemical conversions of biomass-derived platform chemicals over copper–silica nanocomposite catalysts. *ChemSusChem* 2015;**8**:2345–57.
[97] Yang H, Zhang C, Gao P, Wang H, Li X, Zhong L, et al. A review of the catalytic hydrogenation of carbon dioxide into value-added hydrocarbons. *Catal Sci Technol* 2017;**7**:4580–98.
[98] Zaera F. The surface chemistry of metal-based hydrogenation catalysis. *ACS Catal* 2017;**7**:4947–67.
[99] Bratlie KM, Lee H, Komvopoulos K, Yang P, Somorjai GA. Platinum nanoparticle shape effects on benzene hydrogenation selectivity. *Nano Lett* 2007;**7**:3097–101.
[100] Kliewer CJ, Aliaga C, Bieri M, Huang W, Tsung C-K, Wood JB, et al. Furan hydrogenation over Pt (111) and Pt (100) single-crystal surfaces and Pt nanoparticles from 1 to 7 nm: a kinetic and sum frequency generation vibrational spectroscopy study. *J Am Chem Soc* 2010;**132**:13088–95.
[101] Kuhn JN, Huang W, Tsung C-K, Zhang Y, Somorjai GA. Structure sensitivity of carbon–nitrogen ring opening: impact of platinum particle size from below 1 to 5 nm upon pyrrole hydrogenation product selectivity over monodisperse platinum nanoparticles loaded onto mesoporous silica. *J Am Chem Soc* 2008;**130**:14026–7.
[102] Michalak WD, Krier JM, Komvopoulos K, Somorjai GA. Structure sensitivity in Pt nanoparticle catalysts for hydrogenation of 1, 3-butadiene: in situ study of reaction intermediates using SFG vibrational spectroscopy. *J Phys Chem C* 2013;**117**:1809–17.
[103] Pushkarev VV, An K, Alayoglu S, Beaumont SK, Somorjai GA. Hydrogenation of benzene and toluene over size controlled Pt/SBA-15 catalysts: elucidation of the Pt particle size effect on reaction kinetics. *J Catal* 2012;**292**:64–72.
[104] Somorjai GA, Park JY. Colloid science of metal nanoparticle catalysts in 2D and 3D structures. Challenges of nucleation, growth, composition, particle shape, size control and their influence on activity and selectivity. *Top Catal* 2008;**49**:126–35.
[105] Song H, Rioux RM, Hoefelmeyer JD, Komor R, Niesz K, Grass M, et al. Hydrothermal growth of mesoporous SBA-15 silica in the presence of PVP-stabilized Pt nanoparticles: synthesis, characterization, and catalytic properties. *J Am Chem Soc* 2006;**128**:3027–37.
[106] Upare PP, Jeong M-G, Hwang YK, Kim DH, Kim YD, Hwang DW, et al. Nickel-promoted copper–silica nanocomposite catalysts for hydrogenation of levulinic acid to lactones using formic acid as a hydrogen feeder. *Appl Catal Gen* 2015;**491**:127–35.
[107] Upare PP, Lee JM, Hwang YK, Hwang DW, Lee JH, Halligudi SB, et al. Direct hydrocyclization of biomass-derived levulinic acid to 2-methyltetrahydrofuran over nanocomposite copper/silica catalysts. *ChemSusChem* 2011;**4**:1749.
[108] Upare PP, Hwang YK, Hwang DW. An integrated process for the production of 2, 5-dihydroxymethylfuran and its polymer from fructose. *Green Chem* 2018;**20**:879–85.
[109] Lee JM, Upare PP, Chang JS, Hwang YK, Lee JH, Hwang DW, et al. Direct hydrogenation of biomass-derived butyric acid to n-butanol over a ruthenium–tin bimetallic catalyst. *ChemSusChem* 2014;**7**:2998–3001.
[110] Upare PP, Hwang DW, Hwang YK, Lee U-H, Hong D-Y, Chang J-S. An integrated process for the production of 2, 5-dimethylfuran from fructose. *Green Chem* 2015;**17**:3310–3.
[111] Duchesne PN, Li Z, Deming CP, Fung V, Zhao X, Yuan J, et al. Golden single-atomic-site platinum electrocatalysts. *Nat Mater* 2018;**17**:1033–9.
[112] Niu Z, Peng Q, Zhuang Z, He W, Li Y. Evidence of an oxidative-addition-promoted Pd-leaching mechanism in the Suzuki reaction by using a Pd-nanostructure design. *Chem A Euro J* 2012;**18**: 9813–7.

[113] Yin P, Yao T, Wu Y, Zheng L, Lin Y, Liu W, et al. Single cobalt atoms with precise N-coordination as superior oxygen reduction reaction catalysts. *Angew Chem* 2016;**128**:10958−63.
[114] Liu P, Zhao Y, Qin R, Mo S, Chen G, Gu L, et al. Photochemical route for synthesizing atomically dispersed palladium catalysts. *Science* 2016;**352**:797−800.
[115] Paolucci C, Khurana I, Parekh AA, Li S, Shih AJ, Li H, et al. Dynamic multinuclear sites formed by mobilized copper ions in NOx selective catalytic reduction. *Science* 2017;**357**:898−903.
[116] Wei H, Liu X, Wang A, Zhang L, Qiao B, Yang X, et al. FeO x-supported platinum single-atom and pseudo-single-atom catalysts for chemoselective hydrogenation of functionalized nitroarenes. *Nat Commun* 2014;**5**:1−8.
[117] Zhang S, Tang Y, Nguyen L, Zhao Y-F, Wu Z, Goh T-W, et al. Catalysis on singly dispersed Rh atoms anchored on an inert support. *ACS Catal* 2018;**8**:110−21.
[118] Zhang Z, Zhu Y, Asakura H, Zhang B, Zhang J, Zhou M, et al. Thermally stable single atom Pt/m-Al$_2$O$_3$ for selective hydrogenation and CO oxidation. *Nat Commun* 2017;**8**:1−10.
[119] Li Z, Wang D, Wu Y, Li Y. Recent advances in the precise control of isolated single-site catalysts by chemical methods. *Natl Sci Rev* 2018;**5**:673−89.
[120] Liu L, Corma A. Metal catalysts for heterogeneous catalysis: from single atoms to nanoclusters and nanoparticles. *Chem Rev* 2018;**118**:4981−5079.
[121] Cao L, Liu W, Luo Q, Yin R, Wang B, Weissenrieder J, et al. Atomically dispersed iron hydroxide anchored on Pt for preferential oxidation of CO in H 2. *Nature* 2019;**565**:631−5.
[122] Wang C, Gu X-K, Yan H, Lin Y, Li J, Liu D, et al. Water-mediated Mars−van Krevelen mechanism for CO oxidation on ceria-supported single-atom Pt1 catalyst. *ACS Catal* 2017;**7**:887−91.
[123] Wang H, Lu J. Atomic layer deposition: a gas phase route to bottom-up precise synthesis of heterogeneous catalyst. *Acta Phys Chim Sin* 2018;**34**:1334−57.
[124] Yan H, Cheng H, Yi H, Lin Y, Yao T, Wang C, et al. Single-atom Pd1/graphene catalyst achieved by atomic layer deposition: remarkable performance in selective hydrogenation of 1, 3-butadiene. *J Am Chem Soc* 2015;**137**:10484−7.
[125] Fei H, Dong J, Feng Y, Allen CS, Wan C, Volosskiy B, et al. General synthesis and definitive structural identification of MN$_4$C$_4$ single-atom catalysts with tunable electrocatalytic activities. *Nat Catal* 2018;**1**:63−72.
[126] Fei H, Dong J, Arellano-Jiménez MJ, Ye G, Kim ND, Samuel EL, et al. Atomic cobalt on nitrogen-doped graphene for hydrogen generation. *Nat Commun* 2015;**6**:1−8.
[127] Jones J, Xiong H, DeLaRiva AT, Peterson EJ, Pham H, Challa SR, et al. Thermally stable single-atom platinum-on-ceria catalysts via atom trapping. *Science* 2016;**353**:150−4.
[128] Xiong H, Lin S, Goetze J, Pletcher P, Guo H, Kovarik L, et al. Thermally stable and regenerable platinum−tin clusters for propane dehydrogenation prepared by atom trapping on ceria. *Angew Chem* 2017;**129**:9114−9.
[129] Qu Y, Li Z, Chen W, Lin Y, Yuan T, Yang Z, et al. Direct transformation of bulk copper into copper single sites via emitting and trapping of atoms. *Nat Catal* 2018;**1**:781−6.
[130] Yang J, Qiu Z, Zhao C, Wei W, Chen W, Li Z, et al. In situ thermal atomization to convert supported nickel nanoparticles into surface-bound nickel single-atom catalysts. *Angew Chem Int Ed* 2018;**57**:14095−100.
[131] Qu Y, Chen B, Li Z, Duan X, Wang L, Lin Y, et al. Thermal emitting strategy to synthesize atomically dispersed Pt metal sites from bulk Pt metal. *J Am Chem Soc* 2019;**141**:4505−9.
[132] Cui X, Li W, Ryabchuk P, Junge K, Beller M. Bridging homogeneous and heterogeneous catalysis by heterogeneous single-metal-site catalysts. *Nat Catal* 2018;**1**:385−97.
[133] Mitchell S, Vorobyeva E, Pérez-Ramírez J. The multifaceted reactivity of single-atom heterogeneous catalysts. *Angew Chem Int Ed* 2018;**57**:15316−29.
[134] Zhao C, Yu H, Wang J, Che W, Li Z, Yao T, et al. A single palladium site catalyst as a bridge for converting homogeneous to heterogeneous in dimerization of terminal aryl acetylenes. *Mater Chem Front* 2018;**2**:1317−22.
[135] Chen Z, Vorobyeva E, Mitchell S, Fako E, Ortuño MA, López N, et al. A heterogeneous single-atom palladium catalyst surpassing homogeneous systems for Suzuki coupling. *Nat Nanotechnol* 2018;**13**:702−7.

[136] Lee E-K, Park S-A, Woo H, Park KH, Kang DW, Lim H, et al. Platinum single atoms dispersed on carbon nanotubes as reusable catalyst for Suzuki coupling reaction. *J Catal* 2017;**352**:388–93.
[137] Kappes MM, Radi P, Schär M, Schumacher E. Probes for electronic and geometrical shell structure effects in alkali-metal clusters. Photoionization measurements on KxLi, KxMg and KxZn (x< 25). *Chem Phys Lett* 1985;**119**:11–6.
[138] Kulkarni A, Lobo-Lapidus RJ, Gates BC. Metal clusters on supports: synthesis, structure, reactivity, and catalytic properties. *Chem Commun* 2010;**46**:5997–6015.
[139] Vajda S, White MG. Catalysis applications of size-selected cluster deposition. *ACS Catal* 2015;**5**: 7152–76.
[140] He Z, He K, Robertson AW, Kirkland AI, Kim D, Ihm J, et al. Atomic structure and dynamics of metal dopant pairs in graphene. *Nano Lett* 2014;**14**:3766–72.
[141] Yan H, Lin Y, Wu H, Zhang W, Sun Z, Cheng H, et al. Bottom-up precise synthesis of stable platinum dimers on graphene. *Nat Commun* 2017;**8**:1–11.
[142] Mon M, Rivero-Crespo MA, Ferrando-Soria J, Vidal-Moya A, Boronat M, Leyva-Pérez A, et al. Synthesis of densely packaged, ultrasmall Pt02 clusters within a thioether-functionalized MOF: catalytic activity in industrial reactions at low temperature. *Angew Chem* 2018;**130**:6294–9.
[143] Chong L, Wen J, Kubal J, Sen FG, Zou J, Greeley J, et al. Ultralow-loading platinum-cobalt fuel cell catalysts derived from imidazolate frameworks. *Science* 2018;**362**:1276–81.
[144] Wang H, Xu S, Tsai C, Li Y, Liu C, Zhao J, et al. Direct and continuous strain control of catalysts with tunable battery electrode materials. *Science* 2016;**354**:1031–6.
[145] Bogaerts A, Neyts E, Gijbels R, van der Mullen J. Gas discharge plasmas and their applications. *Spectrochimica Acta Part B: Atomic Spectroscopy* 2002;**57**:609–58.
[146] Neyts EC, Ostrikov K, Sunkara MK, Bogaerts A. Plasma catalysis: synergistic effects at the nanoscale. *Chem Rev* 2015;**115**:13408–46.
[147] Neyts EC. Plasma-surface interactions in plasma catalysis. *Plasma Chem Plasma Process* 2016;**36**: 185–212.
[148] Zhang A-J, Zhu A-M, Guo J, Xu Y, Shi C. Conversion of greenhouse gases into syngas via combined effects of discharge activation and catalysis. *Chem Eng J* 2010;**156**:601–6.

Index

Note: 'Page numbers followed by "f" indicate figures and "t" indicate tables'.

A

Acid Orange 7 (AO 7), 309
Acid Red 18 (AR 18), 309
Activated carbon (AC), 134, 230–231
Adsorption mechanism, 76–77
Advanced microscopic techniques, 184–187
    AFM, 184
    cryo-TEM, 184
    TEM and SEM, 186–187
Advanced nanotechnology techniques, 322
Aerosolization system, 297
Aggregated polymer matrix, 46
Alcohols, 132
    CO_2 hydrogenation to, 230–232
Alkaline phosphatase (ALKP), 304–305
Alkyl levulinate (AL), 141–143
Alternaria solani, 302–303
Aluminum (Al), 106
Aminations, 251–252
Amine ligand modification, 280
Aminophenol formaldehyde (APF), 38
3-aminopropyltriethoxysilane (APTES), 36, 38
Ammonia (NH_3), 171–172
Analytic membranes, 174
Anionic polymers, 189–190
Arabidopsis thaliana, 304, 308
Aromatics, 82–83
Artificial intelligence (AI), 14–15
Atom transfer radical polymerization (ATRP), 182
Atomic force microscopy (AFM), 14–15, 183–184
Atomic layer deposition method (ALD method), 335–337
Attenuated Total Reflection-Fourier Transform Infrared spectroscopy (ATR-FTIR spectroscopy), 187

B

Bacillus subtilis, 302–303
Barrett–Joyner–Halenda methods (BJH methods), 14–15
Bentonite, 87–88
Benzimidazoles, 244

Benzylpiperazine, 253
4-Benzylpiperidine, 253
Bifunctional core-shell type magnetic ZSM-5 zeolite structures, 143
Bimetallic Au–Cu@SiO_2 catalyst, 36
Bimetallic Ni–Sn catalysts, 134
Bimetallic particles, 82–83
Bioactive compounds, 252
Biodiesel, 132
    production, 149
Biofuel
    from biomass, 103–104
    CO_2 to, 117–118
    nanocatalysts for biofuel production, 121
Biological oxygen demand (BOD), 103
Biomass
    biomass-derived SA, 137
    to hydrocarbon, 114–117
        gasification, 117
        liquefaction, 116–117
        pyrolysis, 115–116
    refineries, 131–132
Bioremediation, 89
Bisphenol A (BPA), 295
BJH methods. *See* Barrett–Joyner–Halenda methods (BJH methods)
Blending process, 8–9
Bonding energy of surface photon, 181
Boron-doped graphitic carbon nitride (BCN), 233–235
Bottom-up approach, 4–5, 30–31, 53, 205–207
Bromide-treated PdNPs (Br-PdNPs), 290
Bronchoalveolar lavage fluid (BAL), 304–305
Brønsted acidic ionic liquids (BAILs), 149–150
Brunauer–Emmett–Teller method (BET method), 14–15, 55–56
1,4-Butanediol (BDO), 134, 137
    production, 137–140
Butenafine, 253
Butyric acid (BA), 140

C

C-fullerenes, 296

C-SWNTs, 296
CA. *See* Cellulose acetate (CA)
Calcium titanate (CaTiO$_3$), 113
"Capping agents", 6–8
Carbocatalysis, 276
   GO for, 277f
Carbohydrates, 132
Carbon, 289
   carbon-based materials for water treatment, 76–77
   carbon-based NMs, 12–13
   carbon-based NPs, 6–8
   fibers, 84–87
   materials, 84–87, 104
      in nanoparticles stabilization, 38–43
   microspheres, 84–87
Carbon dioxide (CO$_2$), 131–132
   capture, 270–271
   conversion reactions, 224–225
   emission of, 223
   heterogeneous catalyst in CO$_2$ fixation, 233–245
   heterogeneous catalyst in CO$_2$ hydrogenation, 227–233
   to methanol, 117–119
   reduction
      to ethanol, 156
      to formic acid, 158–160
      to methanol, 153–156
      to propanol, 156–158
   sequestration, 189–190
   sequestration, 227
   surface modified carbonaceous nanomaterials, 225–227
   surface-modified nanomaterials for carbon dioxide transformation to liquid fuels, 152–160
   value-added products formed using, 224f
Carbon monoxide (CO), 132, 290–291
   catalytic oxidation of, 330
Carbon nanofibers (CNFs), 113, 231–232, 276–278, 293
Carbon nanomaterials functionalization, 110
Carbon nanotubes (CNTs), 8–9, 291, 293–295
Carbon quantum dots (CQDs), 202
Carbon-based nanocatalysts, 110–112. *See also* Metal oxide nanocatalysts
   functionalized CNT-based nanocatalysts, 111–112

functionalized graphene-based nanocatalysts; also Metal oxide nanocatalysts, 111
   functionalized CNT-based nanocatalysts, 111–112
   functionalized graphene-based nanocatalysts, 111
Carbonaceous materials, 172
Carbonaceous nanomaterials (CNMs), 293
   chemical properties of carboneous NMs catalysts, 298t–299t
Carbonaceous SNMs, 291–293
Carboxylates-based ligands, 279–280
Carboxymethyl cellulose (CMC), 189–190, 307
Catalysis, 270
Catalysts, 1–4, 103–104, 289–290
   economic aspects of NMs used as, 308–310
      CFM@PDA/Pd composite nanocatalyst, 308
      FeOx/C nanocatalyst, 308–309
      FePt–Ag nanocatalysts, 310
      nZVI, 308
      other nanocatalysts, 310
      UV/Ni–TiO2 nanocatalyst, 309–310
   nanoparticles, 321–322
   shape and size dependent catalysts and reactions, 325–334
      hydrogenations, 331–334
      oxidation, 330–331
      shape dependency of activity in water-gas shift, 325–330
   surface modifications of nanocatalysts for catalytic applications, 326t–329t
   synthesis of SNMs with application as, 290–291
Catalytic applications
   of nanomaterial, 323–325
      surface modification of nanoparticles and techniques, 323–325
   surface-modified nanomaterials for
      carbon nanotubes, 293–295
      economic aspects of NMs used as catalysts, 308–310
      fullerene, 295–296
      graphene-based nanomaterials, 296–303
      human health and safety consequences of SNMs, 306–308
      NMs categories and nomenclature, 290, 292f
      other nanomaterials, 303–306
      QDs, 303–306
      synthesis of SNMs with application as catalyst, 290–291
      types of SNMs, 291–293

Index

Catalytic hydrogenations, 252
Catalytic inorganic MR technology, 175—177
Catalytic materials, 174, 289
Catalytic membrane, 174
Catalytic MR, 171—172
   bottlenecks of, 176t
   challenges in SMNs and, 173—174
   types, 174—178
      inorganic MRs, 175—177
      organic MRs, 177—178
Catalytic nanoparticles, surface modification of, 58—59
Catalytic oxidation of CO, 330
Catalytic processes, 320—321
Catalytically active agent, 53—55
Cationic graft-copolymerization, 182
CB. See Conduction band (CB)
C—C Heck coupling reactions, 9—10
CE. See Cellulose esters (CE)
Cellulose acetate (CA), 179
Cellulose esters (CE), 179
Cementitious matrix, 276—278
CFM@PDA/Pd composite nanocatalyst, 308
Chemical hydrolysis, 132
Chemical oxygen demand (COD), 103
Chemical vapor deposition method (CVD method), 183, 207, 229—230
Chemical/electrochemical initiated grafting method, 181—182
Chitosan-based CNFs, 113
Cinnarizine, 253
Clay nanotubes, 306
Click chemistry, 271—272
   modification extension, 272—273
   versatile nature for ligand generation, 273f
Climate change, 223
Clusters, 337—338
CMC. See Carboxymethyl cellulose (CMC)
CN. See Nitrocellulose (CN)
CNFs. See Carbon nanofibers (CNFs)
CNMs. See Carbonaceous nanomaterials (CNMs)
CNTs. See Carbon nanotubes (CNTs)
Cobalt-based heterogeneous catalysts, 258
COD. See Chemical oxygen demand (COD)
Composite NPs, 6—8
Conduction band (CB), 103—104, 197—198
Conventional catalysts, 267—268

Copolymerization of functional organosilanes, 319—320
Copper (Cu)
   Cu-based bimetallic catalyst, 140
   Cu—SiO$_2$ nanocomposite catalyst, 332
Copper-polyaniline (Cu-PANI), 150—152
Core-shell nanoparticles (CSNPs), 134—137, 203—206
Cost-effective protocols, 251—252
Coulombic repulsion, 31
CQDs. See Carbon quantum dots (CQDs)
Cryo-EM. See Cryo-transmission electron microscopy (cryo-TEM)
Cryo-transmission electron microscopy (cryo-TEM), 184
CSNPs. See Core-shell nanoparticles (CSNPs)
CVD method. See Chemical vapor deposition method (CVD method)
Cyclic carbamates, CO$_2$ fixation to, 241—244
Cyclic carbonates, CO$_2$ fixation to, 233—241, 239t—240t

D

Daphnia magna, 307
Dechlorination mechanism, 82—83
Dendritic fibrous nanosilica, 273
Dense nonaqueous phase liquid (DNALP), 88
Density functional theory, 68—69
Depolymerization of biomass, 141—143
Dextran (DEX), 189—190
Dichalcogenides, 210
Diels—Alder reaction, 8—9
Diethylformamide (DFM), 229—230
Differential scanning calorimetry (DSC), 14—15
Dimethylformamide (DMF), 238—241
2,5-dimethylfuran (DMF), 141—143, 146
Direct irradiation method, 180
Disasters, 267—268
Disinfection, 120
DLS. See Dynamic light scattering (DLS)
Double-walled carbon nanotubes (DWCNTs), 293—294
Dry coating strategy, 183
Dry reforming of methane (DRM), 338
Dunaliella tertiolecta, 307
Dyes, photocatalysis of, 309
Dynamic light scattering (DLS), 14—15

E

ECN. *See* Exfoliated graphitic carbon nitride (ECN)
EDA. *See* Ethylenediamine (EDA)
EDS. *See* Energy dispersive X-ray spectroscopy (EDS)
EG. *See* Ethylene glycol (EG)
Electrochemical deposition method, 207
Electrospinning technique, 13–14
Electrostatic stabilization, 31, 81–82
Emulsification method, 88
Emulsified nanoscale zero-valent iron (ENZVI), 88
Endothelial cell (ECs), 296
Energy
　demands, 131
　fuels, 132
Energy dispersive X-ray spectroscopy (EDS), 14–15
Engineered nanoparticle, 171–172
Environmental catalysis, 173
　challenges in SMNs and catalytic MRs, 173–174
　incorporation of SMNs into polymeric membranes, 179–187
　　characterization methods of SMNs based polymer membranes, 183–187
　　methods of SMNs incorporation into polymeric MRs, 180–183
　polymeric MRs, 178–179
　SMNs based polymeric membrane-assisted catalysis, 187–190
　　CO_2 sequestration, 189–190
　　hydrogenation, 188–189
　　pervaporation for esterification, 188
　types of catalytic membrane reactors, 174–178
Escherichia coli, 307
Esterification, 149–150
　pervaporation for, 188
　reaction method, 46–47
Ethanol, 132
　reduction of CO_2 to, 156
5-ethoxy methyl furfural, 141–143
Ethylene glycol (EG), 133
　production, 134–137
Ethylenediamine (EDA), 229–230
Exfoliated graphitic carbon nitride (ECN), 336–337
External/surface modification, 180

F

Facile method, 271–272
Fatty acid methyl ester (FAME), 149
Fe-ZSM-5@ZIF-8 nanocomposite, 153–156
Fenton-type nanocatalyst (FeOx/C nanocatalyst), 308–309
FePt–Ag nanocatalysts, 310
Ferric oxide (Fe_2O_3), 108–109
Few-layered structure (FL), 210
Fibrous nanosilica, 273
Field-effect transistors (FETs), 207
Fischer–Tropsch catalysis (FT catalysis), 229–230
Fischer–Tropsch synthesis (FTS), 117–118, 150–152, 325, 329f
Fixation, 227
Flutamide, 263–264
FOL. *See* Furfuryl alcohol (FOL)
Formic acid, 132
　reduction of CO_2 to, 158–160
Fourier-transform infrared (FT-IR), 14–15
Free radical grafting process, 182
Fullerene (nC_{60}), 295–296
Functional acids, 279–280
Functional nanomaterials, 171–172
Functionalization
　of organic compounds, 319–320
　procedure, 6–9, 203–204
Functionalized CNT-based nanocatalysts, 111–112
Functionalized graphene-based nanocatalysts, 111
Functionalized nanomaterials, 267–268
Functionalized silica nanomaterials, 271
Furfural (FAL), 132, 141–143
　alcohol production and related liquid fuels, 141–144
Furfuryl alcohol (FOL), 141–143
Fusarium oxysporum, 302–303

G

g-butyrolactone (GBL), 137
g-valerolactone (GVL), 141–143
Gamma-glutamyl transpeptidase (GGT), 304–305
Gas chromatography (GC), 14–15
Gas chromatography mass-spectrometry (GC-MS), 14–15
Gas sensors, 198
Gas separation, 187–188
Gas-phase coatings method, 183

Gasification, 117, 132
Glass transition temperatures (Tg), 179
Global warming, 117–118
Global warming, 223
Glutathione (GSH), 295
Graft-copolymerization-based modification, 181
Grafting approach, 63
Grafting-based modifications, 181–183
   chemical/electrochemical initiated grafting, 181–182
   high energy radiation (plasma)induced grafting, 182–183
   photoirradiation-induced grafting, 182
Graphene, 110, 209, 296
   CNTs, 12–13
   graphene-based materials, 235–237, 296–303
      metal and metal oxides, 297–303
   graphene-metal oxides, 210
   modified nanomaterials, 275–278
Graphene oxide (GO), 8–9, 212–213, 275–276
   for carbocatalysis, 277f
Graphene quantum dots doped with nitrogen (NGQDs), 228–229
Graphene-related materials (GRM), 297
Graphite
   graphite-based nanomaterials, 276–278
   nanoplatelets, 276–278
Graphitic carbon nitride (g-C_3N_4), 12–13
Gross domestic product (GDP), 320–321

H

HAP. *See* Hydroxyapatite (HAP)
Hard-templating strategy, 204–205
Hazard, 293
HDTMB. *See* Hexadecyltrimethylammonium bromide (HDTMB)
Heavy metal ion sensors (HMI sensors), 204–205
HEK. *See* Human epidermal keratinocytes (HEK)
HER. *See* Hydrogen evolution reaction (HER)
Heterocypris incongreuens, 307
Heterogeneous catalysis, 2, 320–321
Heterogeneous catalyst, 1–2, 66. *See also* Semiconductor catalysts
   advantages of nanocatalysts in comparison to homo-and heterogeneous catalysts, 3f
   in CO_2 fixation, 233–245
      to cyclic carbamates, 241–244
      to cyclic carbonates, 233–241, 239t–240t
      to other value-added products, 244–245
   in CO_2 hydrogenation, 227–233, 236t
      to alcohols, 230–232
      to hydrocarbons, 228–230
      to value-added products; also Semiconductor catalysts, 232–233
   advantages of nanocatalysts in comparison to homo-and heterogeneous catalysts, 3f
   in CO_2 fixation, 233–245
      to cyclic carbamates, 241–244
      to cyclic carbonates, 233–241, 239t–240t
      to other value-added products, 244–245
   in CO_2 hydrogenation, 227–233, 236t
      to alcohols, 230–232
      to hydrocarbons, 228–230
      to value-added products, 232–233
Heterogeneous nanomaterials, 319–320
Heterogenization method, 268
Hexadecyl-2-hydroxyethyl-dimethyl ammonium dihydrogen phosphate (HHDMA), 65–66
Hexadecyltrimethylammonium bromide (HDTMB), 38
Hexagonal boron nitride (h-BN), 210
Hexatomic clusters, 337–338
High energy radiation (plasma)induced grafting method, 182–183
High performance liquid chromatography (HPLC), 14–15
High-angle annular dark-field scanning transmission electron microscopy (HAADF-STEM), 14–15, 67–68
High-resolution TEM (HRTEM), 14–15
HMF. *See* 5-hydroxymethyl-furfural (HMF)
HMI sensors. *See* Heavy metal ion sensors (HMI sensors)
Hollow spheres, 203–206
Homogeneous catalysts, 1–2, 251, 321–322. *See also* Heterogeneous catalyst
   advantages of nanocatalysts in comparison to homo-and heterogeneous catalysts; also Heterogeneous catalyst, 3f
   advantages of nanocatalysts in comparison to homo-and heterogeneous catalysts, 3f
Hordeum vulgare, 308
HPLC. *See* High performance liquid chromatography (HPLC)
Human epidermal keratinocytes (HEK), 295
Human health and safety (H&S), 289–290
   consequences of SNMs, 306–308
Hybrid nanocatalysts, 64–65

Hybrid nanocomposites, 63
Hydration of CO_2, 189—190
Hydrocarbon
   CO_2 hydrogenation to, 228—230
   production, 103
      biomass to hydrocarbon, 114—117
      CO_2 to methanol and other hydrocarbons, 117—119
      nanocatalysts in production, 114—119
Hydrogen (H_2), 171—172
Hydrogen evolution reaction (HER), 12—13, 83—84
Hydrogenation, 188—189, 227, 331—334
Hydrophilic polymer membranes, 188
Hydrophobic polymers, 188
Hydrothermal growth methods, 207
Hydrothermal method, 44
Hydrothermal-pyrolysis method, 43
Hydroxides, 210, 334
Hydroxyapatite (HAP), 140
5-hydroxymethyl-furfural (HMF), 132
5-hydroxymethylfurfural and related liquid derivatives production, 145—148
3-hydroxypicolinic acid (HPA), 279—280

I
Immobilization process, 280
In situ modification method, 48—49, 49f
In situ polymerization techniques, 8—9
In situ sol—gel-derived silica, 319—320
In-situ water-in-oil microemulsion method, 149—150
INA. *See* Isonicotinic acid (INA)
Incident photon conversion efficiency (IPCE), 282—284
Industrial catalysis, 270—271, 335—337
Industrial catalytic processes, 335—337
Industry applications, SMNs catalysts for
   catalytic applications of nanomaterial, 323—325
   metal-isolated single atoms, 334—338
   nanoparticles catalysts, 321—322
   shape and size dependent catalysts and reactions, 325—334
Infrared spectroscopy, 57
Inorganic agents, 63
Inorganic clays, 87—88
Inorganic core-shell materials, 203—204
Inorganic membrane, 172
Inorganic MRs, 175—177

Inorganic solid materials, 321
Internal/bulk modification approach, 180
Ion exchange reduction method, 44
Ionic grafting process, 182
Iron (Fe), 106, 137—140
   iron-based nanoparticles, 77
Irradiation-based modifications, 180—181
   plasma treatment, 180—181
   UV-irradiation, 181
ISAS fabricated catalysts, 334
Isochrysis galbana, 307
Isolated cluster site catalysts (ICSC), 337—338
Isonicotinic acid (INA), 279—280

K
Kaolinite, 87—88
Khellin, 252

L
Lactase dehydrogenase (LDH), 304—305
Layer-by-layer coating method, 180
Layered double hydroxide (LDH), 87—88, 209
Levulinic acid (LA), 141—143
Ligand
   immobilization techniques to modify catalytic nanoparticles, 63—66
   ligand-stabilized-MNPs, 8—9
   nanoparticle surface modification via, 268f
Lignin, 133
Lignocellulose, 133
Lignocellulosic biomass, 132, 134, 141—143
Linoleic acid (LEA), 279—280
Linolenic acid (LLA), 279—280
Linum usitatissimum, 308
Liquefaction, 116—117
Liquid fuels, 131
   SMNs for biomass conversion to, 133—152
      biodiesel production, 149
      liquid hydrocarbons production, 149—152
      production of 1, 2-propylene glycol and ethylene glycol, 134—137
      production of 1, 4-butanediol, 137—140
      production of 5-hydroxymethylfurfural and related liquid derivatives, 145—148
      production of furfural alcohol and related liquid fuels, 141—144
   surface-modified nanomaterials for carbon dioxide transformation to, 152—160
      reduction of CO_2 to ethanol, 156

reduction of CO_2 to formic acid, 158—160
reduction of CO_2 to methanol, 153—156
reduction of CO_2 to propanol, 156—158
types, 133f
Liquid hydrocarbons production, 149—152
Lolium perenne, 308
Loxapine, 253

M

Machine learning (ML), 14—15
Macromolecules, 49
Maghemite (g-Fe_2O_3), 77
Magnetic surface-modified nanomaterials, 278—281
Magnetically active graphene oxide (MASHGO), 276
Magnetite (Fe_3O_4), 77
Material synthesis methods, 323—324
MDM. *See* Monocyte-derived macrophages (MDM)
MEC. *See* Minimum explosive concentration (MEC)
Medicago sativa, 308
Membrane reactors (MR), 171, 174—178
  inorganic, 175—177
Mercaptopropyl, 268, 279
Mesoporous carbon (MC), 84—87, 214—215
Mesoporous graphitic carbon nitride (mpg-C_3N_4), 68—69
Metal and metal oxides, 297—303
  chemical properties of, 300t—301t
Metal ions, 75
Metal nanoparticles (MNPs), 3—4, 6—8, 46, 53—55, 75
Metal oxide, 334
Metal oxide nanocatalysts, 104—109. *See also* Carbon-based nanocatalysts
  Fe_2O_3, 108—109
  TiO_2, 106—108
  ZnO; also Carbon-based nanocatalysts, 109
  Fe_2O_3, 108—109
  TiO_2, 106—108
  ZnO, 109
Metal-based catalytic systems, 1
Metal-based nanoparticles, 104
Metal-based NPs, 3—4
Metal-isolated single atoms, 334—338
  ALD method, 335—337
  ICSC, 337—338

plasma, 338
Metal-organic chemical vapor deposition (MOCVD), 207
Metal-organic frameworks (MOFs), 210, 273—274
Metal-oxides NPs, 6—8
Metal-supported carbon catalyst, 336—337
Metallic nanoparticles, 205—206, 325—330
Metallic oxide synthesis, 319—320
Methanol, 132
  CO_2 to, 117—119
  economy, 132
  reduction of CO_2 to, 153—156
Methoxsalen, 252
2-methyl furan, 141—143
Methyl orange (MO), 310
Methylbenzylpiperazine, 253
Methylene blue dye, 13—14
6-methylpyridine-2-carboxylic acid (MPCA), 279—280
Microemulsion method, 44
Microscopic NPs, 171—172
Microscopic SMNs, 173
Microwave (MW), 4—5
MIL-125 catalyst, 137
Mineral catalysts, 117
Minimum explosive concentration (MEC), 293
Minimum ignition energy (MIE), 293
Minimum ignition temperature (MIT), 293
ML. *See* Machine learning (ML)
MNPs. *See* Metal nanoparticles (MNPs)
MO. *See* Methyl orange (MO)
MOCVD. *See* Metal-organic chemical vapor deposition (MOCVD)
Modern age materials, surface parameters of, 269
Modern industry applications, SMN-based catalytic materials for
  active role of surface-modified nanomaterials in industry, 269—284
    graphene modified nanomaterials, 275—278
    magnetic surface-modified nanomaterials, 278—281
    silica-modified nanomaterials, 270—275
    TiO_2 surface-modified nanomaterials, 281—284
Modified Hummers' method, 39
MOFs. *See* Metal-organic frameworks (MOFs)
Monocyte-derived macrophages (MDM), 294—295

MPCA. *See* 6-methylpyridine-2-carboxylic acid (MPCA)
MR. *See* Membrane reactors (MR)
Multilayer structure (ML structure), 210
Multiwalled carbon nanotubes (MWCNTs), 84–87, 230–231, 291, 293–294
Multiwalled nanotubes (MWNT), 110–112

N
N-alkylation, 251–252
N-dG. *See* N-doped graphene (N-dG)
N-doped graphene (N-dG), 229
N-methyl-D-aspartate (NMDA), 241
N-methylation, 251–252
Nano scale zerovalent iron, 308
Nanobased catalytic materials
  advancement on modification of nanomaterials, 73–74
  examples of, 75–93
    carbon-based materials for water treatment, 76–77
    NZVI-based materials for water treatment, 77–93
    TiO_2-based materials, 75–76
  nanomaterials for water treatment, 73
Nanobased surface-modified catalysts, 121
Nanobiomaterials, 267–268
Nanobubble-trapped nanoclays, 306
Nanocatalysis, 103
Nanocatalysts (NCs), 1–4, 105, 289–290, 310
  in hydrocarbon production, 114–119
  materials, 103–104
  materials, 104–105
  nanocatalysts bridge between heterogeneous and homogenous catalysis, 322f
  recent advancement and real-time utilization of, 119–121
  surface-functionalized NCs
    characterization of, 14–15
    selective applications of, 9–14
  synthetic strategies of, 4–5, 4f
  technology, 103–104
Nanoclay, 305
  gutbusters, 306
Nanocluster, 203–206
Nanoclusters (NCs), 205–206
Nanodiagnostics, 267–268
Nanoengineering, 267–268
Nanomagnetite, 281

nanomagnetite-supported catalysts, 281
Nanomaterials (NMs), 1–4, 53–55, 131–132, 171–172, 267–269, 289, 305, 319–320
  advancement on modification of, 73–74
  catalytic applications of, 323–325
  categories and nomenclature, 290, 292f
  classification of SMNs based on different parameters, 7f
  economic aspects of NMs used as catalysts, 308–310
  nanomaterial-based photocatalysis, 120
  nanomaterials surface chemistry and Zeta potential, 55–58
  surface modification of nanomaterials and properties for catalytic applications, 320f
  types of NMs and functionalization procedures, 6–9
  for water treatment, 73, 74t
Nanomedicine, 267–268
Nanoparticles (NPs), 29, 38, 53, 171–172, 199, 268, 289
  carbon materials in stabilization of, 38–43
  catalysts, 321–322
  chemical methods of nanoparticle synthesis using various functional surface, 44–46
  ligand immobilization techniques to modify catalytic nanoparticles, 63–66
  nanoparticle-based materials, 319
  silica nanoparticles for nanoparticles stabilization, 32–38
  surface modification
    via ligand, 268f
    of nanoparticles and techniques, 323–325
  surface silica modification methods, 47f
  types of, 30f
Nanoparticles of zero-valent iron (NZVI), 77, 306
  NZVI-based materials for water treatment, 77–93
    bimetallic particles, 82–83
    combined technologies, 88–89
    emulsification, 88
    examples of water-treatment enhancement of NZVI-based materials, 90–93
    modification needs and routes of ZVI modification, 78–81
    NZVI supported on various materials, 84–88, 87f
    sulfidation, 83–84

supported NZVI catalysts onto carbon structures for Cr (VI) degradation, 91t
surface modifiers, 81–82
Nanopharmaceuticals, 267–268
Nanoremediation, 305
Nanoscale zero-valent metal nanocatalysts (nZVM nanocatalysts), 106
Nanosynthesized materials, 73
Nanotechnology, 267–269
  nanotechnology based industrial application in various field, 270f
  for wastewater treatment and pollutant remediation, 119
Nanotransfer printing (nTP), 215
NCs. See Nanocatalysts (NCs); Nanoclusters (NCs)
Newton's laws of motion, 29
NGQDs. See Graphene quantum dots doped with nitrogen (NGQDs)
Nickel (Ni), 106, 189–190
Nickel ferrite (NiFe$_2$O$_4$), 302–303
Nickel oxide (NiO), 145–146
Niclosamide, 263–264
Nilutamide, 263–264
Nimesulide, 263–264
Nimodipine, 263–264
Nitrate (NO$_3$−), 77
Nitrocellulose (CN), 179
Nitrogen (N), 228–229
Nitrogenation, 226
4-nitrophenol (4-NP), 189
NMDA. See N-methyl-D-aspartate (NMDA)
NMs. See Nanomaterials (NMs)
Noble metal, 82–83, 134
  noble metal-based nanoparticles for pharmaceuticals synthesis, 252–253
    synthesis of N, N-dimethylated drug molecules, 254
    synthesis of pharmaceutical molecules by using Pd/SiO$_2$ as catalyst, 255
  noble-metal-based core-shell NMs, 12–13
  noble-metal-free core-shell NMs, 12–13
Noncovalent forces, 76–77
Nonionic poly(vinyl pyrrolidone) (PVP), 189–190
Nonionic PVP, 189–190
Nonmetal-based catalytic systems, 1
Nonnoble metal-based nanoparticles for pharmaceuticals synthesis, 254–264

pharmaceutical compounds
  pharmaceutical compounds prepared by using Co/Melamine-6@TiO$_2$–800–5 catalyst, 259
  preparation of carbon supported Fe$_2$O$_3$-nanoparticles, 263f
  preparation of Fe catalyst, 262
  prepared by using Co-DABCO-TPA@C-800 catalyst, 260
  prepared by using Co$_3$O$_4$eCo/NGr@a-Al$_2$O$_3$ as catalyst, 257
  prepared by using iron catalysts, 258
  synthesis of bioactive compounds using Ni/g-Al$_2$O$_3$, 261
  synthesis of N, N-methylated drug molecules by using Fe$_2$O$_3$/NGr@C catalyst, 264
Nonoxygenated fuel synthesis, 131
Nonphotocatalytic nanocatalysts, 105
Nonsteroidal antiinflammatory drugs (NSAI drugs), 259–261
NPs. See Nanoparticles (NPs)
nTP. See Nanotransfer printing (nTP)
NZVI. See Nanoparticles of zero-valent iron (NZVI)
nZVM nanocatalysts. See Nanoscale zero-valent metal nanocatalysts (nZVM nanocatalysts)

O

o-phenylenediamine (OPD), 244
Occupational Safety and Health Administration (OSHA), 293
One-pot synthesis, 35
One–dimensional nanomaterials (1D nanomaterials), 199, 206–209
  1D NMs-based sensors, 207–209
  synthesis of 1D nanostructures and sensor fabrication, 207
Ordered mesoporous carbon, 84–87
Organic core-shell materials, 203–204
Organic functionalization of nanotubes, 319–320
Organic membrane, 172
Organic MRs, 177–178
Organic water contaminants, 106
Oryzias latipes, 307
Oxidation process, 106, 226, 330–331
Oxidation-reduction potential (ORP), 78
Oxodiperoxomolybdenum complexes, 280
Oxygen evolution reaction (OER), 12–13
Oxygen nanobubble clay (ONB clay), 306
Oyster shells (OS), 153

P

PA. *See* Polyamide (PA)
Palladium (Pd)
    nanoparticles, 32–34, 188
    palladium-loaded TiO$_2$ photocatalyst, 253
PAMAM. *See* Polyamidoamine (PAMAM)
PAN. *See* Polyacrylonitrile (PAN)
PCBs. *See* Polychlorinated biphenyls (PCBs)
PDA. *See* Polydopamine (PDA)
PDMS. *See* Polydimethylsiloxane (PDMS)
PE. *See* Polyethylene (PE)
PEBA. *See* Poly (ether-blockamide) (PEBA)
PEC. *See* Photoelectrocatalytic cell (PEC)
PEI-DTC. *See* Polyethyleneimine dithiocarbamate (PEI-DTC)
Perchlorate (ClO$_4$−), 77
Perovskite, 113
Pervaporation (PV), 187–188
    for esterification, 188
PES. *See* Polyethersulfone (PES)
Pharmaceuticals, 251–252
    noble metal-based nanoparticles for pharmaceuticals synthesis, 252–253
    nonnoble metal-based nanoparticles for pharmaceuticals synthesis, 254–264
    synthesis of pharmaceutical-based products, 252t
Phosphate ester method, 48
Photocatalysis of dyes, 309
Photocatalytic nanocatalysts, 105
Photoelectrocatalytic cell (PEC), 282–284
Photoirradiation-induced grafting method, 182
Physical vapor deposition method (PVD method), 183, 207
Phytoplanktons, 307
PI. *See* Polyimide (PI)
6-(1-piperidinyl)-pyridine-3-carboxylic acid (PPCA), 279–280
Piribedil (anti-Parkinson agents), 253
Plasma catalysis, 338
Plasma polymerization technique, 182–183
Plasma treatment, 180–181
    surface modification of catalytic nanoparticles by and, 58–59
Platinum (Pt)
    Pt nanowire, 158–160
    tungsten-modified platinum, 140
Platinum supported thiolated multiwalled nanotubes (Pt/SMWNT), 336–337
PLD. *See* Pulsed laser deposition (PLD)
PNPs. *See* Polymeric nanoparticles (PNPs)
Point of zero charge (PZC), 57
Pollutant remediation, nanotechnology for, 119, 120t
Pollutants
    sequestration, 78
    in water, 74
Poly (ether-blockamide) (PEBA), 188
Poly (vinyl alcohol), 188
Polyacrylonitrile (PAN), 179
Polyamide (PA), 179, 188
Polyamidoamine (PAMAM), 189
Polychlorinated biphenyls (PCBs), 82–83
Polydimethylsiloxane (PDMS), 188
Polydopamine (PDA), 308
Polyethersulfone (PES), 179
Polyethylene (PE), 179
Polyethyleneimine dithiocarbamate (PEI-DTC), 310
Polyimide (PI), 179, 188
Polymer-based nanocatalysts, 113. *See also* Carbon-based nanocatalysts; Metal oxide nanocatalysts
Polymer-coated nanoparticles, 74
Polymeric ligands, 32
Polymeric MRs, 177–179
    characterization methods of SMNs based polymer membranes, 183–187
        advanced microscopic techniques, 184–187
        X-ray spectroscopic techniques, 187
    incorporation of SMNs into, 179–187
        grafting-based modifications, 181–183
        irradiation-based modifications, 180–181
        methods of, 180–183
        surface coating-based modifications, 183
Polymeric nanoparticles (PNPs), 49
Polymers, 32
Polyol method, 44, 132
Polypropylene (PP), 179
Polysaccharides, 74
Polystyrene (PS), 189, 281–282
Polysulfone (PS), 179
Polytetrafluoroethylene (PTFE), 179
Polyvinyl chloride (PVC), 179
Polyvinylidene fluoride (PVDF), 179
Population growth, 103
Porous carbon-supported NZVI particles, 84–87

Postgrafting protocol method, 272—273
Postmodification method, 180
Powder X-ray diffraction (PXRD), 14—15
PP. See Polypropylene (PP)
PPCA. See 6-(1-piperidinyl)-pyridine-3-carboxylic acid (PPCA)
Precious metals-based nanoparticles, 252
Premodification method, 180
Pristine
  grapheme, 111
  NMs, 289
Propanol, 132
  reduction of CO_2 to, 156—158
1,2-propylene glycol production, 134—137
Pseudomonas syringae, 302—303
Pt/SMWNT. See Platinum supported thiolated multiwalled nanotubes (Pt/SMWNT)
PTFE. See Polytetrafluoroethylene (PTFE)
Pulsed laser deposition (PLD), 207
PV. See Pervaporation (PV)
PVC. See Polyvinyl chloride (PVC)
PVD method. See Physical vapor deposition method (PVD method)
PVDF. See Polyvinylidene fluoride (PVDF)
PVP. See Nonionic poly(vinyl pyrrolidone) (PVP)
PXRD. See Powder X-ray diffraction (PXRD)
Pyridine (PYD), 229—230
  moieties, 279—280
Pyrolysis, 115—116, 132
Pyrrolidine (PYL), 229—230
PZC. See Point of zero charge (PZC)

Q
Quantum dots (QDs), 199—202, 303—306
Quartz crystal microbalance with dissipation monitoring (QCM-D), 297

R
Raman spectroscopy, 14—15
Reactive oxygen species (ROS), 305
Reactive Red 19 (RR), 309
Reactivity enhancement of modified NZVI, 90
Reduced graphene oxide (rGO), 87—88, 111, 212—213, 230—231
Remediation, 305
  bioremediation, 89
  nanoremediation, 305
  nanotechnology for pollutant, 119, 120t
  of wastewater, 103
Renewable biomass feedstock, 132
Reproducible detection system, 207
Repulsive force, 32
Rhodamine B (RhB), 310
Rhodium (Rh)
  Rh-based hydrogenation protocol, 252
  Rh-NP—rGO nanocompositecatalysts, 252
Ru sensitizer-Re catalyst supramolecular metal complex, 282—284

S
SA. See Succinic acid (SA)
SACs. See Single-atom catalysts (SACs)
SAXS. See Small-angle X-ray scattering (SAXS)
Scanning electron microscopy (SEM), 14—15, 183—184, 186—187
Scanning transmission electron microscope (STEM), 134, 187
SCB. See Sugarcane bagasse (SCB)
Schiff base complexes, 280
Selected area (electron) diffraction (SAED), 14—15
Self-templating strategy, 204—205
SEM. See Scanning electron microscopy (SEM)
Semiconductor catalysts, 198—199
  0D nanomaterials, 199—206
  1D nanomaterials, 206—209
  2D nanomaterials, 209—214
  3D nanomaterials, 214—215
  nanomaterials with various dimension, 200t
Sensing mechanism, 202
Sensor technologies, 197
SERS. See Surface-enhanced Raman spectroscopy (SERS)
Shell-isolated nanoparticle—enhanced Raman spectroscopy (SHINERS), 203—204
SI-CRP. See Surface-initiated controlled radical polymerization (SI-CRP)
Silane chemical treatment, 59—63
Silica nanoparticles for nanoparticles stabilization, 32—38
Silica-containing hydroxyl groups, 46
Silica-modified nanomaterials, 270—275
Single-atom anchoring, surface modification via, 66—69
Single-atom catalysts (SACs), 66, 147—148
Single-pot conversion of cellulose, 133
Single-step synthesis protocol, 271—272

Index

Single-walled carbon nanotubes (SWCNTs), 13–14, 293–294
  toxicity of, 294
Single-walled nanotubes (SWNT), 110–112
SLS. See Static light scattering (SLS)
Small-angle X-ray scattering (SAXS), 14–15
Small-scale UV-structure photocatalytic systems, 120
SMNs. See Surface-modified nanomaterials (SMNs)
Soft-templating strategy, 204–205
Sol-gel method, 8–9, 44
Solid-liquid-solid method, 207
Solid-sate nuclear magnetic resonance (ss-NMR), 14–15
Solution-based reduction process, 6–8
Solution-phase growth methods, 207
Solvothermal method, 44, 207
SPR. See Surface plasmon resonance (SPR)
SRFA. See Suwannee River fulvic acids (SRFA)
SRHA. See Suwannee River humic acids (SRHA)
SS. See Stainless steel (SS)
ss-NMR. See Solid-sate nuclear magnetic resonance (ss-NMR)
Stainless steel (SS), 189
Static light scattering (SLS), 14–15
STEM. See Scanning transmission electron microscope (STEM)
Steric stabilization, 32
Stern layer, 31
Succinic acid (SA), 137
Sugarcane bagasse (SCB), 241
Sulfidation, 83–84
  of NZVI, 83–84
Sulfuration, 226
Superoxide, 109
Surface chemistry for nanoparticle modifications
  carbon materials in stabilization of nanoparticles, 38–43
  chemical methods of nanoparticle synthesis using various functional surface, 44–46
  esterification reaction method, 46–47
  phosphate ester method, 48
  in situ modification method, 48–49
  steric stabilization, 32
  types of nanoparticles, 30f
  use of silica nanoparticles for nanoparticles stabilization, 32–38
Surface coating-based modifications, 183
  gas-phase coatings, 183
  wet-phase coatings, 183
Surface functionalization
  of nanomaterials, 172
  need for, 5–6
  types of, 8–9
Surface functionalized NMs, 9
Surface modification, 6
  of catalytic nanoparticles, 58–59
  protocols, 276–278
  via single-atom anchoring, 66–69
  strategy, 291
  of nanomaterials, 269
Surface modified carbonaceous nanomaterials for CO_2, 225–227
Surface modifiers, 81–82
Surface plasmon resonance (SPR), 205–206
Surface-enhanced Raman spectroscopy (SERS), 203–204
Surface-functionalized NCs
  characterization of, 14–15
  selective applications of, 9–14
Surface-initiated controlled radical polymerization (SI-CRP), 181–182
Surface-modified carbon nanocatalysts, 110
Surface-modified nanomaterials (SMNs), 6–8, 131–132, 172, 289, 319
  active role in industry, 269–284
  based polymeric membrane-assisted catalysis, 187–190
  for biomass conversion to liquid fuels, 133–152
  for carbon dioxide transformation to liquid fuels, 152–160
  catalysts for industry applications
    catalytic applications of nanomaterial, 323–325
    metal-isolated single atoms, 334–338
    nanoparticles catalysts, 321–322
    shape and size dependent catalysts and reactions, 325–334
  challenges, 70
  challenges in SMNs and catalytic MRs, 173–174
  human health and safety consequences of, 306–308
  incorporation of SMNs into polymeric membranes, 179–187
  industrial-scale utilization of synthetic methods to preparing, 69

ligand immobilization techniques to modify catalytic nanoparticles, 63–66
nanomaterials surface chemistry and Zeta potential, 55–58
silane chemical treatment, 59–63
surface-modified nanomaterials-based catalytic materials
    nanocatalyst materials, 104–105
    nanocatalysts in hydrocarbon production, 114–119
    recent advancement and real-time utilization of nanocatalysts, 119–121
    wastewater treatment, 105–114
synthesis
    with application as catalyst, 290–291
    methods to produce catalytic nanomaterials and nanoparticle surface structure, 54f
    route for selected nanoparticles used in catalysis, 54t
techniques used to characterize, 56t
types, 291–293, 292f
    carbonaceous SNMs, 291–293
Surface-modified silica NMs, 273
Surfaces modifications techniques, 268
Suwannee River fulvic acids (SRFA), 297
Suwannee River humic acids (SRHA), 297
SWCNTs. *See* Single-walled carbon nanotubes (SWCNTs)
SWNT. *See* Single-walled nanotubes (SWNT)
Synthetic methodologies, 198–199
Synthetic strategies of NCs, 4–5

T
Tetrachroroethylene (PCE), 88
Tetradecyltrimethylammonium bromide (TTAB), 63
Tetrahydrofuran (THF), 137
Tetratomic clusters, 337–338
Thalassiosira pseudonana, 307
Thermal treatment, 59, 291
    surface modification of catalytic nanoparticles by and, 58–59
Thermogravimetric analysis (TGA), 14–15, 243–244
Thiol-capped Au NPs, 8–9
Three-dimensional nanomaterials (3D nanomaterials), 199, 214–215
Titanium dioxide (TiO$_2$), 106–108, 281
    nanocatalyst, 103–104

surface-modified nanomaterials, 281–284
TiO$_2$-based materials, 75–76
Top-down approach, 4–5, 30–31, 53, 205–207
Toxicity
    of nZVI, 306–308
    of QDs, 304
    of SWCNTs, 294
Transesterification, 309–310
Transition metals, 137–140, 251
    transition metal oxide-based nanocatalysts, 104–105
    transition metal oxides, 104–105
    transition metal-based nanomaterials, 131–132
Transition-metal dichalcogenides (TMDs), 210
Transmission electron microscopy (TEM), 14–15, 134, 183–184, 186–187, 332
Transport function of the membrane, 174
Triatomic clusters, 337–338
Triazabicyclodecene (TBD), 149
1,2,3-triazole formation, 272–273
Trichloroethylene (TCE), 88
TTAB. *See* Tetradecyltrimethylammonium bromide (TTAB)
Tungsten-based catalysts, 137
Tungsten-based hydrogenolysis catalysts (Ni–W/M), 137
Turnover frequency (TOF), 143, 308
Two-dimensional nanomaterials (2D nanomaterials), 199, 209–214
Typha latifolia, 308

U
Ultraviolet (UV), 180, 281
    UV-irradiation, 181
    UV/Ni–TiO2 nanocatalyst, 309–310
United Nations Environmental Program (UNEP), 223

V
Valence band (VB), 103–104, 197–198
Value-added products
    CO$_2$ fixation to, 244–245
    CO$_2$ hydrogenation to, 232–233
Vanadium (V), 304–305
Vanadium oxide nanoparticles (VO$_2$NPs), 304–305
Vanadium pentoxide (V$_2$O$_5$), 304–305
Vapor deposition method, 44, 53
Vapor-liquid-solid method, 207

Vapor-phase growth method, 207
Venlafaxine, 253
Visible light, 103–104
Visnagin, 252
Volatile organic compounds (VOCs), 88, 204–205

W

Wastewater treatment, 103, 105–114
    carbon-based nanocatalysts, 110–112
    metal oxide nanocatalysts, 106–109
    miscellaneous nanocatalysts, 113–114
    nanotechnology for, 119, 120t
    polymer-based nanocatalysts, 113
    zero-valent metal nanocatalysts, 106
Water
    purification, 103
    treatment
        carbon-based materials for, 76–77
        nanomaterials for, 73, 74t
        NZVI-based materials for, 77–93
        water-treatment enhancement of NZVI-based materials, 90–93
Water-gas shift reaction (WGS), 325–330
    shape dependency of activity in, 325–330
Wet-phase coatings method, 183
"White grapheme". *See* Hexagonal boron nitride (h-BN)

X

X-ray absorption spectroscopy (XAS), 67–68
X-ray diffraction technique (XRD technique), 187
X-ray photoelectron spectroscopy (XPS), 14–15
X-ray photon correlation spectroscopy (XPS), 187
X-ray spectroscopic techniques, 187

Z

Zeolite, 119, 334
    zeolite-based bifunctional magnetic ZSM-5 catalyst, 143
Zero-dimensional nanomaterials (0D nanomaterials), 199–206
    core-shell nanoparticles, hollow spheres, and nanocluster, 203–206
    QDs, 199–202
Zero-valent metal nanocatalysts, 106
Zeta potential, 55–58
Zinc (Zn), 106
Zinc oxide (ZnO), 13–14, 109
    microfluidic-based in situ synthesis of ZnO nanowires, 209f
    nanoparticles, 303–304
    synthesis route for selected nanoparticles used in catalysis, 54t

Printed in the United States
by Baker & Taylor Publisher Services